College Algebra: A Dream of Fields

Jill A. Dumesnil
University of Alaska Southeast

Colin L. Starr
Willamette University

Typeset using $\LaTeX$.

Contents

College Algebra Overview

This College Algebra text provides the basic knowledge and skills necessary for success in endeavors requiring analytical thought. Our emphasis is on the study of structure and form, the development of analytical reasoning, and the use of algebra as the language of mathematics. College Algebra should encourage higher-level thinking and should involve an investigation of various topics fundamental to the structure of algebra, placing emphasis on understanding rather than on rote manipulation of symbols. These materials do not approach topics in isolation, but integrate them, with the field properties serving as the tie that binds the course together. The use of appropriate technology interwoven throughout the course reflects the belief that technology should be available at all times and provides opportunities for both basic and extended exploration. Students learn best when they are actively involved with the mathematics they are studying, whether it be through explorations with technology, discovery activities, writing assignments, construction of proofs, or extensive exercise sets. Therefore, students who practice communicating about mathematics in a variety of ways develop a more complete grasp of the concepts involved.

Brief List of Topics

1. An Introduction to Logic
    - Basic logical structure
    - Types of arguments
    - Valid and invalid arguments
2. An Introduction to Sets
3. Field Properties
    - Binary operations (closure)
    - Axioms (including equality axioms and order postulates), Theorems, Proofs
    - Sets of numbers (Presented axiomatically as algebraic objects with a variety of applications built into the discussion)
        - Natural numbers
        - Integers
        - Rational numbers
        - Irrational numbers

- Real numbers
- Complex numbers
- Matrices
- Functions
  * Definition, applications, various representations including graphical
  * Polynomial, exponential and logarithmic functions (with applications)
  * Sequences (arithmetic, geometric, and general sequences)

4. Mathematical Induction

Preface to the Student

In order to succeed in this course and with this book, you need to know what our ultimate goals are so that you can maintain focus on them. Here they are:

Firstly, and most importantly, we are working to develop your ability to think critically and abstractly. Re-teaching you high-school algebra would not serve this purpose; instead, we will cover similar topics, but we will also consider the reasons behind what we are doing. This will require you to understand and construct mathematical proofs. We are moving beyond what you already know; it would be both pointless and boring to simply re-hash your high school experience, and we will not do so. Thus, you will need to be prepared to think at a higher level than you were in high school. This is to be expected at a university, and we hope you will welcome this opportunity to stretch your intellectual capacities.

Secondly, we will be studying algebra itself; many of the concepts we discuss will be familiar, but we will discuss them in much greater depth. It is **extremely** important that you not fall into the trap of believing that you already know all about algebra. Because the topics are familiar, many students fall back on old (and bad) habits, and never learn algebra at the level they are capable of. **Do not make this mistake!** We will be building our algebraic systems from the ground up, so you should assume nothing except the common assumptions we will all make together. Please believe us when we tell you that you will not do well this semester unless you can clear your mind and start at the beginning with us.

The point of much of what we do is that, as reasoning and educated adults, you must not accept all that you are told simply on faith. You should probe for yourselves as well. This is a first step along that path to a legitimate liberal arts education.

The class will not be easy. You will need to spend about three (3) hours outside of class for every hour in class; that means about 8 hours per week at a **minimum**. You may find that you need more time; that is perfectly acceptable as long as you take that time. This is why college is a full-time job: a student taking 12 hours should be studying 24 to 36 hours outside of class, for a total of 36 to 48 hours.

If you finish your homework early, you should go over your notes, read the book, and make up more problems for yourself. When you read the book, read it **actively**, not passively. Have a blank piece of paper, and work out all of the details for yourself *even if they are in the book already*. Cover them up and work through them again. I know that this is a lot of work, but if you are engaged at this level, you will do well. Just remember that a math text is not a novel; to read it properly, you must be actively involved. Mathematics is dense with symbols; each one carries a lot of meaning. A great deal of care is required to extract that meaning, but you can do it if you take the time. Commit to it now!

Mathematics is a language. You have studied the alphabet of the language so far; our goal is to raise you to the level where you can read some of the great literature of mathematics. This requires that you acquire the vocabulary and syntax of mathematics, and this is what we will be working on this semester.

A final piece of advice that complements the remarks in the last paragraph: the first step in doing well in this class is to carefully and thoroughly memorize all of the definitions and axioms we come across. These are the foundations we have to work with, and we will not even be able to communicate with each other if you do not learn them immediately.

We welcome you to College Algebra. If you heed the advice above, you will succeed in this class. You are beginning an exciting adventure into the abstract world of mathematics, and we hope very much that you will enjoy it as much as we do.

Dr. Jill Dumesnil and Dr. Colin Starr

May 19, 2017

Chapter 1

Logic

This logic unit provides an introduction to valid mathematical reasoning (through the use of elementary examples to illustrate the differences between valid and invalid reasoning). This will help students understand the fundamental principles of algebra and develop their capacity for logical, rational thought. In all endeavors requiring analytical thought, it is crucial that the processes involved be logically sound; the structure of the argument is as important as the conclusion itself.

Objectives

- To distinguish between valid and invalid arguments
- To motivate the need for proof
- To introduce methods of proof
- To draw appropriate logical conclusions

Terms

- statement
- hypothesis
- conclusion

1.1 The Statement

A **statement** is a sentence that is either true or false, but not both. A statement consists of both a subject and a predicate, and has a truth value. Some sentences are not statements.

There are simple statements whose truth values are relatively easy to discern.

Sentence	Statement?	Truth Value	
SFA's mascot is a lumberjack.	Yes.	True.	
All pencils are yellow.	Yes.	False.	
There was once life on Mars.	Yes.	Unknown.*	
She has black hair.	No.	Not applicable.**	
Apple pie is very tasty.	No.	Not applicable.***	
Reba McIntire has black hair.	Yes.	False.	
$2+3=5$	Yes.	True.	(Identity)
$2x+2y=2(x+y)$	Yes.	True.	(Identity)
$1=4$	Yes.	False.	(Contradiction)
$x+5=13$	No.	Not applicable.	(Conditional equation)

*Even though we do not know the truth value of this sentence, it is a statement because either there was once life on Mars, or there was never life on Mars. The fact that we do not know the truth value does not mean that it does not have one.

**Since we do not know who "she" is, this sentence does not have just one truth value. If we change which "she" we are referring to, we might also change the truth value. Since this would allow the sentence to be sometimes true and sometimes false, it does not fit the definition of a statement. However, in the presence of a context, this same sentence *could* be a statement. For example, at a party, one might point across the room at a particular woman and say, "She has black hair." In that case, the "she" would be known. In this text, we will usually assume that there is some context to avoid such ambiguity.

***This is an opinion; it does not have a particular truth value.

Note that equations are sentences, but may or may not be statements. An equation that is true for all values of the variable(s) involved is called an **identity.** An equation that is true or false depending on the value of the variable(s) is called a **conditional equation;** these are the equations that are "solved" to find the values of the variables making the sentence a true statement. Equations that are false regardless of the value of any involved variable(s) are called **contradictions.**

A sentence with a variable, like "$x+5=13$," is called a **predicate** if for each possible value of x, the sentence becomes a statement. We will not spend much time on this distinction, however.

There are also more complicated statements whose truth values are comparatively difficult to determine.

Statement	Type
I'm at Stephen F. Austin State University.	(simple statement)
I'm in Texas.	(simple statement)
I'm in Texas **and** I'm at SFA .	(conjunction)
I'm in Texas **or** I'm at SFA .	(disjunction)
If I'm at SFA, **then** I'm in Texas.	**conditional**
If I'm in Texas, **then** I'm at SFA.	(converse of conditional)
If I'm not at SFA, **then** I'm not in Texas.	(inverse of conditional)
If I'm not in Texas, **then** I'm not at SFA.	(contrapositive of conditional)

A conditional statement is a statement that can be written in the form "if **hypothesis,** then **conclusion.**" It is **false** if and only if the hypothesis is true and the conclusion is false. For example, the statement "if I'm in Texas, then I'm at SFA" has hypothesis "I'm in Texas" and conclusion "I'm at SFA." (Note that the "if" and the "then" are not part of either the hypothesis or the conclusion.) It is a false statement because one could be in Beaumont, TX and not be at SFA. If there is any case in which a conditional statement could be false, then the statement is declared a false statement. If a conditional statement is true, the hypothesis is said to **imply** the conclusion.

Note that when we say, "If $x = 2$, then $2x = 4$," we are really saying, "If we *assume* that $x = 2$ is true, then we can conclude that $2x = 4$ is also true."

Two statements are **logically equivalent** if each implies the other. For example, $x = 2$ implies that $2x = 4$, and $2x = 4$ implies that $x = 2$, so these two statements are logically equivalent. If two statements are logically equivalent, we will usually say "Statement 1 if and only if Statement 2." In the preceding example, we would say, "$x = 2$ if and only if $2x = 4$." Any conditional statement is logically equivalent to its contrapositive.

Section 1.1 Exercises

Determine which of the following are statements. If a sentence is a statement, give its truth value (if possible).

1. George Washington had wooden teeth.

2. Audrey Hepburn was very graceful.

3. All college presidents are men.

4. Some roses are red.

5. Some roses are not red.

6. Chocolate is delicious.

Identify the hypothesis and conclusion of each conditional statement. You may need to rewrite the statement in "if-then" form first.

7. If $x = 3$, then $x^2 - 1 = 8$.

8. John will use his umbrella if it rains.

9. If I stay up too late tonight, then I won't feel well tomorrow.

10. If I study appropriately, then I will pass this course.

11. My car stalls whenever it runs out of gas.

12. Every good boy deserves fudge.

Determine whether each equation is an identity, a conditional equation, or a contradiction.

13. $x^3 = 8$

14. $2x+3 = 2x-1$

15. $3(x+y) = 3x+3y$

16. $x+5 = 7$

17. $5+3 = 15$

18. $4x+5y = 5y+4x$

19. Consider the two statements, "Oranges contain Vitamin C" and, "Plastic is made from iron." Is the conjunction of these two statements true? Is the disjunction of these two statements true?

20. Consider the two statements, "Humans need oxygen to live" and, "Humans need water to live." Is the conjunction of these two statements true? Is the disjunction of these two statements true?

Write the converse and the contrapositive of each conditional statement. Assuming that the given conditional statements are true, determine the truth values of the the converse and the contrapositive, if possible.

21. If $x = 2$, then $x^2 = 4$.

22. If it snows, then school is closed.

23. If I study appropriately, then I will pass this course.

24. If this book is too long, then I won't finish it before the test.

Consider the conditional statement "If I stay up too late tonight, then I won't feel well tomorrow."

25. Identify the hypothesis and the conclusion in the statement.

26. What, if anything, do you know is true if I don't feel well tomorrow?

27. What, if anything, do you know is true if I stay up too late tonight?

28. What, if anything, do you know is true if I do feel well tomorrow?

Consider the conditional statement "If $x = 2$, then $(x-2)(x+3)(x-1) = 0$."

29. Identify the hypothesis and the conclusion in the statement.

30. What, if anything, do you know is true if $x = 2$?

31. What, if anything, do you know is true if $(x-2)(x+3)(x-1) \neq 0$?

32. What, if anything, do you know is true if $(x-2)(x+3)(x-1) = 0$?

1.2 The Argument

An **argument** is a series of assumptions, known as **hypotheses,** leading to a **conclusion.** There are two types of arguments: **inductive** and **deductive.** An inductive argument is an argument in which a conclusion (a conjecture) is formed based on observed data; such arguments attempt to infer a general result from specific observations. Inductive conclusions are never guaranteed although evidence for their likelihood may be quite strong.

A deductive argument is an argument in which a specific conclusion is derived from a general principle or rule. Such conclusions are completely guaranteed provided the deductive reasoning is **valid** and their hypotheses are true. (It is possible for an argument to be valid with false or nonsensical hypotheses, but for present purposes, assume that hypotheses are true and not absurd.) A deductive argument is deemed **invalid** if it is possible to draw conflicting conclusions, none of which violate the hypotheses.

Example 1.2.1. (Deductive) Some professors are men. Pat is a professor. Therefore, Pat is a man.

In this example, we could conclude that Pat is a woman without violating the hypotheses. Therefore, this argument is invalid.

□

Example 1.2.2. (Inductive) The sun has risen every morning throughout recorded history. Therefore, the sun will rise tomorrow.

While the evidence is very strong, the conclusion that the sun will rise tomorrow is **not** guaranteed.

□

In college mathematics courses, many arguments consist of "proving" conditional statements. It is helpful to think of conditional statements as promises kept or promises broken.

Example 1.2.3. If Mary Beth attends every class meeting, then Mary Beth will pass the course.

□

Under what conditions is the above statement true or false? If the promise is kept, the statement is true; if the promise is broken, the statement is false. There are four (cases three and four are combined) cases to consider:

1. Mary Beth attends every class AND passes the course. The promise was kept, so the conditional statement is true.

2. Mary Beth attends every class BUT receives a failing grade. The promise was broken, so the conditional statement is false.

3. Mary Beth does not attend every class. Therefore, regardless of the course grade, the promise has NOT been broken. Hence, whether the conclusion is true or false, the conditional statement is true.

Note that a conditional statement is false exactly when the hypothesis is true and the conclusion is false. A false hypothesis can imply a true conclusion or a false conclusion. (For the most part, we will not consider the validity of conclusions derived from hypotheses we know to be false.)

Example 1.2.4. The product of any two even integers is even.

□

How does one prove such a statement? Notice first that the statement can be rewritten as a conditional statement:

If x and y are (arbitrary) even integers, then xy is even.

The original statement contains a hidden "universal quantifier"; that is, the phrase "any two even integers" refers to "all possible pairs of (not necessarily related) even integers."

Recalling that a conditional statement is true when the conclusion follows directly from the hypotheses (using legitimate laws of mathematics), assume that x and y are even integers. This means that x and y are each a multiple of 2; say $x = 2s$ and $y = 2t$, where s and t are integers. Then we have that the product $xy = (2s)(2t) = 2(s \cdot 2 \cdot t)$ is even since $(s \cdot 2 \cdot t)$ is an integer, making xy an integer multiple of 2.

Section 1.2 Exercises

Determine which arguments below are valid. Assume in each case that the hypotheses are true.

1. Every mathematician likes numbers. Everyone who likes numbers is weird. Therefore, all mathematicians are weird.

2. If John gets too hot, he will pass out. Yesterday, John passed out. Therefore, John got too hot yesterday.

3. All dogs go to heaven. Buffy is not a dog. Therefore, Buffy will not go to heaven.

4. Apples are good for you. Fruit is good for you. Therefore, apples are fruit.

5. Argue that if x and y are arbitrary even integers, then $x + y$ is an even integer.

6. When an investment company gives a prospectus for some fund it manages, there is usually a disclaimer of the form, "The Fund's past performance does not necessarily indicate how it will perform in the future." What kind of argument are they being wary of?

Determine what can be concluded (if anything) if the given statements are assumed to be true.

7. Every person who likes to read, will read. Tameka likes to read.

8. Some children like to play video games. Everyone who likes to play video games understands electronics.

9. Consider the following argument: Some cars made by Company X are still in use after 20 years. No cars made by Company Y have been in use for even 15 years. Therefore, cars made by Company X are more dependable than cars made by Company Y. Determine whether this argument is valid.

10. Consider the following argument: College students should expect to spend two to three hours studying for every hour they are in class. Janet is a college student and is in class 15 hours per week. Therefore, Janet should expect to spend 30 to 45 hours studying per week. Determine whether this argument is valid.

11. Consider the following argument: All birds have wings. All chickens have wings. Therefore, all chickens are birds. Determine whether this argument is valid.

12. Consider the following argument: All rainy days are cloudy. Today is not cloudy. Therefore, today is not rainy. Determine whether this argument is valid.

13. Consider the following argument: All rainy days are cloudy. Today is not rainy. Therefore, today is not cloudy. Determine whether this argument is valid.

Suppose that you are given a collection of objects as described below. Determine whether the given object is or is not a part of the collection, based on valid logical reasoning. Note: a statement of the form, "The collection consists of precisely those things that weigh exactly four pounds" can be translated into a statement such as, "A thing is part of the collection **if and only if** it weighs exactly four pounds."

14. The collection consists of precisely those animals that have (or had) four legs; Mr. Ed, the talking horse.

15. The collection consists of precisely those numbers x satisfying $x^2 > 3$; $x = -2$.

16. The collection consists precisely of large numbers; 1000.

Chapter 2

Sets

This unit provides an introduction to the basic principles of sets, set notation, operations on sets, and the fundamental role sets play in all branches of mathematics. The basic notion of set is so fundamental to mathematics that it is impossible to give a precise definition of set in terms of simpler concepts.

Objectives

- To understand the idea of and notation for sets
- To introduce operations on sets and study their properties
- To introduce addition and multiplication of natural numbers in their set theoretic form
- To begin a study of "special" sets of numbers

Terms

- set
- element
- empty set
- universal set
- subset
- equality of sets
- disjoint sets
- complement
- intersection
- union
- cardinality
- one-to-one correspondence
- equivalent sets

2.1 Sets: Basic Notions

A **set** is a collection of objects described in such a way that it is discernible whether or not a particular object does or does not belong to the set. The objects in a set are called the **elements** of the set. Consider, for example, the following collections of objects.

Collection	Set?	
All students enrolled in this course	Yes	(Context assumed)
All integers greater than 25	Yes	
All large numbers	No	(What is "large"?)
All lines in a plane through a given point	Yes	
All rational numbers	Yes	
All x such that $x^2 - 4x - 5 = 0$	Yes	
All words that describe themselves	No	(Non-self-descriptive)
The collection $\mathscr{S}$ of all sets	No	(Classic paradox)

Sets are typically named with upper-case letters, and the elements of sets are denoted by lower-case letters. If A is a set and a is an element of A, we denote this membership by $a \in A$, pronounced, "a is an element of A" or, "a is in A." Similarly, if the object a is not an element of the set A, then we write $a \notin A$, pronounced, "a is not an element of A."

Two sets of particular importance are the **empty set** and the **universal set**. The **empty set** is the set containing no elements; it is denoted by $\emptyset$. It may help to think of a set as a box; in this sense, the empty set is just a box with nothing in it. It is *not* the same as nothing! (Note: the symbol $\emptyset$ is the last letter of the Danish-Norwegian alphabet, *not* the Greek letter ϕ, to which it bears some resemblance.) The **universal set,** U, consists of all elements under consideration in a given discussion. Obviously, different problems may involve different universal sets. For example, the set of all integers greater than 25 might be considered in the "universe" of all integers, while the set of all students enrolled in this course might be considered in the "universe" of all students at your university or the "universe" of all college students or the "universe" of all people, whichever is most appropriate in the given context. (Note: There is no universal set for all sets.)

There are two basic methods for defining or describing sets. These are the **roster method** and the **rule method.** The roster (or tabulation) method consists of either listing the elements of the set in braces ({ and }), or indicating a partial list of the elements forming a pattern to be continued either indefinitely or to a specified end.

Example 2.1.1. The set S_1 of "counting numbers" from 5 to 10 (inclusive) is given in roster notation as $S_1 = \{5,6,7,8,9,10\}$. This notation is pronounced, "the set S_1 is the set containing the elements 5, 6, 7, 8, 9, and 10." Note that the sets $\{5,6,7,8,9,10\}$ and $\{5,5,6,7,8,9,10\}$ contain precisely the same elements; generally, we do not list an element more than once since that would not add any new information about what is in the set. (We already knew that $5 \in S_1$.)

□

Example 2.1.2. The set S_2 of "counting numbers" from 1 to 100 (inclusive) is given in roster notation as $S_2 = \{1,2,3,4,\ldots 100\}$. The $\ldots$ is to indicate that the established pattern should be continued until 100 is reached.

□

Example 2.1.3. The set $\mathbb{Z}$ of integers is given in roster notation as $\mathbb{Z} = \{\ldots, -3, -2, -1, 0, 1, 2, 3, \ldots\}$. Here, the $\ldots$ on both sides indicates that the established pattern is to be continued in both directions without stopping.

□

The **rule** (or **set-builder**) **method** consists of indicating in braces the type of element along with a description of the elements in the set. The type of element is separated from the description by a vertical bar $|$, which is pronounced "such that."

Example 2.1.4. The set S_2 above can be expressed in set-builder notation as

$$S_2 = \{x \in \mathbb{Z} | 1 \leq x \leq 100\}.$$

This is pronounced, "The set of all x in $\mathbb{Z}$ (or integers x) such that $1 \leq x$ and $x \leq 100$." The x is referred to as a "dummy variable." It represents any element satisfying the given conditions.

□

Example 2.1.5. The set $\mathbb{Q}$ of rational numbers is difficult to express in roster notation; $\mathbb{Q}$ is given in set-builder notation as $\mathbb{Q} = \left\{ \frac{a}{b} \middle| a, b \in \mathbb{Z}, b \neq 0 \right\}$. This is pronounced, "The set of all $\frac{a}{b}$ such that a and b are elements of $\mathbb{Z}$ and b does not equal zero."

□

Example 2.1.6. The set S_5 of all nonvertical lines in the plane passing through the point (x_1, y_1) is given in set builder notation by $S_5 = \{y - y_1 = m(x - x_1) | m \text{ is a real number }\}$.

□

Relationships among sets may often be clarified by an illustration. Within a rectangle representing the universal set, any set A may be represented by a closed region.

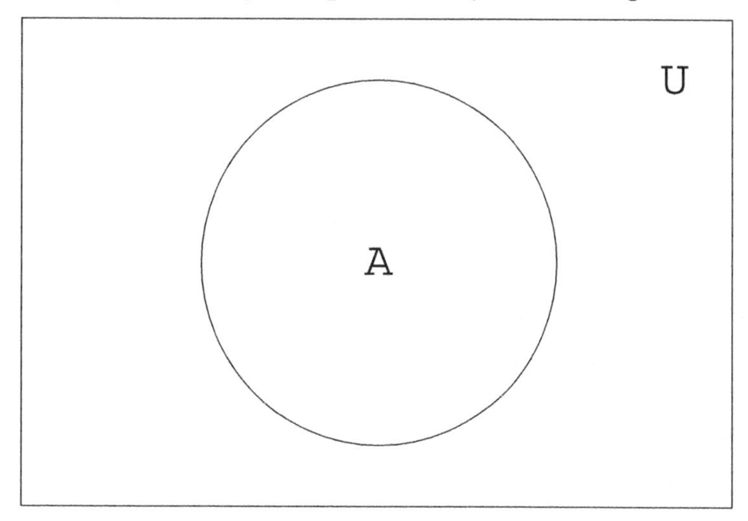

Such diagrams are called **Venn diagrams** after the English logician John Venn (1834-1883), and are useful in many applications involving relationships among sets. Among other things, Venn diagrams are useful in

1. justifying set equality,
2. understanding the definitions and properties of set operations,
3. making conjectures about sets, and
4. modeling information to solve problems.

Example 2.1.7. When surveyed, 110 college freshmen said they took the following courses as high school seniors: 25 took physics, 45 took biology, 48 took mathematics, 10 took physics and mathematics, 8 took biology and mathematics, 6 took physics and biology, and 5 took all three subjects. How many students interviewed took biology, but neither physics nor mathematics? How many students interviewed did not take any of the three mentioned subjects?

□

Let U denote the universe of the 110 students interviewed. Let P, B, and M denote the sets of interviewed students who took physics, biology, and mathematics, respectively. Each of these is represented by its own closed set, as shown. Beginning in the center, we put a 5 in the region common to all three circles (since 5 students took all three subjects). Knowing that 6 took both physics and biology, of which 5 are already accounted for, there remains only 1 student who took physics and biology, but not mathematics. We fill in the rest of the numbers with similar reasoning.

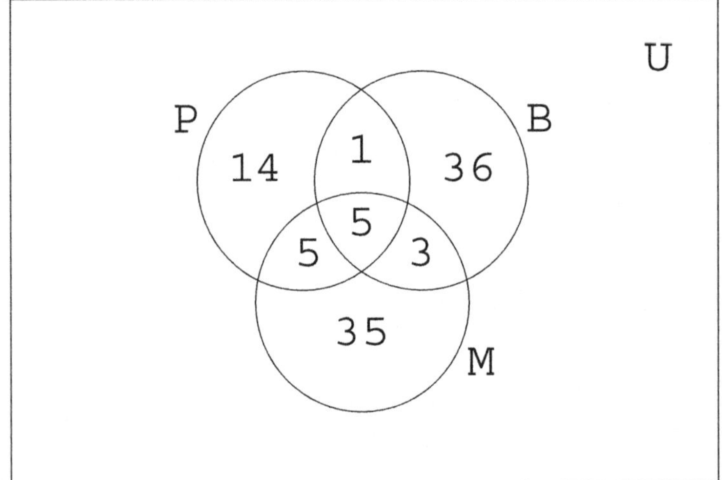

Adding, we see that of the 110 students interviewed, 36 took biology *only* and 11 took none of the mentioned subjects.

Example 2.1.8. Draw a Venn diagram accurately depicting the following sets.

U is the Universe of all air-breathing creatures.
A is the set of all birds.
B is the set of all dogs.
C is the set of all bird dogs.

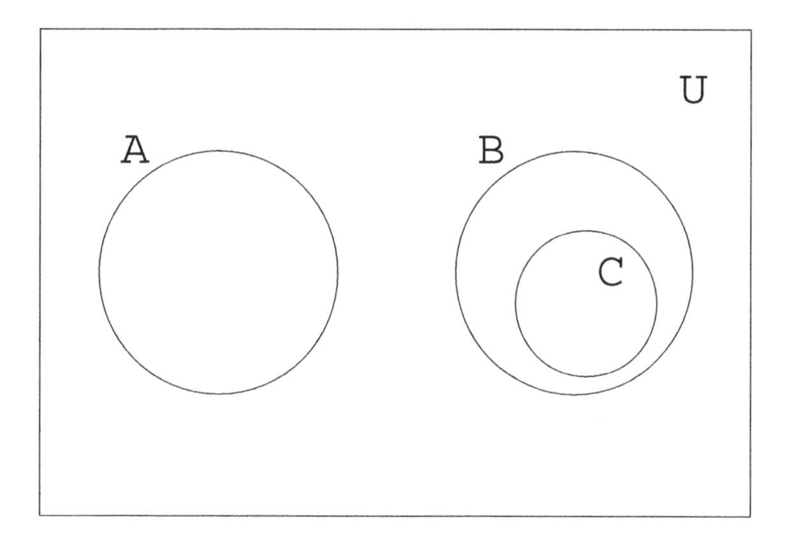

We know that no air-breathing creatures are simultaneously birds and dogs.

□

Example 2.1.9. Decide whether or not the following argument is valid.

- All chickens have wings.
- All birds have wings.
- Therefore, all chickens are birds.

Letting W denote the set of all winged creatures, C denote the set of all chickens, and B denote the set of all birds, we have the following (accurate) Venn diagram .

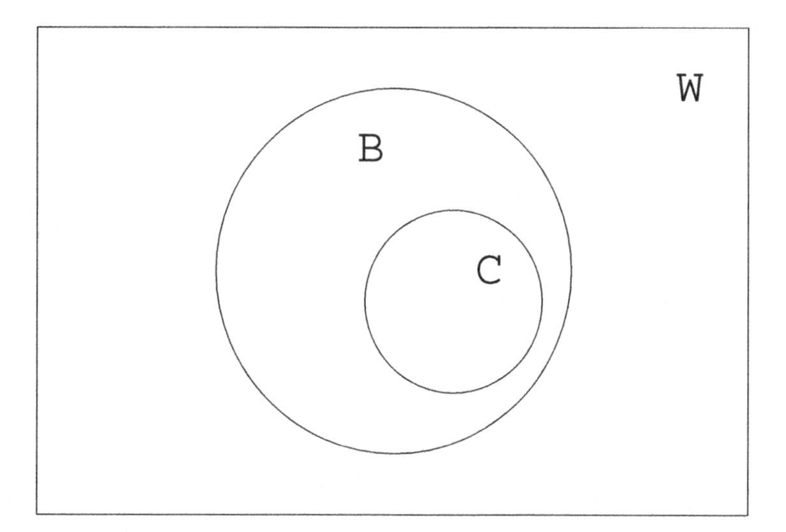

Notice that the conclusion, "all chickens are birds," is *true*, but the argument is *invalid* because the conclusion doesn't follow from the hypotheses. All flies have wings, too, but flies aren't birds. (See the set F in the diagram below.) Simply because two sets reside inside a third doesn't *guarantee* that the three sets are nested.

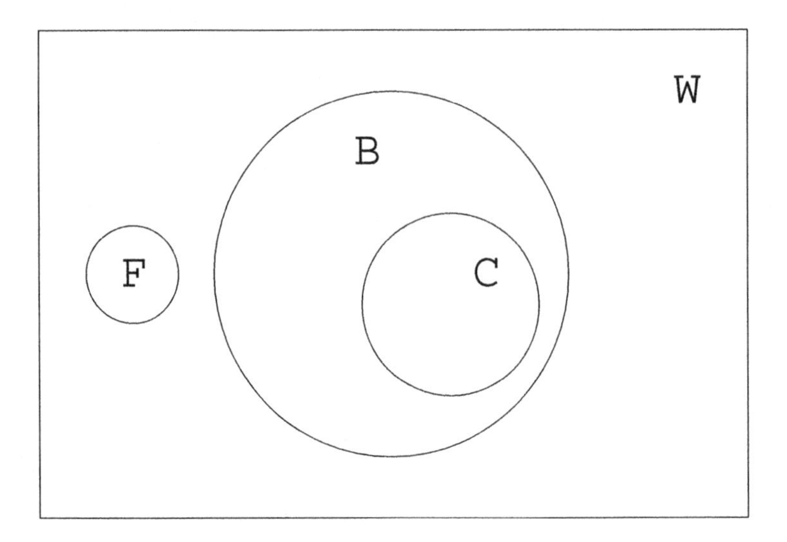

□

Let A and B be sets. If every element of A is also an element of B, then A is called a **subset** of B, denoted $A \subseteq B$. If A is a subset of B but B contains elements that are not in A, then A is called a **proper subset** of B, denoted $A \subset B$. One way to think of subsets is to recall that a set may be thought of as a box with elements in it; we can form a subset by removing some of those elements. Two sets are **equal** if each is a subset of the other; in this case, we write $A = B$. That is, $A = B$ if and only if $A \subset B$ and $B \subseteq A$. Intuitively, this means that A and B must have exactly the same elements.

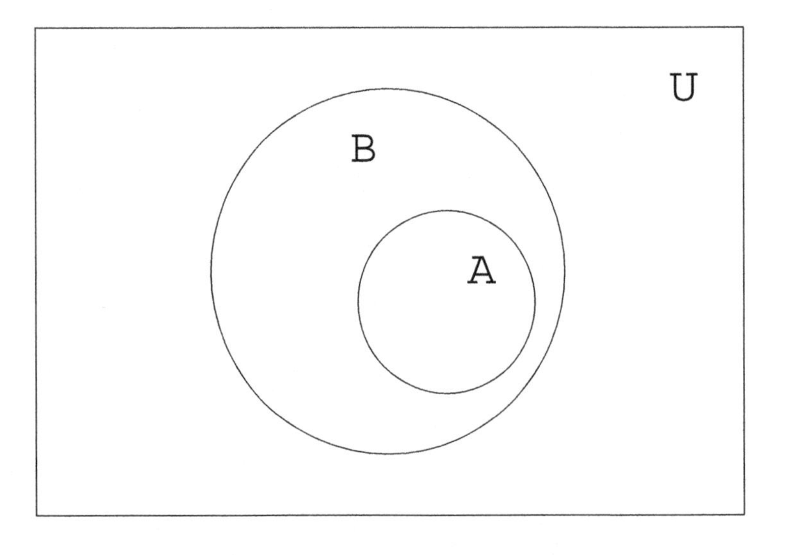

$A \subseteq B$

Example 2.1.10. The set of all roses is a proper subset of the set of all flowers.

□

Example 2.1.11. The set of "counting numbers" is a proper subset of the set of integers.

□

Example 2.1.12. The set $\{x|x \text{ is an even integer}\}$ is a subset of the set $\{2n|n \text{ is an integer}\}$, but not a proper subset.

□

To prove that a given set A is a subset of a given set B, it is useful to rewrite the definition of subset in conditional form: "$A \subseteq B$" is equivalent to "if $x \in A$, then $x \in B$." Therefore, to prove $A \subseteq B$, assume the hypothesis $x \in A$ is true and logically deduce that $x \in B$.

Example 2.1.13. Let A, B, and C be sets. If $A \subseteq B$ and $B \subseteq C$, then $A \subseteq C$. That is, subset is a **transitive** relation.

How would one prove that the notion of "subset" is transitive? Since we have a conditional statement, we assume that the hypotheses (H_1) $A \subseteq B$ and (H_2) $B \subseteq C$ are true and try to deduce that $A \subseteq C$. Thus, we need to verify that a conditional statement is true, so we assume $x \in A$, and show (using our assumptions) that $x \in C$.

Since $x \in A$ and $A \subseteq B$, we may conclude that $x \in B$, using our definition above. Now, since $x \in B$ and $B \subseteq C$, we may conclude that $x \in C$, as desired. Therefore, we've established our objective.

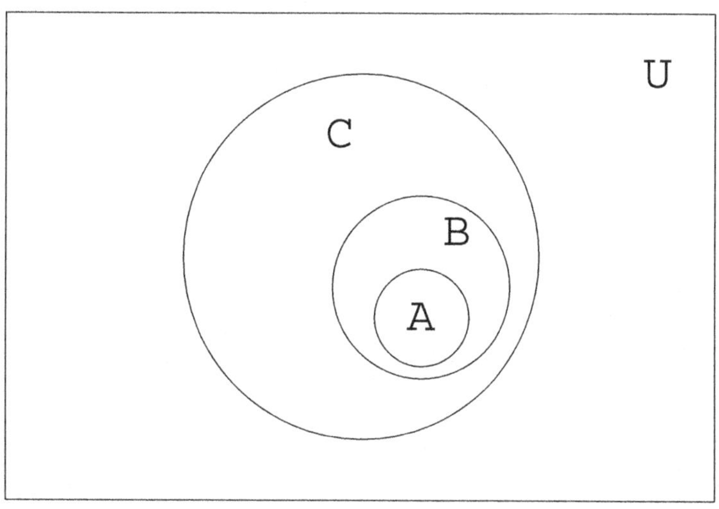

□

Section 2.1 Exercises

Determine whether the collection described is a set.

1. The collection of all cats in the world.

2. The collection of all big houses in the United States.

3. The collection of all fast computers in the world.

4. The collection of all smart students at SFASU.

5. The collection of all rich people in the world.

6. The collection of all trucks in Texas.

Write each set in roster notation. What is a reasonable universal set?

7. The set of all living former presidents of the United States.

8. The set of all solutions to the equation $(x-2)(x+5)(x-1)=0$.

9. The set of all positive even integers less than 21.

10. The set of all odd whole numbers less than 12.

Let $A=\{1,2,3,4,5\}$. Mark each statement as true or false, and give an explanation.

11. $2 \in A$.

12. $-2 \in A$.

13. $4 \notin A$.

14. $7 \notin A$.

Write each set in set-builder notation.

15. $\{2,4,6,8,10,12,\ldots\}$

16. The set of all integers greater than 12.

17. The set of even integers less than 15.

18. The set of odd whole numbers less than 100.

Determine which sets contain the same number of elements.

19. $A=\{x|x-5=0\}$; $B=\{t|(t+2)(t-2)=0\}$; $C=\{P|P$ is a point of intersection of two given distinct lines in a plane$\}$; $D=\{-2,2\}$.

20. $A=\{5,10,15,20\}$; $B=\{P|P$ is a vertex of the square $WXYZ\}$; $C=\{x|(x-1)(x+1)=0\}$; $D=\{1,2,3,4\}$.

Determine which sets contain exactly the same elements.

21. $A=\{x|x-5=0\}$; $B=\{t|(t+2)(t-2)=0\}$; $C=\{P|P$ is a point of intersection of two given distinct lines in a plane$\}$; $D=\{-2,2\}$.

22. $A=\{1,2,2,3\}$; $B=\{1,2,3\}$; $C=\{x|x$ is an integer greater than zero and less than $4\}$; $D=\{0,1,2,3\}$.

23. Let $U = \{1,2,3,4,5,6,7,8,9,10\}$. Let $A = \{2,4,6,8,10\}, B = \{5,6,7,8,9,10\}$, and $C = \{2,3,4,5\}$. Draw a Venn diagram illustrating these sets.

24. A class of 30 third-graders had an ice-cream party. When asked about their preferences, 22 said they liked vanilla, 15 said they liked chocolate, and 9 said they liked both. Draw an accurate Venn diagram illustrating this situation. Be sure to specify any sets that you name, including the universal set. How many did not say they liked either chocolate or vanilla?

In David's bicycle shop, 45 bicycles are inspected. If 20 needed new tires and 30 needed gear repairs, answer the following questions.

25. What is the greatest number of bikes that could have needed both?

26. What is the least number of bikes that could have needed both?

27. What is the greatest number of bikes that could have needed neither?

28. Suppose that all of the elements of a set A are also in the set B. Draw a Venn diagram that would illustrate this idea.

Draw a Venn diagram to illustrate each argument, and determine whether the argument is valid.

29. All computers are machines. All machines are made by people. Therefore, all computers are made by people.

30. Whenever it rains, Tammy uses her umbrella. Tammy used her umbrella last Tuesday. Therefore, it rained last Tuesday. (Hint: consider using the set of all days as your universal set.)

31. Let $U = \{a,b,c,\ldots,z\}$. Let $S = \{x|x \text{ is a vowel}\}$ and $T = \{a,b,c,d,e\}$.

    (a) Draw a Venn diagram illustrating this situation.

    (b) Let $V = \{x|x \in S \text{ or } x \in T\}$. How does V fit into the Venn diagram?

    (c) Let $R = \{x|x \in S \text{ and } x \in T\}$. How does R fit into the Venn diagram?

    (d) What is the relationship between S and V? Between T and V? Between R and S? Are there any similarities in these relationships?

32. Let $U = \mathbb{Z}$. Let $A = \{2x|x \in \mathbb{Z}\}$, $B = \{6x|x \in \mathbb{Z}\}$, and $C = \{3x|x \in \mathbb{Z}\}$. Draw a Venn diagram illustrating this situation. What can be said about the set B?

33. Explain why $\mathbb{Z} \subseteq \mathbb{Q}$.

34. A pollster interviewed 500 college seniors who owned credit cards. She reported that 240 owned Goldcard, 290 had Blackcard, and 270 had Silvercard. Of these seniors, the report said that 80 owned only a Goldcard and a Blackcard, 70 owned only a Goldcard and a Silvercard, 60 owned only a Blackcard and a Silvercard, and 50 owned all three cards. When the report was submitted to the local campus newspaper for publication, the editor refused to publish it, claiming that the poll was not accurate. Who was correct - the editor or the pollster? Explain.

True or False:

35. $\{1,2,3\} \subseteq \{1,2,3\}$.

36. $\{1,2,3\} \subset \{1,2,3\}$.

37. $\{1,3\} \subset \{1,2,3\}$

38. $\{1,2,3\} \in \{1,2,3\}$.

39. $\{1,2,3,4,5,6\} \subseteq \{1,2,3\}$.

40. $\{-2,5,12\} \subseteq \{7x-2 | x \in \mathbb{Z}\}$.

41. $\{x | x \text{ is a horse}\} \subset \{x | x \text{ is a four-legged animal}\}$.

42. Prove that $\subseteq$ is a **reflexive** relation. That is, show that for every set A, $A \subseteq A$.

43. Show that $\subseteq$ is **not** a **symmetric** relation. That is, find sets A and B such that $A \subseteq B$, but $B \not\subseteq A$.

44. Suppose that A and B are given sets.

    (a))Let $C = \{x | x \in A \text{ and } x \in B\}$. Show that $C \subseteq A$ and $C \subseteq B$. (You may want to draw a Venn diagram to help you.)

    (b) Let $D = \{x | x \in A \text{ or } x \in B\}$. Show that $A \subseteq D$ and $B \subseteq D$.

45. Use the definition of set equality from this section to show that $\{x | x^2 = 1\} = \{-1,1\}$.

46. Explain why the empty set is a subset of every set. (Hint: it may help here to think of the "box" analogy.)

2.2 Operations on Sets

It is useful to define a variety of operations on sets; some of these operations have already shown up in the exercises. We will consider **set complement, set difference, set intersection, set union,** and the **Cartesian product** of sets. In the following discussion, assume A and B are sets in a universal set U.

The **complement** of A is the set defined by $A' = \{x \in U | x \notin A\}$.

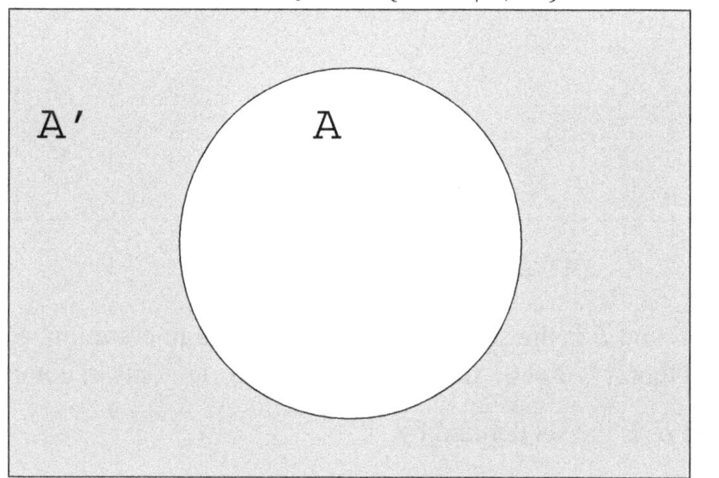

A' is shaded grey

The complement of a set A is simply the set of all elements (in U) that are *not* in A.

The **complement of** A **relative to** B or the **set difference of** B **and** A is the set

$$B - A = \{x \in U | x \in B \text{ and } x \notin A.\}$$

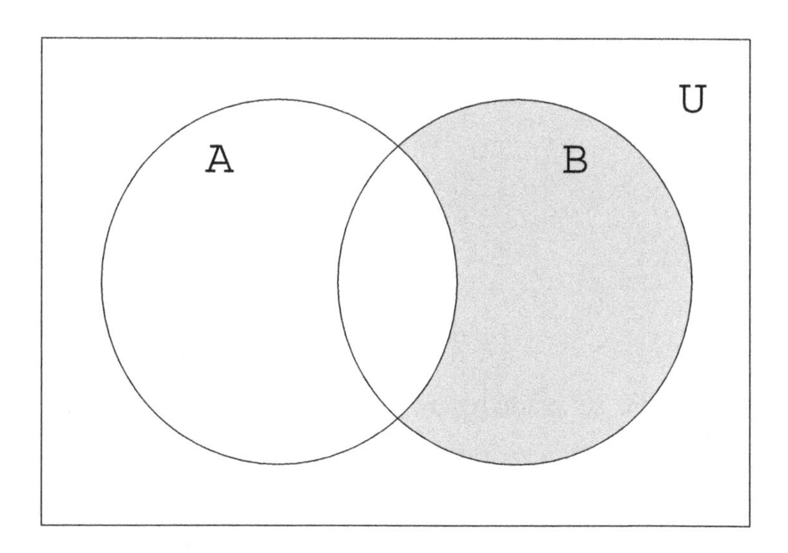

$B - A$ is shaded grey

The complement of A relative to B is the set of all elements *in* B that are not in A.

The **intersection of** A **and** B is the set defined by $A \cap B = \{x \in U | x \in A \text{ and } x \in B\}$.

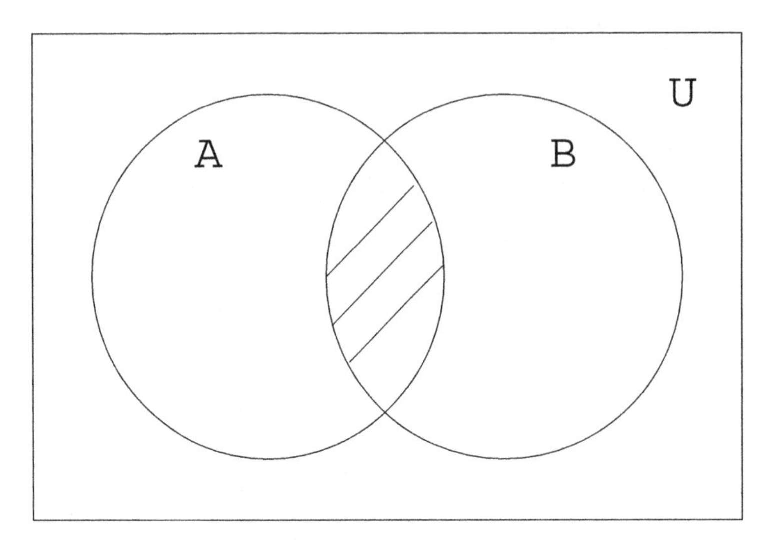

$A \cap B$ is shaded with slanted lines

The intersection of A and B is the set of elements they have in common. Sets A and B are said to be **disjoint** provided that $A \cap B = \emptyset$; that is, they have no elements in common.

The **union of** A **and** B is the set defined by

$$A \cup B = \{x \in U | x \in A \text{ or } x \in B\}.$$

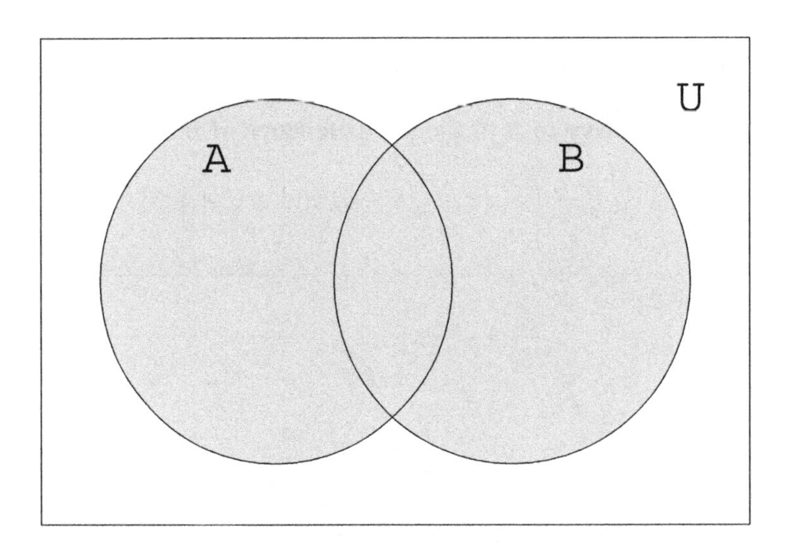

$A \cup B$ is shaded grey

The union of A and B is the set made by combining the elements of A with those of B.

The **Cartesian product of** A **and** B is the set

$$A \times B = \{(a,b) | a \in A \text{ and } b \in B\}.$$

That is, the Cartesian product $A \times B$ is the set made up of all **ordered pairs** in which the first element comes from the set A and the second element comes from the set B.

Example 2.2.1. Let $U = \{0,1,2,3,4,5,6,7,8,9\}$ denote the universal set, and let $A = \{1,2,3\}$ and $B = \{1,3,5,7,9\}$. Then

$$\begin{aligned} A' &= \{x \in U | x \notin A\} = \{0,4,5,6,7,8,9\}, \\ B - A &= \{x \in U | x \in B \text{ and } x \notin A\} = \{5,7,9\}, \\ A - B &= \{x \in U | x \in A \text{ and } x \notin B\} = \{2\}, \\ A \cap B &= \{x \in U | x \in A \text{ and } x \in B\} = \{1,3\}, \\ A \cup B &= \{x \in U | x \in A \text{ or } x \in B\} = \{1,2,3,5,7,9\}, \\ (A \cup B)' &= \{x \in U | x \notin A \cup B\} = \{0,4,6,8\}, \\ \emptyset' &= \{x \in U | x \notin \emptyset\} = U, \\ U' &= \{x \in U | x \notin U\} = \emptyset. \end{aligned}$$

□

Example 2.2.2. Let $A = \{1,2,3\}$ and $B = \{p,q\}$; then

$$A \times B = \{(a,b) | a \in A \text{ and } b \in B\} = \{(1,p),(1,q),(2,p),(2,q),(3,p),(3,q)\}.$$

□

Example 2.2.3. We will show that $A \cap B = B \cap A$. To do this, we must show that $A \cap B \subseteq B \cap A$ and $B \cap A \subseteq A \cap B$. (See the section on Subsets.)

We first show that $A \cap B \subseteq B \cap A$; this is equivalent to saying that if $x \in A \cap B$, then $x \in B \cap A$. Thus, we first let $x \in A \cap B$. Then $x \in A$ and $x \in B$. Therefore, $x \in B$ and $x \in A$, so $x \in B \cap A$. Similarly, we may show that if $x \in B \cap A$, then $x \in A \cap B$, so $A \cap B = B \cap A$.

□

Section 2.2 Exercises

Let $U = \{-10,-9,-8,-7,-6,-5,-4,-3,-2,-1,0,1,2,3,4,5,6,7,8,9,10\}$, and let $A = \{x \in U | x^2 \geq 4\}$, $B = \{-2,-1,0,1,2\}$, and $C = \{x \in U | x \text{ is even}\}$. Determine each set below.

1. $A \cap B$
2. $B \cap C$
3. $B \cup C$
4. $C - B$
5. A'
6. $(A \cap B)'$

7. $(A \cup B) \cap C$

8. Let $A = \{c, p\}$ and let $B = \{at, ut, ot\}$. Determine $A \times B$.

9. Show that $A \cap B \subseteq A$ and $A \cap B \subseteq B$.

10. Show that $A \subseteq A \cup B$ and $B \subseteq A \cup B$.

11. Show that $A \cap B = B \cap A$.

12. Show that $A \cup B = B \cup A$.

13. Find sets to illustrate the fact that $A \times B \neq B \times A$.

14. Show that $(A \cap B)' = A' \cup B'$. (One of DeMorgan's laws.)

15. Show that $(A \cup B)' = A' \cap B'$. (One of DeMorgan's laws.)

16. The **power set** of a set A is the set $\mathscr{P}(A)$ of all subsets of A. For example, if $A = \{1,2,3\}$, then $\mathscr{P}(A) = \{\emptyset, \{1\}, \{2\}, \{3\}, \{1,2\}, \{1,3\}, \{2,3\}, \{1,2,3\}\}$.

    (a) If $A = \{1,2\}$, find $\mathscr{P}(A)$.
    (b) If $A = \{1,2,3,4\}$, find $\mathscr{P}(A)$.
    (c) In general, if A contains n elements, how many elements will $\mathscr{P}(A)$ contain?

17. Maria has three hats and two scarves. Let A be the set of hats Maria owns, and let B be the set of scarves Maria owns. Determine $A \times B$, and also discover how many different hat-and-scarf combinations she could wear. (You will need to find suitable elements for A and B.)

2.3 Set Equivalence and Equality

As is evident from the earlier examples, there are many different "sizes" of sets. The number of elements in a set S is called the **cardinality** or **cardinal number** of S, denoted $n(S)$ or $|S|$. Intuitively, a **finite** set is a set containing only a finite number of elements. In the examples from section 2.1 , $S_1 = \{5,6,7,8,9,10\}$ and $S_2 = \{1,2,3,\ldots,100\}$ are finite sets with $n(S_1) = 6$ and $n(S_2) = 100$. Any set that is not finite is called an **infinite** set. The sets $\mathbb{Q}$, $\mathbb{Z}$, and $S_5 = \{y - y_1 = m(x - x_1) | m$ is a real number $\}$ defined previously are infinite sets.

Sets A and B are said to be in **one-to-one correspondence** provided that each element of A can be paired with exactly one element of B in such a way that every element of B belongs to exactly one pairing. Two sets are said to be **equivalent** or **of the same size** provided there exists a one-to-one correspondence between them. When sets A and B are equivalent, it is denoted as $A \sim B$, read, "A is equivalent to B."

Example 2.3.1. The set $A = \{x | x$ is a letter of the English alphabet$\}$ is equivalent to the set $B = \{n | n$ is an integer and $1 \leq n \leq 26\}$; the pairing $(A,1),(B,2),(C,3),\ldots,(Z,26)$ is a specific one-to-one correspondence between the sets. One can see that each letter appears in exactly one pair, and so does one number. Furthermore, this pairing illustrates exactly what we do when we count!

□

Example 2.3.2. The notion of one-to-one correspondence is essential for the concept of cardinal number. All sets equivalent to the set $\{1\}$ are said to have cardinal number *one*; all sets equivalent to the set $\{1,2\}$ are said to have cardinal number *two*; all sets equivalent to the set $\{1,2,3,\ldots,n\}$ are said to have cardinal number n. The empty set has cardinal number 0.

□

It should be noted that a set S is **finite** when S can be put into one-to-one correspondence with a set of the form $\{1,2,3,\ldots,n\}$, where n is natural number. A set T is **infinite** when T can be put into one-to-one correspondence with a proper subset of itself.

Example 2.3.3. The set $\mathbb{N} = \{1,2,3,4,\ldots\}$ contains $E = \{2,4,6,8,\ldots\}$ as a proper subset, and the correspondence $n \leftrightarrow 2n$ is a one-to-one correspondence. Therefore, $\mathbb{N}$ is an infinite set.

□

Two sets A and B are **equal,** denoted $A = B$, provided each element of A is an element of B and each element of B is an element of A; that is, provided A and B consist of precisely the same elements. We may also describe set equality in terms of subsets: two sets A and B are equal if and only if $A \subseteq B$ and $B \subseteq A$.

Example 2.3.4. The sets $B = \{n | n$ is an integer and $1 \leq n \leq 26\}$ and $C = \{1,2,3,4,\ldots,26\}$ are equal. However, the set $A = \{x | x$ is a letter of the English alphabet$\}$ is *not* equal to B, even though they are equivalent.

□

Set equality satisfies the properties of an **equivalence relation;** that is, set equality is **reflexive, symmetric,** and **transitive.**

1. Reflexive: For any set A, $A = A$.
2. Symmetric : If A and B are sets with $A = B$, then $B = A$.
3. Transitive : If A, B, and C are sets with $A = B$ and $B = C$, then $A = C$.

Section 2.3 Exercises

1. Determine the cardinality of each set.
   (a) $A = \{2,4,6,8\}$
   (b) $V = \{x|x \text{ is a vowel in the English alphabet}\}$.
2. For each set in exercise 1, find a set of the form $\{1,2,\ldots,n\}$ that is equivalent to the given set, and give an explicit one-to-one correspondence between the sets.
3. Explain why equal sets are automatically equivalent .
4. Find an example (different from the one given in the text!) of two sets that are equivalent but not equal.
5. Find an explicit one-to-one correspondence between the infinite sets $\{0,1,2,3,4,5,\ldots\}$ and $\{2,4,6,8,10,12,\ldots\}$.
6. Explain why the sets $\{1\}$ and $\{1,2\}$ **cannot** be put in one-to-one correspondence. (What happens if you try to put them in one-to-one correspondence?)
7. Determine whether the given sets are equal.
   (a) $A = \{1,2,2,3,3\}$; $B = \{1,2,3\}$.
   (b) $A = \{0,1,2,3,4,\ldots\}$; $B = \mathbb{Z}$.
   (c) $S = \{1,2,3,\ldots\}$; $T = \{1,2,3,4\ldots\}$.
   (d) $A = \{x|x^2 = 4\}$; $B = \{-2,0,2\}$.
8. If $A = \{2,5,7\}$ and $B = \{-1,2,5\}$, what set C is formed by combining the elements from A and B? What set D is formed by taking only those elements A and B have in common?
9. Give examples of finite sets A and B such that
   (a) $n(A \cup B) = n(A) + n(B)$
   (b) $n(A \cup B) \neq n(A) + n(B)$

Chapter 3

Special Sets of Numbers

This unit provides an introduction to special sets of "numbers." By "numbers" we include such generalized number notions as natural numbers, integers, rational numbers, real numbers, complex numbers, and matrices. We present all of these as sets with associated binary operations whose properties we wish to study.

Objectives

- To distinguish among various sets of numbers
- To introduce addition and multiplication of natural numbers in set theoretic form
- To begin a study of the field properties
- To extend the definitions of addition and multiplication to the "larger" sets as necessary to have the field properties satisfied
- To introduce the technique of proof by contradiction

Terms

- natural numbers
- integers
- rational numbers
- irrational numbers
- real numbers
- field properties

3.1 "Special" Sets of Real Numbers

Real numbers are numbers that express quantity; they are one-dimensional numbers. If we were to draw a "picture" of the real numbers, we could use a one-dimensional number line – every real number would have exactly one location on the line. The set of real numbers is denoted by $\mathbb{R}$.

The first set of numbers we typically encounter is the set of **natural numbers** or "counting numbers." The set of natural numbers is defined by

$$\mathbb{N} = \{1, 2, 3, 4, \ldots\}.$$

The set of **whole numbers** is the set

$$W = \{0, 1, 2, 3, \ldots\} = \mathbb{N} \cup \{0\}.$$

The set of **integers**, traditionally denoted by $\mathbb{Z}$ (standing for the German word "zahlen," meaning numbers), is the set

$$\mathbb{Z} = \{\ldots, -3, -2, -1, 0, 1, 2, 3, \ldots\}.$$

The set of **rational numbers** is the set

$$\mathbb{Q} = \left\{ \frac{a}{b} \middle| a, b \in \mathbb{Z}, b \neq 0 \right\}.$$

Real numbers that are not rational are called **irrational numbers,** the set of all irrational numbers is denoted by J. Each of these sets will be defined in more detail later. For now, it suffices to consider that $\mathbb{R} = \mathbb{Q} \cup J$, and the other sets mentioned above may be depicted as in the following Venn Diagram.

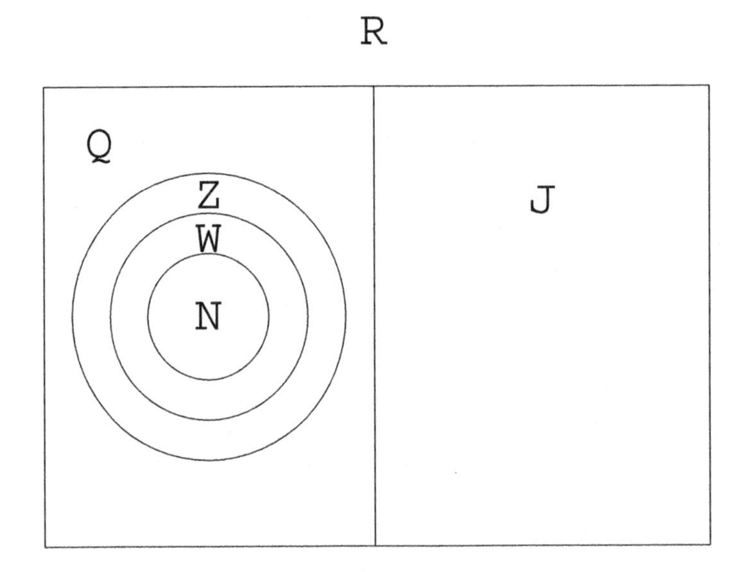

3.2 Natural Numbers

As mentioned earlier, the set $\mathbb{N}$ of natural numbers (or counting numbers) is the first set of numbers one typically encounters. For the most part, our intuitive notion of natural numbers is sufficient. The set of **natural numbers** is given by

$$\mathbb{N} = \{1,2,3,4,\ldots\}.$$

Notice that these numbers represent counting numbers or *cardinalities*.

Example 3.2.1. The natural number $3 = n(A)$, where the set $A \sim \{a,b,c\}$.

□

Suppose you wanted to teach a young child how to add; how would you go about it? (Assume that the child already knows how to count.) Take as an example the addition problem $2+3$. You might first find two objects, say blocks, and put them to one side, and then find three other blocks and put them to the other side. The child counts two objects and three objects. Then you could combine your two sets of blocks into one larger set containing five elements, which the child counts to get the sum. This is precisely the idea we present below for addition; multiplication requires some more motivation.

The **binary operations** $+$ (addition) and $\cdot$ (multiplication) can be defined on $\mathbb{N}$ in terms of set cardinality: let a and b denote arbitrary natural numbers. Then there exist disjoint sets A and B with cardinalities a and b, respectively. That is, $n(A) = a$ and $n(B) = b$, with $A \cap B = \emptyset$.

The **sum of** a **and** b is defined by

$$a + b = n(A \cup B);$$

the **product of** a **and** b is defined by

$$a \cdot b = n(A \times B).$$

Example 3.2.2. Let $a = 2$ and $b = 3$. To compute $a+b$ and $a \cdot b$ **by definition**, one needs two disjoint sets A and B with cardinalities a and b, respectively. Let $A = \{1,2\}$ and $B = \{\clubsuit, \diamondsuit, \spadesuit\}$. (Notice that $A \cap B = \emptyset$ – the piles of blocks need to be separate!) Then $n(A) = 2$ and $n(B) = 3$; also, $A \cup B = \{1,2,\clubsuit,\diamondsuit,\spadesuit\}$ (this is where we combine the blocks), so $n(A \cup B) = 5$. By definition, we have

$$2+3 = a+b = n(A \cup B) = 5.$$

Furthermore, $A \times B = \{(1,\clubsuit),(2,\clubsuit),(1,\diamondsuit),(2,\diamondsuit),(1,\spadesuit),(2,\spadesuit)\}$, so $n(A \times B) = 6$. Therefore,

$$2 \cdot 3 = a \cdot b = n(A \times B) = 6.$$

□

Note that sets A and B must be disjoint for the sum to be accurately defined, but the product would be as accurately defined without the stipulation that $A \cap B = \emptyset$. For convenience, we will continue to use disjoint sets A and B for both sums and products of natural numbers.

For acceptable binary operations to be defined on particular sets of numbers, the set must be **closed** with respect to the operation. This means that given two elements of the set, the element resulting from the binary operation on the two elements must also be a member of the set. By the definitions given above in terms of cardinalities, it is apparent that the set of natural numbers is closed under addition and multiplication; that is, if $a, b \in \mathbb{N}$, then $a+b \in \mathbb{N}$ and $a \cdot b \in \mathbb{N}$. This is because each is defined as a cardinality of a specific set, and that is exactly what natural numbers are.

Example 3.2.3. The set $\mathbb{N}$ is closed under the operation $x \circ y = x+y+1$. $\mathbb{N}$ is not closed under subtraction: $3-5=-2 \notin \mathbb{N}$, even though $3 \in \mathbb{N}$ and $5 \in \mathbb{N}$.

□

Theorem 1. *The sum and product of natural numbers satisfy the following properties.*

1. *The sum of natural numbers is* **associative***; i.e., for natural numbers a, b, and c,* $(a+b)+c = a+(b+c)$.
2. *The sum of natural numbers is* **commutative***; i.e., for natural numbers a and b,* $a+b=b+a$.
3. *The product of natural numbers is* **associative***; i.e., for natural numbers a, b, and c,* $(a \cdot b) \cdot c = a \cdot (b \cdot c)$.
4. *The product of natural numbers is* **commutative***; i.e., for natural numbers a and b,* $a \cdot b = b \cdot a$.
5. *For natural numbers, multiplication* **distributes** *over addition; i.e., for natural numbers a, b, and c,* $a \cdot (b+c) = a \cdot b + a \cdot c$.

The above properties 1 and 2 are consequences of the corresponding properties for set union, and may be justified by drawing the appropriate Venn diagrams. We will prove property 2 directly.

Let $a, b \in \mathbb{N}$. Then there exist disjoint sets A and B such that $n(A) = a$ and $n(B) = b$. Now $a+b = n(A \cup B)$ by definition, and $b+a = n(B \cup A)$, also by definition. We have already observed that $A \cup B = B \cup A$ and that equal sets are automatically equivalent; therefore $n(A \cup B) = n(B \cup A)$, so $a+b=b+a$.

The proofs of properties 3, 4, and 5 require proving the appropriate conditional statements and are beyond the scope of what is intended at this point.

Section 3.2 Exercises

Use sets to illustrate each computation.

1. $4+3=7$
2. $4 \cdot 3 = 12$
3. $2 \cdot (3+4) = (2 \cdot 3)+(2 \cdot 4)$ (Multiplication distributes over addition.)

4. $(2+3)+4 = 2+(3+4)$ (Addition is associative.)

5. $2 \cdot 4 = 4 \cdot 2$ (Multiplication is commutative.)

6. Use Venn diagrams to show that $(A \cup B) \cup C = A \cup (B \cup C)$. What property of natural number addition can this be used to illustrate?

7. Let $A = \{1,2,3\}$ and $B = \{a,b,c\}$.

    (a) Find $A \times B$.

    (b) Find $B \times A$.

    (c) Find an explicit one-to-one correspondence between $A \times B$ and $B \times A$. What theorem can this be used to illustrate?

8. Find two examples of sets with operations such that the sets are not closed with respect to those operations.

9. Explain why there is no natural number a such that $2 \cdot a = 1$.

10. Explain why we cannot use the sets $A = \{1,2,3,4,5\}$ and $B = \{4,5,6\}$ to illustrate the computation $5+3 = 8$. What goes wrong? Be explicit.

11. Show that for any set A, $\mathscr{P}(A)$ is closed under taking unions. (Recall that $\mathscr{P}(A)$ is the power set of A; that is, the set of all subsets of A.)

12. Show that for any set A, $\mathscr{P}(A)$ is closed under taking intersections.

3.3 The Field Properties

Let S denote any set of "numbers" with binary operations addition and multiplication. The following properties may or may not hold for the set S; we have seen that many of them hold for $\mathbb{N}$, but others do not. As we encounter more sets of numbers with their own operations throughout the course, we will investigate which of these properties hold and which do not.

Closure

- *Additive Closure:* For all a and b in S, $a+b$ is also in S.
- *Multiplicative Closure:* For all a and b in S, $a \cdot b$ is also in S.

Associative Laws

- *Associative Law of Addition:* $(a+b)+c = a+(b+c)$ for all a, b, c in S.
- *Associative Law of Multiplication:* $(a \cdot b) \cdot c = a \cdot (b \cdot c)$ for all a, b, c in S.

Commutative Laws

- *Commutative Law of Addition:* $a+b = b+a$ for all a, b in S.
- *Commutative Law of Multiplication:* $a \cdot b = b \cdot a$ for all a, b in S.

Existence of Identities

- *Additive Identity Property:* There is a number, 0, in S such that $a+0 = a$ and $0+a = a$ for all a in S. A number 0 with this property is called an **additive identity.**
- *Multiplicative Identity Property:* There is a number, 1, in S such that $a \cdot 1 = a$ and $1 \cdot a = a$ for all a in S. A number 1 with this property is called a **multiplicative identity.**

Existence of Inverses

- *Additive Inverse Property:* For each a in S, there is a number b in S such that $a+b = 0$ and $b+a = 0$. This number b is called an **additive inverse of** a and is denoted by $-a$. Notice that $b = -a$ and $a = -b$.
- *Multiplicative Inverse Property:* For each a in S with $a \neq 0$, there is a number b in S such that $a \cdot b = 1$ and $b \cdot a = 1$. This number b is called a **multiplicative inverse of** a and is denoted by a^{-1}. Notice that $b = a^{-1}$ and $a = b^{-1}$.

Distributive Law

- *Distributive Law of Multiplication over Addition:* For all elements a, b, c in S, $a \cdot (b+c) = a \cdot b + a \cdot c$.

The above properties are called the **field properties** or **field axioms**, and sets with two binary operations satisfying all of these properties are known as **fields**.

These properties are fundamental to the study of algebra; the sooner you learn to recognize and apply them, the more successful you will be in this course. **Knowing the field properties is a must.**

Section 3.3 Exercises

Assume that the set F is a field Identify which property above is used to justify each statement.

1. $-3.751+0=-3.751$
2. $(6\cdot 5)\cdot 2=6\cdot(5\cdot 2)$
3. $3x+6x=(3+6)x$
4. $(42)(42^{-1})=1$
5. $7+(-7)=0$
6. $0+y=y$
7. $1\cdot x=x$
8. $(x+y)\cdot(a+b)=(a+b)\cdot(x+y)$
9. $(x+y)a+(x+y)b=(x+y)(a+b)$
10. $a+b=b+a$
11. $a+(-a)=0$
12. $(x+2)(x+2)^{-1}=1$
13. The set $\mathbb{Z}$ is not a field. Which field property does not hold in $\mathbb{Z}$?
14. Identify which field property is used in each step below.

$(x+2)(x+3)$	$=$	$[(x+2)\cdot x]+[(x+2)\cdot 3]$	__________________
	$=$	$[x\cdot(x+2)]+[3(x+2)]$	__________________
	$=$	$[x\cdot x+x\cdot 2]+[3\cdot x+3\cdot 2]$	__________________
	$=$	$[x^2+2x]+[3x+6]$	__________________
	$=$	$[(x^2+2x)+3x]+6$	__________________
	$=$	$[x^2+(2x+3x)]+6$	__________________
	$=$	$[x^2+(2+3)x]+6$	__________________
	$=$	$[x^2+5x]+6$	3+2=5
	$=$	x^2+5x+6	__________________

15. A customer in a back aisle of your store has seven items worth \$1.52 each. She tells you that her partner will soon be arriving with three more of the same item. What field property allows you to quickly compute the total value of the items, $7 \cdot 1.52 + 3 \cdot 1.52$?

16. If the sales tax on a dress worth \$78 dollars is 8.25%, the cost of the dress including sales tax is $1.0825 \cdot 78$ dollars. If the dress is part of a 25% off sale, its cost becomes $78 \cdot 0.75$ dollars.

    (a) As the customer, would you rather apply the sales tax to the sale price, or take 25% off of the after-tax price of the dress?

    (b) Compute the cost of the dress if the sales tax applies only to the sale price.

    (c) Compute the cost of the dress if the sales tax is computed before taking 25% of the price of the dress.

    (d) What field property guarantees that the two answers will be the same? This means that it doesn't matter which you compute first!

17. Show that if a, b, and c are elements of a field, then $(b+c) \cdot a = b \cdot a + c \cdot a$.

3.4 "Extending" the Number System

The set $\mathbb{N} = \{1,2,3,4,\dots\}$ of natural numbers satisfy field properties 1, $1'$, 2, $2'$, and 5. However, the set $\mathbb{N}$ fails to satisfy the other field properties. Consequently, we wish to "extend" the set of natural numbers to sets of numbers that will satisfy the remaining field properties.

Dealing first with the additive properties, we obtain an additive identity by simply extending $\mathbb{N}$ to the set $W = \mathbb{N} \cup \{0\}$ of whole numbers. Recall that $0 = n(\emptyset)$, and if $a \in \mathbb{N}$, then $a = n(A)$ for some nonempty set A. Thus, we see that

$$0 + a = n(\emptyset \cup A) = n(A) = a \text{ and } a + 0 = n(A \cup \emptyset) = n(A) = a.$$

Therefore, the element $0 \in W$ satisfies the requirements to be an additive identity on W. However, additive inverses are still missing for all whole numbers except 0. We need to "extend" to the set $\mathbb{Z} = \{\dots,-3,-2,-1,0,1,2,3,\dots\}$ of integers by including the additive inverse of each natural number. Conceptually, this is a reasonable thing to do: if natural numbers represent steps forward, then their additive inverses represent steps backward. Thus, 4 steps forward followed by 4 steps backward puts you right where you started, as though you had taken zero steps forward: $4+(-4) = 0$.

The set $\mathbb{Z}$ of integers is closed under addition; *i.e.*, for all $a,b \in \mathbb{Z}$, $a+b \in \mathbb{Z}$. The addition on $\mathbb{Z}$ satisfies field properties 1, 2, 3 and 4. Thus, addition on $\mathbb{Z}$ is associative and commutative, $\mathbb{Z}$ contains an additive identity, and every element of $\mathbb{Z}$ has an additive inverse in $\mathbb{Z}$.

When a set S has an "addition" satisfying the above properties, we can define **subtraction** on the set as the **inverse of addition**:

$$\text{For } a,b \in S, \quad a - b \equiv a + (-b).$$

The symbol $\equiv$ is used to indicate a definition; thus, $a - b = a + (-b)$.

Considering now the multiplicative properties, it is true that the "extended" multiplication defined on $\mathbb{Z}$ is associative, commutative, and distributes over the addition. [Note: In a future section on consequences of the field properties, we'll prove mathematically that the extended multiplication on $\mathbb{Z}$ does indeed satisfy the commonly known properties of integer multiplication, *e.g.*, for $a,b \in \mathbb{Z}$, $(-a)(b) = -(ab)$.]

Also, $\mathbb{Z}$ contains a multiplicative identity element, 1. However, for most nonzero elements in $\mathbb{Z}$, the set of integers does not contain a multiplicative inverse. In fact, the integers 1 and -1 are the only integers with *integer* multiplicative inverses. Hence, we have the need to "extend" the number system further to the set $\mathbb{Q}$ of rational numbers.

Once we have a set S with an associative, commutative "multiplication" with a multiplicative identity element and such that all nonzero elements have multiplicative inverses, we can define **division** on the set as the **inverse of multiplication**:

$$\text{For } a,b \in S, \quad a \div b \equiv a \cdot b^{-1}.$$

The set $\mathbb{Q}$ of rational numbers satisfies all of the field axioms; therefore, $\mathbb{Q}$ is a field.

We adopt the usual conventions for order of operations. Grouped items receive the highest "priority". Since subtraction is really just addition (of an additive inverse), addition and subtraction have the same priority in computations. Similarly, since division is really just multiplication (by a multiplicative inverse), multiplication and division have the same priority in computations. Multiplication and division are given a higher priority than addition and subtraction. Operations at the same priority are performed from left to right. Thus, when we write $2+3+4$, it means $(2+3)+4$; similarly, $3 \div 7 \cdot 12$ means $(3 \div 7) \cdot 12$.

Example 3.4.1. Include grouping symbols to indicate the order of operations for $2+3 \cdot 6-4$.

$2+3 \cdot 6-4=[2+(3 \cdot 6)]-4.$

□

If we had wanted the addition first, we would have had to have specifically said so by using grouping symbols: $(2+3) \cdot 6-4$ is different from the given computation.

Note that to apply associative laws, we must first rewrite the operations in terms of addition and/or multiplication.

Section 3.4 Exercises

Determine the additive inverse of each element.

1. -3.14159
2. $\frac{52}{71}$
3. $-a$
4. $(x+y)$
5. $-12?$
6. 0

Determine the multiplicative inverse of each element, if it has one.

7. $4x+7$
8. 0
9. $\frac{13}{4}$
10. $(x+y)$
11. $3+4$
12. 1

Write each subtraction in terms of addition.

13. $13-9$
14. $-3-(-6)$
15. $x-y$

Include grouping symbols to indicate the order of operations for each computation.

16. $4 \cdot 5+9$
17. $2-3+6$
18. $12 \cdot 3+2 \div 4 \cdot 20$
19. $a \cdot b+c \div d-e \cdot f$

3.5 Rational Numbers

Definition 3.5.1. The set $\mathbb{Q}$ of **rational numbers** is defined by

$$\mathbb{Q} = \left\{ \frac{a}{b} \middle| a, b \in \mathbb{Z}, b \neq 0 \right\}.$$

The number a is the **numerator**, and b is the **denominator**.

This set together with the associated addition and multiplication satisfies all the properties of a field. Recall that

$$\frac{a}{b} + \frac{c}{d} = \frac{ad + bc}{bd} \quad \text{and} \quad \frac{a}{b} \cdot \frac{c}{d} = \frac{ac}{bd}$$

where $\frac{a}{b}, \frac{c}{d} \in \mathbb{Q}$. (These formulas for addition and multiplication will be proven later.)

An equivalent definition of the set of rational numbers is "the set of real numbers whose decimal expansion either terminates or repeats."

Example 3.5.1. The numbers $\frac{1}{2}$, $0.\overline{3} = 0.3333\ldots$, 0.25, $-\frac{5}{2}$, $0.11\overline{2}$, and $1.8\overline{23}$ are rational numbers.

□

Since the two definitions of rational numbers are equivalent, every rational number can be expressed in both forms: as a ratio of integers (a fraction) and as a terminating or repeating decimal. To convert from a fractional representation to a decimal representation, one divides numerator by denominator. It's useful to know in advance whether the decimal representation will be terminating or repeating.

Theorem 2. *Let $\frac{a}{b}$ represent a rational number in lowest terms. The decimal representation of $\frac{a}{b}$ is terminating if and only if the prime factorization of b contains only the prime numbers 2 and/or 5.*

A momentary digression: the number 432 stands for "four hundreds, three tens, and two ones." Each position is worth 10 times as much as the position to its immediate right; this is the case in a "base 10" system. The prime numbers 2 and 5 happen to be the only factors of 10; it is no coincidence that those are the factors of the denominator that will cause a terminating decimal expansion! Think about why that should be the case.

Example 3.5.2. The decimal representation of $\frac{1}{8}$ is terminating (0.125), while the decimal representation of $\frac{3}{11}$ is repeating ($0.\overline{27}$).

□

Example 3.5.3. The decimal representation of $\frac{3}{20}$ is terminating (0.15), while the decimal representation of $\frac{1}{7}$ is repeating ($0.\overline{142857}$).

□

Given that the decimal representation will repeat, it is also useful to know in advance the length of the string of repeating digits. (The string of repeating digits is called the **repetend**.)

Theorem 3. *If the rational number $\frac{a}{b}$ has a repeating decimal representation, then the repetend is at most $b-1$ digits in length.*

Example 3.5.4. The decimal representation of $\frac{1}{7}$ has a repetend of maximal length 6 digits.

□

Example 3.5.5. The decimal representation of $\frac{1}{17}$ is repeating and of maximal length 16 digits.

□

Since few hand-held calculators have a 16-digit display, how would one discern the *exact* decimal representation of $\frac{1}{17}$? One, perhaps unsatisfying, option would be to use pencil-and-paper long-division. Another, perhaps more satisfying, option is to use a combination of electronic and pencil-and-paper calculation. The division of 1 by 17 on an inexpensive eight-digit-display calculator will show a quotient of 0.05882353. It isn't clear whether or not the last "3" is accurate or rounded; temporarily truncate the last digit.

```
      0.0588235
  17)1.0000000000000000
      0.9999995
      0.0000005000000000
```

Now divide the "remainder" 5 by 17 electronically and obtain 0.29411765. Again, it isn't clear whether or not the last "5" is accurate or rounded; temporarily truncate. (Notice that the 3 we truncated in the previous step was rounded up from 0.29411765.)

```
      0.05882352941176
  17)1.0000000000000000
      0.9999995
      0.0000005000000000
      0.00000049999992
      0.0000000000000800
```

Finally, divide 8 by 17 to obtain .47588235 to finish the problem.

```
      0.0588235294117647058
  17)1.0000000000000000
      0.9999995
      0.0000005000000000
      0.00000049999992
      0.0000000000000800
      0.0000000000000799
```

Therefore, we get that $\frac{1}{17} = 0.\overline{0588235294117647}$. (We can conclude that we have finished since we have 16 digits, and that is as many as possible before repetition begins.)

Consider, now, the problem of converting a rational number from decimal representation to a ratio of integers. There are basically two cases: the decimal is terminating or the decimal is repeating. The conversion of a terminating decimal to a ratio of integers consists of writing the terminating string of digits over the appropriate denominator (and reducing to lowest terms, if desired). See the examples below.

Example 3.5.6. The terminating decimal 0.25 corresponds to the fraction $\frac{25}{100}$, which may be reduced to $\frac{1}{4}$.

□

Example 3.5.7. The terminating decimal 0.3 corresponds to the fraction $\frac{3}{10}$, which is in lowest terms.

□

Example 3.5.8. The terminating decimal 0.333 corresponds to the fraction $\frac{333}{1000}$.

□

The conversion of a repeating decimal to a ratio of integers often requires a degree of mathematical finesse. Consider the following examples.

Example 3.5.9. Many people simply recognize the repeating decimal $0.\overline{3}$ as being equivalent to the fraction $\frac{1}{3}$.

□

Example 3.5.10. To convert the repeating decimal $0.\overline{6}$ to a ratio of integers, we may either recognize from experience that $0.\overline{6} = \frac{2}{3}$, or multiply both sides of the equality $0.\overline{3} = \frac{1}{3}$ by 2.

□

Example 3.5.11. To convert the repeating decimal $0.\overline{9}$ to a ratio of integers, we might multiply both sides of the equality $0.\overline{3} = \frac{1}{3}$ by 3, which yields that $0.\overline{9} = 1$. (Think about that!)

□

Example 3.5.12. To convert the repeating decimal $0.\overline{2}$ to a ratio of integers requires the solution of an appropriate system of linear equations. Say $n = 0.\overline{2}$. Multiplying both sides by 10 yields $10n = 2.\overline{2}$. This gives us a linear system of equations, and subtraction yields that

$$\begin{array}{rcl} 10n & = & 2.\overline{2} \\ n & = & 0.\overline{2} \\ \hline 9n & = & 2 \end{array}$$

Division of the resulting equation yields the rational number $n = \frac{2}{9}$.

Notice the strategy we used: we needed a multiple of n to be equal to an *integer* so that we could express n as a ratio of integers. Accordingly, we found a multiple of n (namely, $10n$) that matched exactly with n to the right of the decimal. What makes this desirable is that subtraction will lead to complete cancellation of everything after the decimal!

□

Example 3.5.13. To convert the repeating decimal $0.\overline{23}$ to a ratio of integers also requires the solution of an appropriate system of linear equations. Say $n = 0.\overline{23}$. Multiplying both sides by 100 yields $100n = 23.\overline{23}$. Thus, we have the linear system of equations, and subtraction yields that

$$\begin{array}{rcl} 100n &=& 23.\overline{23} \\ n &=& 0.\overline{23} \\ \hline 99n &=& 23.00 \end{array}$$

Division of the resulting equation yields the rational number $n = \frac{23}{99}$. The strategy was the same as above; only the numbers have changed.

□

Example 3.5.14. To convert the repeating decimal $0.2\overline{34}$ to a ratio of integers, write $n = 0.2\overline{34}$. Multiplying both sides by 10 yields $10n = 2.\overline{34}$, and multiplying both sides by 1000 yields $1000n = 234.\overline{34}$. Thus, we have the following linear system of equations, and subtraction yields that

$$\begin{array}{rcl} 1000n &=& 234.\overline{34} \\ 10n &=& 2.\overline{34} \\ \hline 990n &=& 232.00 \end{array}$$

Division of the resulting equation yields the rational number $n = \frac{232}{990}$. Notice that this time we had to find two different multiples of n in order to "line up" the decimal parts for cancellation.

□

Depending on the given situation, it is sometimes advantageous to express rational numbers as a ratio of integers rather than in decimal form, as in the following example.

Theorem 4. *The rational numbers are* **dense***; i.e., between any two distinct rational numbers there is another rational number.*

Proof. Let $a, b \in \mathbb{Q}$ with $a < b$. A reasonable guess for the rational number we seek is the average of a and b: we claim that

$$1)\quad \frac{a+b}{2} \in \mathbb{Q} \quad \text{and} \quad 2)\quad a < \frac{a+b}{2} < b.$$

Together, these say that the number $\dfrac{a+b}{2}$ is the rational number we need between our two given rational numbers. We'll prove the first claim now and defer the second claim until a later section on inequalities.

Since $a, b \in \mathbb{Q}$, we can express $a = \dfrac{x}{y}$ and $b = \dfrac{p}{q}$, where $x, y, p, q \in \mathbb{Z}$, $y \neq 0$, and $q \neq 0$. Then

$$\frac{a+b}{2} = (a+b)\cdot\frac{1}{2} = \left(\frac{x}{y}+\frac{p}{q}\right)\cdot\frac{1}{2} = \left(\frac{xq+yp}{yq}\right)\cdot\frac{1}{2} = \frac{xq+yp}{2yq},$$

which is rational since both the numerator, $xq+yp$, and the nonzero (why?) denominator, $2yq$, are integers. □

Section 3.5 Exercises

Perform the following computations.

1. $\frac{1}{2}+\frac{5}{6}$

2. $\frac{3}{10}-\frac{4}{5}$

3. $\frac{3}{4}+\frac{5}{6}$

4. $\frac{1}{7}+\frac{-2}{8}$

5. $\left(\frac{5}{7}+\frac{3}{8}\right)+\frac{1}{3}$

6. $\left(\frac{1}{2}-\frac{2}{3}\right)+\frac{3}{4}$

7. $\frac{8}{27}-\frac{-1}{12}$

8. $\frac{2}{3}\cdot\frac{5}{7}$

9. $\frac{12}{-25}\cdot\frac{15}{4}$

10. $\frac{13}{17}\cdot\frac{-1}{2}$

11. $\frac{24}{28}\cdot\frac{3}{5}$

12. $\frac{1}{2}\cdot\left(\frac{4}{3}+\frac{4}{7}\right)$

13. $\frac{2}{3}\cdot\left(\frac{8}{9}-\frac{4}{3}\right)$

14. $\frac{1}{6}\div\frac{2}{3}$

15. $\frac{12}{17}\div\frac{2}{3}$

16. $\frac{1}{2}\div\left(\frac{4}{3}+\frac{4}{7}\right)$

17. $\frac{2}{3}-\left(\frac{4}{3}+\frac{4}{7}\right)$

18. $\frac{-1}{4}\div\left(\frac{4}{3}+\frac{4}{7}\right)$

19. $\frac{2}{3}-\left(\frac{1}{9}+\frac{2}{3}\right)$

20. $\frac{3}{7}\div\left(\frac{4}{3}\div\frac{4}{7}\right)$

21. $\frac{-1}{4}\cdot\left(\frac{4}{9}\cdot\frac{2}{-7}\right)$

22. $3+\frac{5}{6}$

Represent each repeating decimal as a ratio of integers.

23. $0.\overline{3}$

24. $1.\overline{3}$

25. $9.\overline{9}$

26. $0.\overline{19}$

27. $0.\overline{248}$

28. $0.\overline{471}$

29. $0.1\overline{9}$

30. $2.\overline{34}$

31. $21.3\overline{67}$

32. $0.5\overline{464}$

33. $0.\overline{2891}$

34. $2.34\overline{5678}$

Determine whether each fraction's decimal expansion terminates or repeats.

35. $\frac{3}{20}$

36. $\frac{3}{51}$

37. $\frac{57}{120}$

38. $\frac{173}{310}$

39. $\frac{41356}{64000}$

40. $\frac{455}{250}$

41. Compute $0.\overline{4}-0.444$.

42. Find a rational number between $\frac{17}{19}$ and $\frac{18}{19}$.

43. Find a rational number between $\frac{37}{15}$ and $\frac{38}{17}$.

44. Find a rational number between 0.23 and 0.24.

45. Find THREE rational numbers between 5.71 and 5.72.

Recall that a set S is **closed** with respect to an operation provided every time the operation is applied to any two elements of the set S the result (or answer) is also a member of the set S.

46. Verify that $\frac{12}{19} \cdot \frac{9}{5}$ is a rational number.

47. Verify that $\frac{5}{8} + \frac{3}{7}$ is a rational number.

48. Show that $\mathbb{Q}$ is closed under addition.

49. Show that $\mathbb{Q}$ is closed under multiplication.

50. If **subtraction** on $\mathbb{Q}$ is defined as $\frac{a}{b} - \frac{c}{d} = \frac{ad - bc}{bd}$, show that $\mathbb{Q}$ is closed under subtraction.

51. If **division** on $\mathbb{Q}$ is defined as $\frac{a}{b} \div \frac{c}{d} = \frac{ad}{bc}$, show that $\mathbb{Q} - \{0\}$ is closed under division.

Given the definitions of addition and multiplication on $\mathbb{Q}$, verify each of the following. Assume a, b, c, d, e, and f are integers, assume denominators are not zero, and use any of the properties we've discussed for operations on *integers.*

52. $\frac{a}{b} + \frac{c}{d} = \frac{c}{d} + \frac{a}{b}$

53. $\frac{a}{b} \cdot \frac{c}{d} = \frac{c}{d} \cdot \frac{a}{b}$

54. $\left(\frac{a}{b} + \frac{c}{d}\right) + \frac{e}{f} = \frac{a}{b} + \left(\frac{c}{d} + \frac{e}{f}\right)$

55. $\left(\frac{a}{b} \cdot \frac{c}{d}\right) \cdot \frac{e}{f} = \frac{a}{b} \cdot \left(\frac{c}{d} \cdot \frac{e}{f}\right)$

56. $\frac{0}{1} + \frac{a}{b} = \frac{a}{b}$

57. $\frac{1}{1} \cdot \frac{a}{b} = \frac{a}{b}$

58. $\frac{a}{b} + \frac{-a}{b} = \frac{0}{1}$ (Hint: Assume $\frac{0}{x} = \frac{0}{1}$ when $x \neq 0$. This will be proven later.)

59. $\frac{a}{b} \cdot \frac{b}{a} = \frac{1}{1}$ (Hint: Assume $\frac{x}{x} = \frac{1}{1}$ whenever $x \neq 0$. This will be proven later.)

60. What do exercises 46 through 59 prove about the set $\mathbb{Q}$?

3.6 Irrational Numbers

The set J of **irrational numbers** is defined by $J = \mathbb{R} - \mathbb{Q}$, or, equivalently, as the set of real numbers whose decimal expansion neither terminates nor repeats.

Example 3.6.1. The numbers π, $\sqrt{2}$, e, and $0.12122122212222\ldots$ are irrational numbers.

□

That $0.12122122212222\ldots$ is irrational is apparent provided one recognizes that the pattern of an increasing string of the digit "2" followed by the single digit "1" has no repetend and is not terminating. The proofs that π (the ratio of the circumference of any circle to its diameter) and e (the base for the natural logarithm) are irrational are quite beyond our scope at present. However, the proof that $\sqrt{2}$ is irrational is possible at this stage.

In order to prove that $\sqrt{2}$ is irrational, we'll employ a proof technique known as **proof by contradiction**. In proof by contradiction, we assume the theorem we wish to prove is *false* and logically arrive at a contradiction (something that cannot possibly be true given our hypotheses). Since the only questionable "fact" is the original assumption, and all of our steps are logically valid, we conclude that our assumption – that the theorem is false – is incorrect, and the theorem is therefore *true*.

Theorem 5. *$\sqrt{2}$ is irrational.*

Proof. We will assume that $\sqrt{2}$ is *rational*, and find that something impossible follows.

Suppose $\sqrt{2} = \frac{a}{b}$, where $a, b \in \mathbb{Z}$, and $\frac{a}{b}$ is in lowest terms. That means that a and b have no common factors; in particular, a and b cannot both be even.

$$\begin{aligned} \sqrt{2} &= \frac{a}{b} && \text{Given} \\ \sqrt{2}b &= a && \text{Multiply both sides by } b \\ \left(\sqrt{2}b\right)^2 &= a^2 && \text{Square both sides} \\ 2b^2 &= a^2 && \text{Simplification} \end{aligned}$$

Thus $a^2 = 2b^2$. Since a^2 is 2 times another integer, a^2 is even. This can only happen if a itself is even, so a is 2 times some integer, and we may write $a = 2k$ for some $k \in \mathbb{Z}$. Now compute some more:

$$\begin{aligned} a^2 &= 2b^2 && \text{Continuation from above} \\ (2k)^2 &= 2b^2 && \text{Substitution} \\ 4k^2 &= 2b^2 && \text{Simplification} \\ 2k^2 &= b^2 && \text{Divide both sides by 2} \end{aligned}$$

We see that b^2 must also be even, forcing b to be even as well. This means that we have a and b both even, contradicting the fact that a and b have no common factors! Since this is impossible, we must have made an error somewhere. However, we can check that all of our steps were valid, so the error must be in the original assumption that $\sqrt{2}$ is rational; therefore, $\sqrt{2}$ is irrational. □

Corollary 3.6.1. *$\sqrt{2} + 5$ is irrational.*

Proof. We will again argue by contradiction. If $\sqrt{2}+5$ is rational, then, since $\mathbb{Q}$ is closed under subtraction, $(\sqrt{2}+5)-5=\sqrt{2}$ is also rational. But this is false by the preceding theorem! Again, we must have made an error, and again, the only place we could have made an error is in our initial assumption that $\sqrt{2}+5$ is rational. Therefore, $\sqrt{2}+5$ is irrational. □

The grid below summarizes our findings for the subsets of real numbers we have discussed. Note that it does not make sense to talk about additive inverses if there is no additive identity.

Set	Clos.	Assoc.	Comm.	Dist.	Add. Id.	Mult. Id.	Add. Inv.	Mult. Inv.	Field?
$\mathbb{N}$	$+,\cdot$	$+,\cdot$	$+,\cdot$	Yes	No	1	NA	No	No
W	$+,\cdot$	$+,\cdot$	$+,\cdot$	Yes	0	1	No	No	No
$\mathbb{Z}$	$+,\cdot$	$+,\cdot$	$+,\cdot$	Yes	0	1	Yes	No	No
$\mathbb{Q}$	$+,\cdot$	$+,\cdot$	$+,\cdot$	Yes	0	1	Yes	Yes	Yes
J	No	NA	NA	Yes	No	No	NA	NA	No
$\mathbb{R}$	$+,\cdot$	$+,\cdot$	$+,\cdot$	Yes	0	1	Yes	Yes	Yes

Section 3.6 Exercises

Identify each number as rational or irrational.

1. $\dfrac{17}{31}$

2. $.121221222$

3. $(\sqrt{2}+13)-(1+\sqrt{2})$

4. $\dfrac{\sqrt{2}+1}{\sqrt{2}}$

5. Show that J is not closed under multiplication by finding two irrational numbers whose product is rational.

6. Show that J is not closed under addition by finding two irrational numbers whose sum is rational.

7. Prove that $\sqrt{3}$ is irrational.

8. Show that $-\sqrt{2}$ is irrational. [Hint: think about the proof of the corollary.]

9. Explain why $0.101001000100001000000\ldots$ is irrational.

10. Given that π is irrational, show that $\pi-\dfrac{2}{5}$ is irrational.

Use proof by contradiction to prove each of the following.

11. If $\sqrt{2}\notin\mathbb{Q}$, then $\sqrt{2}+3\notin\mathbb{Q}$.

12. If $\sqrt{2} \notin \mathbb{Q}$, then $3\sqrt{2} \notin \mathbb{Q}$.

13. If $a \notin \mathbb{Q}$ and $b \in \mathbb{Q}$, then $a+b \notin \mathbb{Q}$.

14. If $a \notin \mathbb{Q}$ and $b \in \mathbb{Q}$ with $b \neq 0$, then $ab \notin \mathbb{Q}$.

15. Write a short paragraph explaining (in your own words!) the idea behind a proof by contradiction.

Chapter 4

Theorems and Proofs

This unit provides basic understanding of the field properties, the equality axioms and their consequences. Students will begin to acquire analytical reasoning skills and focus on fundamental understanding of algebraic structure and concepts. In particular, we derive all the usual rules of algebra using only judicious application of the field properties and the equality axioms. We also introduce the principle of mathematical induction as a proof technique and use it to obtain the usual laws of exponents.

Objectives

- To present the rules of algebra as natural consequences of the field properties and equality axioms
- To introduce the notion of a deductive logical system
- To introduce the principle of mathematical induction
- To convey some basic ideas and methods universal to mathematics

Terms

- field properties
  - closure
  - associative law(s)
  - commutative law(s)
  - identity element(s)
  - inverse element(s)
  - distributive law
- equality axioms
  - reflexive
  - symmetric
  - transitive
  - substitution
- Principle of Mathematical Induction

4.1 The Field Properties

Let S denote any set of "numbers" with binary operations addition and multiplication. The following properties may or may not hold for the set S.

Closure

- *Additive Closure:* For all a and b in S, $a+b$ is also in S.
- *Multiplicative Closure:* For all a and b in S, $a \cdot b$ is also in S.

Associative Laws

- *Associative Law of Addition:* $(a+b)+c = a+(b+c)$ for all a,b,c in S.
- *Associative Law of Multiplication:* $(a \cdot b) \cdot c = a \cdot (b \cdot c)$ for all a,b,c in S.

Commutative Laws

- *Commutative Law of Addition:* $a+b = b+a$ for all a,b in S.
- *Commutative Law of Multiplication:* $a \cdot b = b \cdot a$ for all a,b in S.

Existence of Identities

- *Additive Identity Property:* There is a number, 0, in S such that $a+0 = a$ and $0+a = a$ for all a in S. A number 0 with this property is called an **additive identity.**
- *Multiplicative Identity Property:* There is a number, 1, in S such that $a \cdot 1 = a$ and $1 \cdot a = a$ for all a in S. A number 1 with this property is called a **multiplicative identity.**

Existence of Inverses

- *Additive Inverse Property:* For each a in S, there is a number b in S such that $a+b = 0$ and $b+a = 0$. This number b is called an **additive inverse of** a and is denoted by $-a$. Notice that $b = -a$ and $a = -b$.
- *Multiplicative Inverse Property:* For each a in S with $a \neq 0$, there is a number b in S such that $a \cdot b = 1$ and $b \cdot a = 1$. This number b is called a **multiplicative inverse of** a and is denoted by a^{-1}. Notice that $b = a^{-1}$ and $a = b^{-1}$.

Distributive Law

- *Distributive Law of Multiplication over Addition:* For all elements a,b,c in S, $a \cdot (b+c) = a \cdot b + a \cdot c$.

The above properties are called the **field properties** or **field axioms**, and sets with two binary operations satisfying all of these properties are known as **fields**.

Recall the following definitions. Let $a,b \in S$ where S is any set in which addition and multiplication are defined. Then we define **subtraction** and **division** as follows:

$$a - b \equiv a + (-b) \text{ provided } -b \text{ exists, and}$$
$$a \div b \equiv a \cdot b^{-1} \text{ provided } b^{-1} \text{ exists.}$$

We have an understanding of rational numbers from our experience of the world. In algebra, we examine the properties displayed by the rational numbers (the Commutative Law, for example), and then try to generalize them. This is what led us to the Field Axioms stated previously. From these, we can deduce all of the familiar "rules" of algebra; we will see that they are neither arbitrary nor unreasonable. Every one of them follows from the axioms, which were very reasonably modeled after the rational numbers. Below are two computations that we are taught to perform in high school algebra courses; here, we see that these computations are nothing more than *direct consequences* of the field axioms.

Example 4.1.1. Show that $(x+1)6+(x+1)y = (y+6)(x+1)$.

$$\begin{aligned}
(x+1)6+(x+1)y &= (x+1)(6+y) && \text{Distributive Law of Multiplication over Addition} \\
&= (x+1)(y+6) && \text{Commutative Law of Addition} \\
&= (y+6)(x+1) && \text{Commutative Law of Multiplication}
\end{aligned}$$

□

Example 4.1.2. Show that $(x+2)(x+3) = x^2+5x+6$. (Note: x^2 means $x \cdot x$.)

$$\begin{aligned}
(x+2)(x+3) &= [x(x+3)]+[2(x+3)] && \text{Distributive Law of Multiplication over Addition} \\
&= [x \cdot x + x \cdot 3]+[2 \cdot x + 2 \cdot 3] && \text{Distributive Law of Multiplication over Addition} \\
&= [x^2+3x]+[2x+6] && \text{Commutative Law of Multiplication; Definition of } x^2 \\
&= [(x^2+3x)+2x]+6 && \text{Associative Law of Addition} \\
&= [x^2+(3x+2x)]+6 && \text{Associative Law of Addition} \\
&= [x^2+(3+2)x]+6 && \text{Distributive Law of Multiplication over Addition} \\
&= [x^2+5x]+6 && 3+2=5 \\
&= x^2+5x+6 && \text{Order of Operations}
\end{aligned}$$

□

This second example illustrates *by valid reasoning from the axioms* what is usually passed down as "FOIL"ing in most high school algebra courses (without any justification). The mystery is revealed!

Theorem 6 (Factor Permutation). *For field elements a, b, c, and d,*

$$(ab)(cd) = (ac)(bd) = (ad)(bc).$$

Theorem 7 (Summand Permutation). *For field elements a, b, c, and d,*

$$(a+b)+(c+d) = (a+c)+(b+d) = (a+d)+(b+c).$$

These are just several applications of the Commutative and Associative Laws, as seen in the proof above. In fact, all permutations of $abcd$ are equal, and all permutations of $a+b+c+d$ are equal, and we usually write them without parentheses.

Section 4.1 Exercises

1. Find two examples of a set with an operation that is *not* commutative.
2. Find two examples of a set with an operation that is *not* associative.
3. Show that "squaring" does *not* distribute over addition. That is, find numbers a and b such that $(a+b)^2 \neq a^2+b^2$

Indicate a specific reason why each of the following is an identity.

4. $a \div b = a \cdot b^{-1}$
5. $0+3=3$
6. $\pi + (-\pi) = 0$
7. $(x+y)\cdot(a+b) = (a+b)\cdot(x+y)$
8. $a+b=b+a$
9. $2-5=2+(-5)$
10. $(a+x)(a+x)^{-1}=1$
11. $(x+y)\div(a+b)=(x+y)\cdot(a+b)^{-1}$
12. $(\alpha+\beta)+\gamma = \alpha+(\beta+\gamma)$
13. $\alpha(\delta+\varepsilon) = \alpha\delta+\alpha\varepsilon$
14. Show that $(x-2)(x+2)=x^2-4$ by using the field properties. You will need to use the fact that $x-a=x+(-a)$.
15. Show that $(x+1)(x+2)=x^2+3x+2$ by using the field properties.
16. Show that $x^2+6x+5=(x+1)(x+5)$ by using the field properties.
17. In algebra, we often have a need to "combine like terms." For example, $7x+5x=12x$. What field property allows us to combine like terms? Show that $7x+5x=12x$ using the field properties.
18. Assume that F is a field. Show that for all $a,b,c \in F$, $(b+c)\cdot a = b\cdot a + c\cdot a$.
19. Explain why $0\cdot 1 = 0$.
20. Prove the Factor Permutation Theorem.
21. Prove the Summand Permutation Theorem.

4.2 Equality Axioms

We assume that the **equality relation** on any set S satisfies the following axioms.

1. Equality is **reflexive**: $a = a$ for each $a \in S$.
2. Equality is **symmetric**: If $a = b$, then $b = a$.
3. Equality is **transitive**: If $a = b$ and $b = c$, then $a = c$.
4. Equality satisfies the **substitution axiom**: If $a = b$, then b may be substituted for a and a may be substituted for b in statements concerning a and/or b.

The equality axioms and the field properties may be used to prove theorems concerning sets of "numbers." Of particular importance to us at this time are **real numbers**, but we will state many of the theorems in greater generality, recalling that the set $\mathbb{R}$ of real numbers is a field.

Theorem 8. *For any a, b, and c in a field, if $a = b$, then $a + c = b + c$ and $a \cdot c = b \cdot c$.*

Note: You are to supply a justification wherever there is a blank line in a proof. Look to the field axioms! For many of these proofs, we will use a two-column proof format that may be familiar from geometry.

Proof. Let a, b, c be elements of a field.

Statement	Justification
$a = b$	Hypothesis (Given)
$a+c = a+c$	Equality is Reflexive
$a+c = b+c$	Substitution Axiom

Similarly, we may establish that $a \cdot c = b \cdot c$. □

Theorem 9 (Additive Cancellation). *Let a, b, c be elements of a field. If $a + c = b + c$, then $a = b$.*

Proof. Let a, b, c be elements of a field.

$a+c = b+c$	Given
$(a+c)+(-c) = (b+c)+(-c)$	Existence of additive inverse; Theorem 1 above
$a+[c+(-c)] = b+[c+(-c)]$	____________
$a+0 = b+0$	Additive Inverse; Substitution Axiom $(c+(-c)=0)$
$a = b$	Additive Identity; Substitution Axiom

Therefore, $a = b$ whenever $a + c = b + c$. □

Notice the use of a *previously proved theorem* in the proof above. Also, we will usually use the substitution axiom without explicitly mentioning it. Above, we saw that $c + (-c) = 0$, and substituted 0 for $c + (-c)$ (on both sides of the equation). We will generally just refer to this as "Additive Inverse" instead of "Additive Inverse; Substitution Axiom." Similarly, the last justification will be "Additive Identity." Just remember that there is a substitution going on!

Theorem 10 (Multiplicative Cancellation). *Let a, b and c be elements of a field with $c \neq 0$. If $ac = bc$, then $a = b$.*

Proof. Let a, b, c be elements of a field.

$$
\begin{array}{rcll}
a \cdot c &=& b \cdot c & \text{Given} \\
(a \cdot c) \cdot c^{-1} &=& (b \cdot c) \cdot c^{-1} & \text{Existence of multiplicative inverse of } c \text{ since } c \neq 0; \\
&&& \quad \text{equals may be multiplied by equals} \\
a \cdot [c \cdot c^{-1}] &=& b \cdot [c \cdot c^{-1}] & \underline{\hspace{3cm}} \\
a \cdot 1 &=& b \cdot 1 & \underline{\hspace{3cm}} \\
a &=& b & \underline{\hspace{3cm}}
\end{array}
$$

Therefore, $a = b$ whenever $ac = bc$ and $c \neq 0$. □

Section 4.2 Exercises

Identify which equality axiom is being used in each case.

1. $-4 = -4$
2. If $2 = x - 1$, then $x - 1 = 2$.
3. If $x = 4$, then $x^2 - 7 = 4^2 - 7$
4. If $x = 2y - 1$ and $2y - 1 = 4z$, then $x = 4z$.
5. Argue from the equality axioms that if $a = b$ and $a = c$, then $b = c$.
6. Show that if $2x + 4 = 6 + 4$, then $x = 3$, using *only* the results of this section.
7. Show that if $a = b$ and $c = d$, then $a + c = b + d$. This property of equality is often stated in the form, "Equals may be added to equals."
8. Suppose you know that the sticker price of your new car (before tax) is the same as the sticker price of your neighbor's car (also before tax). What theorem guarantees that you paid the same price (after tax)?
9. Explain the differences between the ideas of *commutativity of addition* and *symmetry of equality*.
10. Prove that if a, b, and c are elements of a field and $a = b$, then $ac = bc$.
11. Show that if $a = b$ and $c = d$, then $a \cdot c = b \cdot d$. This property of equality is often stated in the form, "Equals may be multiplied by equals."

4.3 The Notion of Uniqueness

An important concept in mathematics is that of uniqueness. Many times it is implicitly assumed that an element with a particular property is unique. Here we investigate uniqueness of identities and inverses in a field.

Definition 4.3.1. An element satisfying a certain property is called **unique** if it is the only element satisfying that property.

Theorem 11. *In a field F, the identity element for addition is unique.*

Proof. Suppose there is another additive identity z such that $a = z + a$ for all $a \in F$. Then in particular, $0 = z + 0$. Since 0 also is an additive identity, $z + 0 = z$. Thus $z = 0$ since equality is transitive. (This means that our "other" additive identity is actually 0 also.) □

Theorem 12. *The additive inverse of any field element is unique.*

Proof. Suppose that $a + b = 0$ and $a + c = 0$ for some a, b, and c in the field. That is, suppose that b and c are both additive inverses for a. Then $0 = a + c$ by symmetry of equality, so

$$
\begin{array}{rcll}
a+b & = & a+c & \text{Transitivity of equality} \\
b+(a+b) & = & b+(a+c) & \text{Equals may be added to equals} \\
(b+a)+b & = & (b+a)+c & \text{Associative Law of Addition} \\
0+b & = & 0+c & \text{Additive Inverse } (b+a=0) \\
b & = & c & \text{Additive Identity}
\end{array}
$$

Thus b and c must in fact be the same element. □

The proofs of the following multiplicative analogs are exercises:

Theorem 13. *In a field, the identity element for multiplication is unique.*

Theorem 14. *The multiplicative inverse of any nonzero field element is unique.*

Since identities and inverses are unique, we are justified in referring to "the" additive identity, etc.

The idea of uniqueness is very powerful, as we shall see. If we can show that a given element has a property that we know only one element has, then our given element must be that one!

Section 4.3 Exercises

1. Prove that in a field, the identity element for multiplication is unique.
2. Prove that the multiplicative inverse of any nonzero field element is unique.
3. Suppose that $6x = 1$. What theorem guarantees that $x = 6^{-1}$?
4. Prove the following stronger uniqueness theorem for the additive identity: If $z + x = x$ for some element $x \in F$, then $z = 0$. (This is "stronger" in the sense that we are only assuming that $z + a = a$ for *some* a, instead of for *all* a. That means we cannot choose the a to be 0 as we did above.)
5. If $x + 2 = 0$, what theorem guarantees that $x = -2$?

4.4 Multiplication with Additive Identities and Inverses

Now we will work to prove some of the basic theorems we ordinarily take for granted. Throughout this section, a, b, c, and d are elements of a field F.

Theorem 15 (Zero-Factor Theorem). *For all field elements a, $a \cdot 0 = 0$.*

Proof. Let a be an arbitrary field element. Then

$$\begin{array}{rcll} a \cdot 0 &=& a \cdot (0+0) & \text{Additive Identity; } 0+0=0 \\ a \cdot 0 &=& a \cdot 0 + a \cdot 0 & \text{Distributive Law} \\ -(a \cdot 0) + a \cdot 0 &=& -(a \cdot 0) + (a \cdot 0 + a \cdot 0) & \text{Existence of } -(a \cdot 0)\text{; add } -(a \cdot 0) \text{ to both sides} \\ -(a \cdot 0) + a \cdot 0 &=& [-(a \cdot 0) + a \cdot 0] + a \cdot 0 & \text{Associative Law of Addition} \\ 0 &=& 0 + a \cdot 0 & \text{Additive Inverse} \\ 0 &=& a \cdot 0 & \text{Additive Identity} \\ a \cdot 0 &=& 0 & \text{Symmetry of Equality} \end{array}$$

□

Thus, the familiar "anything times zero is zero" is justified in a field, and this explains why 0 does not have a multiplicative inverse: If a were a multiplicative inverse for 0, then $a \cdot 0 = 1$, but we know from the theorem that $a \cdot 0 = 0$, giving us the ridiculous conclusion that $0 = 1$ by transitivity of equality. Therefore, no element of the field can be a multiplicative inverse for 0, and we see that *division by 0 is never allowed*, since division is really multiplication by the multiplicative inverse.

Example 4.4.1. For which real numbers x does $x+2$ have a multiplicative inverse?

Solution: In order for $\frac{1}{x+2}$ to be defined, we must have $x+2 \neq 0$. Therefore, $x+2$ will have an inverse if and only if $x \in \{x \in \mathbb{R} | x \neq -2\}$.

□

The next theorem is the converse of the previous, and it is very important in solving equations. Notice that the Zero-Factor Theorem says that when zero is a **factor** the product is zero. The Zero-Product Theorem says when zero is the **product** at least one of the factors must be zero.

Theorem 16 (Zero-Product Theorem). *Let a and b be elements of a field. If $ab = 0$, then $a = 0$ or $b = 0$.*

Proof. If $a = 0$, then we already have $a = 0$ or $b = 0$. If $a \neq 0$, then a^{-1} exists. Thus

$$\begin{array}{rcll} ab &=& 0 & \text{Given} \\ a^{-1}(ab) &=& a^{-1} \cdot 0 & \text{Property of equality} \\ (a^{-1}a)b &=& 0 & \text{Associative Law of Multiplication; Theorem 15} \\ 1 \cdot b &=& 0 & \text{Multiplicative Inverse} \\ b &=& 0 & \text{Multiplicative Identity} \end{array}$$

Thus if $a \neq 0$, then $b = 0$. □

Why is this important to equation solving? Consider the following example.

Example 4.4.2. Solve $(x+2)(x-1)=0$ for x.

Solution: We have $a=x+2$ and $b=x-1$, and $ab=0$. From the Zero-Product Theorem, we know that either $a=0$ or $b=0$, so $x+2=0$ or $x-1=0$. If $x+2=0$, then x must be the (unique) additive inverse of 2, so $x=-2$. If $x-1=0$, then $x+(-1)=0$, so x must be the (unique) additive inverse of -1; we get $x=1$. Therefore, either $x=-2$ or $x=1$. Notice that we *could not* have made our first step without the Zero-Product Theorem. Also, we needed uniqueness of additive inverses to get $x=-2$ or $x=1$.

□

Combining the previous two theorems we have the following theorem.

Theorem 17. *A product of two factors is zero if and only if at least one of the factors is zero.*

Theorem 18. *For any field element a, $(-1)a=-a$.*

Proof. We will show that $(-1)a$ is the unique additive inverse for a; therefore $(-1)a=-a$.

$$\begin{array}{rcll} a+(-1)a & = & (1)a+(-1)a & \text{Multiplicative Identity} \\ & = & (1+(-1))a & \text{Distributive Law} \\ & = & (0)a & 1+(-1)=0 \\ & = & 0 & \underline{\hspace{4cm}} \end{array}$$

Thus $a+(-1)a=0$ by transitivity of equality (applied several times). Since $-a$ is the only field element satisfying $a+(-a)=0$, it must be that $-a=(-1)a$. □

In a string of equalities such as the one above, transitivity is applied repeatedly. For example, if we have

$$\begin{array}{rcl} a & = & b \\ & = & c\ , \\ & = & d \end{array}$$

then $a=b$ and $a=c$, so $a=c$ by transitivity. Now $a=c$ and $c=d$, so $a=d$ by transitivity.

Theorem 19. *For field elements a and b, $(-a)b=-(ab)=a(-b)$.*

Proof. We will first give one proof using the above theorem that $(-1)a=-a$.

$$\begin{array}{rcll} (-a)b & = & ((-1)a)b & \underline{\hspace{4cm}} \\ & = & (-1)(ab) & \underline{\hspace{4cm}} \\ & = & -(ab) & \underline{\hspace{4cm}} \end{array}$$

Thus $(-a)b=-(ab)$ by transitivity of equality.

Now $a(-b)=(-b)a$ by the Commutative Law, and $(-b)a=-(ba)$ by the first part of the theorem. Finally, $-(ba)=-(ab)$ by the Commutative Law, so $a(-b)=-(ab)$ by transitivity of equality. □

Proof. An alternate proof may be given using the uniqueness of additive inverses.

$$\begin{array}{rcll} (-a)b+ab & = & ((-a)+a)b & \text{Distributive Law} \\ & = & 0\cdot b & \text{Additive Inverse} \\ & = & 0 & \underline{\hspace{4cm}} \end{array}$$

Thus $(-a)b+ab=0$ by transitivity of equality. Hence, by uniqueness of additive inverses, $(-a)b=-(ab)$. Similarly, $a(-b)=-(ab)$. ☐

Theorem 20. *For field elements a and b, $(-a)(-b)=ab$.*

Proof. This first proof uses two applications of the previous theorem.

$$\begin{array}{rcll} (-a)(-b) & = & -(a(-b)) & \text{Previous theorem } (-x)(y)=-(xy) \\ & = & -(-(ab)) & \text{Previous theorem } (x)(-y)=-(xy) \\ & = & ab & \underline{\hspace{5cm}} \end{array}$$

Thus $(-a)(-b)=ab$ by transitivity of equality. ☐

Proof. The second proof uses the fact that $(-1)a=-a$ (several times).

$$\begin{array}{rcll} (-a)(-b) & = & [(-1)a][(-1)b] & \underline{\hspace{5cm}} \\ & = & ((-1)(-1))(ab) & \underline{\text{Factor Permutation}} \\ & = & (-(-1))(ab) & \underline{\hspace{5cm}} \\ & = & (1)(ab) & \text{Additive Inverse of } -1 \\ & = & ab & \underline{\hspace{5cm}} \end{array}$$

Thus $(-a)(-b)=ab$ by transitivity of equality. ☐

Theorem 21 (Generalized Distributive Law)**.** *Let a,b,c be elements of a field. Then $a(b-c)=ab-ac$, $-(a+b)=-a-b$, and $-(a-b)=-a+b$.*

Proof.

$$\begin{array}{rcll} a(b-c) & = & a[b+(-c)] & \text{Definition of } b-c \\ & = & ab+a(-c) & \underline{\hspace{5cm}} \\ & = & ab+(-(ac)) & \underline{\hspace{5cm}} \\ & = & ab-ac & \underline{\hspace{5cm}} \end{array}$$

Thus $a(b-c)=ab-ac$ by __________________________.

The other parts may be proven either in a similar fashion or by using the uniqueness of additive inverses. These proofs are left as exercises. ☐

Section 4.4 Exercises

Solve each equation. Name the axioms or theorems you use.

1. $(x+1)(x-3)=0$
2. $(2x+3)(2.3x-5)=0$
3. $4(x+2)=0$
4. $(5x-2)(2x+3)(15x-1)=0$
5. $(x-\frac{1}{2})(2x-\frac{2}{3})(5x+\frac{5}{4})=0$
6. Prove that $(a+b)(c+d)=ac+ad+bc+bd$ by using the field axioms. You may use the permutation theorems as needed.
7. Use the previous problem to show that $(x+a)(x+b)=x^2+(a+b)x+ab$. This shows how to factor a quadratic polynomial with leading term x^2 : find factors of the constant term whose sum is the coefficient of x.
8. Prove that $(x-a)(x+a)=x^2-a^2$. (Difference of squares)
9. Prove that $(x+a)(x+a)=x^2+2ax+a^2$. (Perfect square)

Factor each quadratic polynomial.

10. x^2-4
11. $x^2-8x+12$
12. x^2+6x+9
13. x^2-64
14. x^2-3
15. $x^2+2x-15$

16. Show that $(ax+b)(cx+d)=acx^2+(ad+bc)x+bd$. This shows how to factor an arbitrary quadratic polynomial: multiply the constant term and the leading coefficient, and then find factors whose sum is the coefficient of x.

Factor each quadratic.

17. $2x^2-x-3$
18. $2x^2+9x+4$
19. $3x^2-13x+4$
20. $6x^2+11x-10$
21. $12x^2+4x-5$
22. $9x^2+14x-8$

Solve each equation for x.

23. $x^2-3x+2=0$
24. $x^2+6x+6=1$

25. $x^2+2x-15=0$

26. $x^2+x-12=0$

27. $6x^2+11x+4=0$

28. $9x^2+14x=8$

29. $12x^2-5=4x$

30. $6x^2+15=19x$

31. Prove that $-(a-b)=-a+b$.

32. Prove that $-(a+b)=-a-b$.

33. Prove that if $abc=0$, then $a=0$ or $b=0$ or $c=0$.

34. A rock falls from a cliff 80 feet high. Its distance from the ground after t seconds is given by $h(t)=-16t^2+80$. Determine at what time the rock hits the ground. [Hint: what is the height of a rock on the ground?]

35. The product of one more than a whole number and two more than the same number is equal to 2. What is the number?

36. If the dimensions of a warehouse must be such that its length is 10 feet longer than its width and its floor space must consist of 3000 square feet, find its length and width.

4.5 Fractions

Throughout this section we assume that F is a field.

Definition 4.5.1. Let $a, b \in F$, where F is a field. We define the **fraction** $\frac{a}{b}$ by

$$\frac{a}{b} = a \cdot b^{-1},$$

where $b \neq 0$. The element a is the **numerator** of the fraction, and b is the **denominator** of the fraction.

Theorem 22 (Fraction Identities.). *Let a and b be field elements, $b \neq 0$. Then*

$$b\left(\frac{a}{b}\right) = a, \frac{a}{1} = a, \frac{0}{b} = 0, \frac{b}{b} = 1, \text{ and } \frac{1}{b} = b^{-1}.$$

Proof. Since $b \neq 0$, b^{-1} exists.

$b\left(\frac{a}{b}\right)$	$=$	$b(ab^{-1})$	Definition of $\frac{a}{b}$
	$=$	$b(b^{-1}a)$	____________
	$=$	__________	Associative Law
	$=$	$1 \cdot a$	Multiplicative Inverse
	$=$	__________	____________

Thus $b\left(\frac{a}{b}\right) = a$ by transitivity of equality.

Also, $1 = \frac{1}{1}$, since $1 \cdot 1^{-1} = 1$. Thus $\frac{a}{1} = a \cdot \left(\frac{1}{1}\right) = a \cdot 1 = a$. The proofs of the last three parts are left as exercises. □

Theorem 23 (Fraction Multiplication). *For field elements $a, b \neq 0, c, d \neq 0$, $\frac{a}{b}\frac{c}{d} = \frac{ac}{bd}$.*

Proof.

$\frac{ac}{bd}$	$=$	$ac(bd)^{-1}$	Definition of fraction
	$=$	$(ac)\frac{1}{bd}$	____________
	$=$	$(ac)\frac{1}{bd} \cdot 1 \cdot 1$	Multiplicative Identity (twice)
	$=$	$(ac)\frac{1}{bd}(bb^{-1})(dd^{-1})$	Substitution: $bb^{-1} = 1, dd^{-1} = 1$
	$=$	$(ac)\frac{1}{bd}(bd)(b^{-1}d^{-1})$	Factor Permutation
	$=$	$(ac) \cdot 1 \cdot (b^{-1}d^{-1})$	Multiplicative Inverse
	$=$	$(ac)(b^{-1}d^{-1})$	Multiplicative Identity
	$=$	$(ab^{-1})(cd^{-1})$	Factor Permutation
	$=$	$\frac{a}{b}\frac{c}{d}$	Definition

Thus $\frac{ac}{bd} = \frac{a}{b}\frac{c}{d}$ by transitivity of equality, so $\frac{a}{b}\frac{c}{d} = \frac{ac}{bd}$ by symmetry of equality. □

Notice that the fraction bar acts as a grouping symbol: $\frac{ac}{bd}$ means $(ac)(bd)^{-1}$. Similarly, $\frac{a+b}{c}$ means $(a+b)\cdot c^{-1}$; the numerator is always thought of as being grouped, and the denominator is always thought of as being grouped. Furthermore, the fraction itself is all grouped together. Thus, the symbol $\frac{a}{b}$ should be thought of as $\left(\frac{(a)}{(b)}\right)$.

Theorem 24 (Fraction Simplification). *For field elements a,b,c, where $b,c \neq 0$, $\frac{ac}{bc} = \frac{a}{b}$.*

Proof.

$\frac{ac}{bc}$	$= \frac{a}{b}\frac{c}{c}$	Fraction Multiplication
	$= \frac{a}{b}(c \cdot c^{-1})$	____________________
	$= \frac{a}{b} \cdot 1$	Multiplicative Inverse
	$=$ ____________	Multiplicative Identity

The conclusion follows by the transitive property of equality. Similarly, $\frac{ca}{cb} = \frac{a}{b}$. □

Example 4.5.1. Simplify $\frac{x^2-4}{x^2-3x+2} \cdot \frac{x^2-1}{x^2+4x+4}$.

Solution: In order to use the fraction simplification theorem, we must first factor the numerators and denominators. Notice that this is the *only* method we have right now to simplify fractions. **If you haven't factored, you aren't ready to simplify.**

$$\begin{aligned}
\frac{x^2-4}{x^2-3x+2} \cdot \frac{x^2-1}{x^2+4x+4} &= \frac{(x-2)(x+2)}{(x-2)(x-1)} \cdot \frac{(x-1)(x+1)}{(x+2)(x+2)} \\
&= \frac{x+2}{x-1} \cdot \frac{(x-1)(x+1)}{(x+2)(x+2)} \\
&= \frac{(x+2)(x-1)(x+1)}{(x-1)(x+2)(x+2)} \\
&= \frac{x+1}{x+2}.
\end{aligned}$$

Note that we added parentheses as needed to keep the numerators and denominators grouped appropriately. We also used the fraction simplification theorem several times, as well as the factor permutation theorem.

□

Theorem 25 (Fraction Equivalence). $\frac{a}{b} = \frac{c}{d}$ *if and only if $ad = bc$, where $b,d \neq 0$.*

Proof. Remember, "p if and only if q" means "if p, then q" *and* "if q, then p." We must prove both statements.

First suppose that $ad = bc$ is true. Since $b, d \neq 0$, b^{-1} and d^{-1} exist.

$$\begin{array}{rcll} ad &=& bc & \text{Given} \\ (ad)d^{-1} &=& (bc)d^{-1} & \text{Property of equality} \\ a(dd^{-1}) &=& b(cd^{-1}) & \text{Associative Law} \\ a \cdot 1 &=& (cd^{-1})b & \text{Multiplicative Inverse; Commutative Law} \\ a &=& (cd^{-1})b & \text{Multiplicative Identity} \\ ab^{-1} &=& (cd^{-1})bb^{-1} & \text{Property of equality} \\ \frac{a}{b} &=& \frac{c}{d} \cdot 1 & \text{Definition; Multiplicative Inverse} \\ \frac{a}{b} &=& \frac{c}{d} & \text{Multiplicative Identity} \end{array}$$

Thus, if $ad = bc$ is true (with $b, d \neq 0$), then $\frac{a}{b} = \frac{c}{d}$ must be true.

Now suppose that $\frac{a}{b} = \frac{c}{d}$.

$$\begin{array}{rcll} \frac{a}{b} &=& \frac{c}{d} & \text{Given} \\ ab^{-1} &=& cd^{-1} & \text{Definition} \\ ab^{-1}bd &=& cd^{-1}bd & \text{Property of equality; Associative Law} \\ a \cdot 1 \cdot d &=& cbd^{-1}d & \text{Multiplicative Inverse; Factor Permutation} \\ ad &=& cb \cdot 1 & \text{Multiplicative Identity; Multiplicative Inverse} \\ ad &=& bc & \text{Multiplicative Identity; Commutative Law} \end{array}$$

Thus, if $\frac{a}{b} = \frac{c}{d}$, then $ad = bc$. Note the use of the Associative Law – since the factors can be grouped in any way, we omitted the parentheses. □

Now we concern ourselves with addition of fractions within a field, beginning with fractions having a common denominator.

Theorem 26 (Fraction Addition I)**.** *For field elements a, b, c, if $c \neq 0$, then* $\frac{a+b}{c} = \frac{a}{c} + \frac{b}{c}$.

Proof.

$$\begin{array}{rcll} \frac{a+b}{c} &=& (a+b)c^{-1} & \text{Definition – note } \textbf{grouping} \\ &=& ac^{-1} + bc^{-1} & \underline{\hspace{6cm}} \\ &=& \frac{a}{c} + \frac{b}{c} & \underline{\hspace{6cm}}. \end{array}$$

The conclusion follows by the transitive property of equality. □

The following theorem shows that in any field, addition of fractions mimics addition of rational numbers.

Theorem 27 (Fraction Addition II)**.** *For field elements a, b, c, d, with $b, d \neq 0$,*
$\frac{a}{b} + \frac{c}{d} = \frac{ad+bc}{bd}$.

Proof.

$$\begin{aligned}
\frac{a}{b}+\frac{c}{d} &= \frac{a}{b}\cdot 1+1\cdot\frac{c}{d} && \text{Multiplicative Identity (twice)}\\
&= \frac{a}{b}\frac{d}{d}+\frac{b}{b}\frac{c}{d} && \text{Theorem: } \frac{x}{x}=1 \text{ if } x\neq 0\\
&= \frac{ad}{bd}+\frac{bc}{bd} && \text{Multiplication of Fractions}\\
&= \frac{ad+bc}{bd} && \text{Fraction Addition I}
\end{aligned}$$

The conclusion follows by the transitive property of equality. □

It is quite likely that bd is not the "least common denominator" of the two given fractions; the theorem above says that one need not find the least common denominator in order to add fractions. It is often convenient to do so (as it can save simplification later), but it is by no means necessary.

Example 4.5.2. Add $x+2$ and $\frac{3}{x}$.

Solution 1:

$$\begin{aligned}
(x+2)+\frac{3}{x} &= \frac{x+2}{1}+\frac{3}{x} && \frac{a}{1}=a\\
&= \frac{x+2}{1}\cdot 1+\frac{3}{x} && \text{Multiplicative Identity}\\
&= \frac{x+2}{1}\frac{x}{x}+\frac{3}{x} && \frac{x}{x}=1\\
&= \frac{(x+2)x}{1\cdot x}+\frac{3}{x} && \text{Fraction Multiplication}\\
&= \frac{x^2+2x}{x}+\frac{3}{x} && \text{Distributive Law}\\
&= \frac{x^2+2x+3}{x} && \text{Fraction Addition I}
\end{aligned}$$

Solution 2:

$$\begin{aligned}
(x+2)+\frac{3}{x} &= \frac{x+2}{1}+\frac{3}{x} && \text{Previous Theorem: } \frac{a}{1}=a\\
&= \frac{(x+2)x+1\cdot 3}{1\cdot x} && \text{Fraction Addition II}\\
&= \frac{x^2+2x+3}{x} && \text{Distributive Law, Multiplicative Identity}
\end{aligned}$$

□

Example 4.5.3. Compute $\dfrac{2x+1}{x-1} - \dfrac{x-4}{(x-1)(x+1)}$.

Solution:

$$
\begin{aligned}
\frac{2x+1}{x-1} - \frac{x-4}{(x-1)(x+1)} \\
&= \frac{2x+1}{x-1} + \left(-\frac{x-4}{(x-1)(x+1)}\right) && \rule{6cm}{0.4pt} \\
&= \frac{2x+1}{x-1} + \frac{-(x-4)}{(x-1)(x+1)} && \rule{6cm}{0.4pt} \\
&= \frac{2x+1}{x-1} + \frac{-x+4}{(x-1)(x+1)} && \rule{6cm}{0.4pt} \\
&= \frac{(2x+1)(x-1)(x+1)+(-x+4)(x-1)}{(x-1)(x-1)(x+1)} && \rule{6cm}{0.4pt} \\
&= \frac{(x-1)[(2x+1)(x+1)+(-x+4)]}{(x-1)(x-1)(x+1)} && \rule{6cm}{0.4pt} \\
&= \frac{2x^2+3x+1+(-x)+4}{(x-1)(x+1)} && \rule{6cm}{0.4pt} \\
&= \frac{2x^2+2x+5}{(x-1)(x+1)} && \rule{6cm}{0.4pt}
\end{aligned}
$$

□

Theorem 28 (Fraction Division). *For field elements a, b, c, d, with $b, c, d \neq 0$,* $\dfrac{a}{b} \div \dfrac{c}{d} = \dfrac{a}{b}\dfrac{d}{c}$.

Proof. Note that we need $c \neq 0$ in order for division by $\dfrac{c}{d}$ to be defined. Also, notice that $\dfrac{c}{d}\dfrac{d}{c} = 1$, so that $\dfrac{d}{c} = \left(\dfrac{c}{d}\right)^{-1}$.

$$
\begin{aligned}
\frac{a}{b} \div \frac{c}{d} &= \frac{a}{b} \cdot \left(\frac{c}{d}\right)^{-1} && \text{Definition of Division} \\
&= \frac{a}{b}\frac{d}{c} && \text{Observation above}
\end{aligned}
$$

The conclusion follows by the transitive property of equality. □

Section 4.5 Exercises

1. Prove that if $b \neq 0$, then $\frac{1}{b} = b^{-1}$.

2. Prove that if $b \neq 0$, then $\frac{0}{b} = 0$.

3. Prove that $\frac{b}{b} = 1$ for any $b \neq 0$.

4. Prove that $\left(\frac{a}{b}\right)^{-1} = \frac{b}{a}$ if $a \neq 0$. (Be sure to explain why we need the condition $a \neq 0$.)

5. Prove that $x - (a - b) = x - a + b$.

6. Prove that $x - (a + b) = x - a - b$.

7. Prove that $-\frac{1}{b} = \frac{1}{-b}$.

8. Prove that $-\frac{a}{b} = \frac{-a}{b} = \frac{a}{-b}$.

9. Prove that if $\frac{a}{b} = 0$ (and $b \neq 0$), then $a = 0$.

Compute and simplify. Cite the theorems or axioms you use.

10. $\frac{x-3}{3x-2} + \frac{2x+5}{x+2}$

11. $\frac{3}{x} + \frac{2x+3}{5}$

12. $\frac{x-2}{2-x}$

13. $x - \frac{3-x}{2}$

14. $\frac{2x+1}{3x^2-7x-6} + \frac{x+4}{2x^2-x-15}$

15. $\frac{x^2-9x+20}{x-6} \cdot \frac{x-3}{x^2-7x+12}$

16. $\frac{4x+3}{x^2-1} \cdot \frac{x+1}{4x^2-3x-1}$

17. $\frac{4x+3}{2x^2-2x-1} - \frac{x-1}{16x^2-1}$

18. $\frac{x^2-7x-8}{x^2+6x+8} \div \frac{x^2-1}{x^2+x-2}$

19. $\frac{4x^2+20x+25}{x^2+7x+10} \div \frac{4x^2+4x-15}{x^2+10x+25}$

20. $\frac{\left(\frac{x}{4}+\frac{4}{x}\right)}{\left(\frac{4}{x}-\frac{x}{4}\right)}$

21. $\frac{\left(\frac{x}{x-1}+\frac{x-1}{x+2}\right)}{\left(\frac{x-1}{x+2}+\frac{x+1}{x-1}\right)}$

22. $\frac{\left(\frac{x+3}{x-2}+\frac{x-1}{x+2}\right)}{\left(\frac{x+3}{x+2}+\frac{x+2}{x-2}\right)}$

23. Show that $\frac{x^2+1}{x^2+2} \neq \frac{1}{2}$. [Hint: find a value for x that makes the two expressions different.]

Solve for x: [This is an excellent place to remember that fractions have a lot of hidden grouping symbols!]

24. $\dfrac{x}{2}+\dfrac{5x}{7}=0$

25. $\dfrac{x-2}{3x+5}=0$

26. $\dfrac{x-2}{x+1}+3=5$

27. $\dfrac{x+2}{x}-\dfrac{x-1}{x+3}=0$

28. $\dfrac{\frac{x+3}{x-2}+\frac{x-1}{x+2}}{\frac{x+3}{x+2}+\frac{x+2}{x-2}}=0$

4.6 Exponents and Induction

We are led by our experiences to infer certain conclusions from long observation. We expect that a few hours after a meal, we get hungry; we expect that the news will come on a 6:00; we expect that when we flip the light switch, the light will come on. These are all examples of inductive reasoning. However, no matter how many times these things happen, the next occurrence *is never guaranteed*: we may become ill and lose our appetites; the news may be preempted by a basketball game; the light bulb may be burned out.

In mathematics, the case is a little cleaner. We have a method for guaranteeing that "the next occurrence" *will* happen: it is the **principle of mathematical induction.**

Definition 4.6.1. Principle of Mathematical Induction: Let $S_1, S_2, S_3, \ldots$ represent statements, one statement for each natural number. Suppose that

1. S_1 is known to be true (the **base case**), and

2. if S_k is known to be true for some $k \in \mathbb{N}$, then S_{k+1} is also true (the **induction step**).

Then for every natural number n, S_n is true.

The statement "S_k is true for some $k \in \mathbb{N}$" is known as the **induction hypothesis**.

This is a fairly reasonable principle. We are told that S_1 is true (this is the "some value of k"), and that if S_1 is true, so is S_2; thus, S_2 is true. Also, if S_2 is true, then so is S_3; thus, S_3 is true. If S_3 is true, then so is S_4, and so on. This is what we meant above by guaranteeing the "next" occurrence.

We will illustrate the use of this principle in the following definition and theorems, and we will see throughout the text that induction is a useful tool.

Definition 4.6.2. Let a be a field element, and $n \in \mathbb{N}$. Define

$$a^1 = a, a^2 = a^1 \cdot a, a^3 = a^2 \cdot a.$$

In general, if a^n has already been defined, then

$$a^{n+1} = a^n \cdot a.$$

In an expression of the form a^n, a is called the **base**, and n is called the **exponent** or **power**.

The symbol a^n is pronounced, "a to the n" or "a to the nth power." Intuitively,

$$a^n = \underbrace{a \cdot a \cdot a \cdots a}_{n \text{ factors}}.$$

Theorem 29. *For any* $n \in \mathbb{N}$, $1^n = 1$.

Proof. We have the statements $S_1 : 1^1 = 1, S_2 : 1^2 = 1, S_3 : 1^3 = 1$, and so on. We already know that S_1 is true, so we assume that for some natural number k, S_k is true. This means that $1^k = 1$ for that particular natural number k.

Now $1^{k+1} = 1^k \cdot 1$ by definition. Since 1 is the multiplicative identity, $1^k \cdot 1 = 1^k$.

Our induction hypothesis now guarantees us that $1^k = 1$, so by transitivity of equality, $1^{k+1} = 1$ also. Therefore, by the principle of mathematical induction, we have that $1^n = 1$ for all $n \in \mathbb{N}$. □

Notice how the proof worked. We used the *known* result that $1^k = 1$, and achieved the desired result by multiplying both sides of the equation by 1.

Theorem 30 (Multiplication of Like Bases). *For natural numbers m and n and for any field element a,* $a^m \cdot a^n = a^{m+n}$.

Proof. Let a be an element of a field, and let $m \in \mathbb{N}$. The statements are as follows:

$$\begin{array}{ll} S_1: & a^m \cdot a^1 = a^{m+1} \\ S_2: & a^m \cdot a^2 = a^{m+2} \\ S_3: & a^m \cdot a^3 = a^{m+3} \\ \vdots & \quad \vdots \\ S_n: & a^m \cdot a^n = a^{m+n} \\ \vdots & \quad \vdots \end{array}$$

In accordance with the principle of mathematical induction, we first establish that S_1 is true:

$$\begin{array}{rcll} a^m \cdot a^1 & = & a^m \cdot a & \text{Definition of } a^1 \\ & = & a^{m+1} & \text{Definition of } a^{m+1}. \end{array}$$

Therefore, S_1 is true. Next, suppose that we know that S_k is true for some $k \in \mathbb{N}$. This means that $a^m \cdot a^k = a^{m+k}$ *for that particular k*. Our job is to show that S_{k+1} is also true; that is, we must show that $a^m \cdot a^{k+1} = a^{m+k+1}$.

$$\begin{array}{rcll} a^m \cdot a^k & = & a^{m+k} & \text{Induction hypothesis} \\ (a^m \cdot a^k) \cdot a^1 & = & a^{m+k} \cdot a^1 & \text{Property of equality} \\ a^m \cdot (a^k \cdot a^1) & = & a^{(m+k)+1} & \text{Associativity; Base Case} \\ a^m \cdot a^{k+1} & = & a^{m+k+1} & \text{Base Case} \end{array}.$$

This shows that if S_k is true, then so is S_{k+1}; therefore, by the principle of mathematical induction, S_n is true for all $n \in \mathbb{N}$. That is, $a^m \cdot a^n = a^{m+n}$ for all $n \in \mathbb{N}$. ☐

We also offer the following intuitive argument:

Proof.

$$\begin{array}{rcll} a^n \cdot a^m & = & (\underbrace{a \cdot a \cdot a \cdots a}_{n \text{ factors}})(\underbrace{a \cdot a \cdot a \cdots a}_{m \text{ factors}}) & \text{Definition} \\ & = & \underbrace{a \cdot a \cdot a \cdots a}_{n+m \text{ factors}} & \text{Associative Law} \\ & = & a^{n+m} & \text{Definition} \end{array}$$

☐

Example 4.6.1. We have that $x^5 \cdot x^7 = x^{5+7} = x^{12}$.

☐

Theorem 31 (Raising a Power to a Power). *For natural numbers m and n and for any field element a,* $(a^n)^m = a^{nm}$.

Proof. We omit the induction argument for this theorem, leaving it as an exercise. Instead, we have the following intuitive argument:

Let a be an element of a field, and let m and n be natural numbers. Then

$$\begin{aligned}
(a^n)^m &= \underbrace{a^n \cdot a^n \cdot a^n \cdots a^n}_{m \text{ factors}} && \text{Definition} \\
&= \underbrace{\underbrace{(a \cdot a \cdot a \cdots a)}_{n \text{ factors}}\underbrace{(a \cdot a \cdot a \cdots a)}_{n \text{ factors}} \cdots \underbrace{(a \cdot a \cdot a \cdots a)}_{n \text{ factors}}}_{m \text{ factors of } n \text{ factors each}} && \text{Definition} \\
&= \underbrace{a \cdot a \cdot a \cdots a}_{nm \text{ factors}} && \text{Multiplication of natural numbers} \\
&= a^{nm} && \text{Definition}
\end{aligned}$$

□

Example 4.6.2. We have that $(w^5)^3 = w^{5 \cdot 3} = w^{15}$.

□

Theorem 32 (Power of a Product)**.** *Let a and b be elements of a field. For any natural number n, $(ab)^n = a^n b^n$.*

Proof. (Using Mathematical Induction) Let S_n represent the statement $(ab)^n = a^n b^n$. Then S_1 is true, since $(ab)^1 = ab$ (1 factor of ab). We must show that **if** S_k is true for some $k \in \mathbb{N}$, then so is S_{k+1}. Assume S_k is true: $(ab)^k = a^k b^k$. Compute $(ab)^{k+1}$:

$$\begin{aligned}
(ab)^{k+1} &= (ab)^k(ab) && \text{Definition of } x^{k+1} \\
&= (a^k b^k)(ab) && \text{Induction hypothesis} \\
&= (a^k \cdot a)(b^k \cdot b) && \text{Factor Permutation} \\
&= a^{k+1} b^{k+1} && \text{Definition of } x^{k+1}
\end{aligned}$$

Therefore, $(ab)^{k+1} = a^{k+1} b^{k+1}$ by transitivity of equality. We now know that S_1 is true, and that if S_k is true for some value of k, then so is S_{k+1}. By the principle of mathematical induction, S_n is true for every natural number n. □

Example 4.6.3. Application of the Power of a Product Theorem yields $(2x)^5 = 2^5 x^5 = 32x^5$.

□

Definition 4.6.3. If a is a nonzero field element and $n \in \mathbb{N}$, then $a^{-n} = (a^{-1})^n$, and $a^0 = 1$.

Note that this is consistent with the use of -1 as an exponent to mean multiplicative inverse, and this permits us to extend our idea of exponents to negative integers.

Theorem 33. *If $a \neq 0$ and $n \in \mathbb{N}$, then $(a^{-1})^n = (a^n)^{-1}$.*

This says, "A power of an inverse is the inverse of the power."

Proof. We will again use mathematical induction. Let S_n represent the statement that $(a^{-1})^n = (a^n)^{-1}$. Then S_1 says that $(a^{-1})^1 = (a^1)^{-1}$, which is true since both sides are just a^{-1}.

Now assume that S_k is true for some $k \in \mathbb{N}$; that is, assume that for some $k \in \mathbb{N}$, $(a^{-1})^k = (a^k)^{-1}$. Then

$$\begin{array}{rcll} a^{k+1}(a^{-1})^{k+1} & = & (a^k \cdot a)((a^{-1})^k(a^{-1})^1) & \text{Multiplication of like bases} \\ & = & (a^k \cdot a)((a^k)^{-1}a^{-1}) & \text{Induction hypothesis} \\ & = & (a^k \cdot (a^k)^{-1})(a \cdot a^{-1}) & \text{Factor permutation} \\ & = & 1 \cdot 1 & \text{Multiplicative inverse} \\ & = & 1 & \text{Multiplicative identity} \end{array}$$

That is, $a^{k+1}(a^{-1})^{k+1} = 1$. We already know that $a^{k+1}(a^{k+1})^{-1} = 1$ by the definition of multiplicative inverse, so by the uniqueness of multiplicative inverses, $(a^{k+1})^{-1} = (a^{-1})^{k+1}$. That is, S_{k+1} is true. By the principle of mathematical induction, S_n is true for all $n \in \mathbb{N}$.

Here is an alternative proof: $a^n(a^{-1})^n = (a \cdot a^{-1})^n$ by the previous theorem. This is just $1^n = 1$ (as we proved earlier). Since the multiplicative inverse of a^n is unique, we must have $(a^n)^{-1} = (a^{-1})^n$. □

Notice that for $a \neq 0$, $(a^n)^{-1} = (a^{-1})^n$ may also be written in reciprocal notation as $\frac{1}{a^n} = \left(\frac{1}{a}\right)^n$. Also, for any **integer** n, $\frac{1}{a^n} = a^{-n}$ and $a^n = \frac{1}{a^{-n}}$.

Example 4.6.4. We have $3^{-4} = \frac{1}{3^4} = \frac{1}{81}$.

□

These results allow us to use negative integers as exponents as summarized in the next theorem.

Theorem 34 (Division of Like Bases). *If $a \neq 0$ is a field element and n and m are positive integers, then $\frac{a^n}{a^m} = a^{n-m}$.*

Proof. We will consider three cases: $n = m$, $n > m$, and $n < m$.

Note that $\frac{a^n}{a^m} = a^n(a^m)^{-1}$ by the definition of fraction.

If $n = m$, this is equal to 1.

If $n > m$, then $a^n = a^m \cdot a^{n-m}$ by multiplication of like bases. (Check that this works.) Now $\frac{a^n}{a^m} = \frac{a^m \cdot a^{n-m}}{a^m \cdot 1} = \frac{a^{n-m}}{1} = a^{n-m}$, where we have made use of the fraction simplification theorem.

If $m > n$, then $a^m = a^n \cdot a^{m-n}$ by multiplication of like bases, and, again by the fraction simplification theorem, we get that $\frac{a^n}{a^m} = \frac{a^n \cdot 1}{a^n \cdot a^{m-n}} = \frac{1}{a^{m-n}} = a^{-(m-n)} = a^{n-m}$. □

Example 4.6.5. Observe that $\frac{1}{a^{-2}} = a^{-(-2)} = a^2$.

□

The following theorem amplifies our previous theorems about exponents. Notice that the exponents in the theorem below may be **any** integers; the earlier theorems allowed only natural numbers.

Theorem 35. *Assume that a and b are field elements and* $m, n \in \mathbb{Z}$. *Also, assume that if* $n = 0$, *then* $ab \neq 0$. *Then*

1. $a^n a^m = a^{n+m}$ *To multiply like bases, add the exponents*

2. $(ab)^n = a^n b^n$ *A power of a product is the product of the powers*

3. $(a^n)^m = a^{nm}$ *To raise a power (of a) to a power, multiply the exponents*

4. $\dfrac{a^n}{a^m} = a^{n-m}$ *(for* $a \neq 0$*) To divide like bases, subtract the exponents*

Proof. The proofs are not difficult, but they are tedious, and we have seen the ideas required in proving the initial versions. We will prove part 1.

Suppose that $n = 0$. Then $a^n a^m = a^0 a^m = 1 \cdot a^m = a^m = a^{m+0}$. Similarly, if $m = 0$, part (a) holds. Henceforth, we may assume that neither m nor n is 0.

If $n, m \in \mathbb{N}$, then this is a theorem we have proved before.

If $n \in \mathbb{N}$ and $m < 0$, then $a^n a^m = \dfrac{a^n}{a^{-m}}$, where $-m > 0$.

- If $n > -m$, then this becomes $a^{n-(-m)} = a^{n+m}$ (by a previous theorem), as desired.
- If $n = -m$, this becomes $1 = a^0 = a^{n-n} = a^{n+m}$, again as desired.
- Finally, if $n < -m$, then this becomes $\dfrac{1}{a^{-m-n}} = \dfrac{1}{a^{-(n+m)}} = a^{n+m}$.

The case $m \in \mathbb{N}$ and $n < 0$ is similar.

The final case is $n < 0$ and $m < 0$. Then $a^n a^m = \dfrac{1}{a^{-n}} \dfrac{1}{a^{-m}}$, with $-n, -m \in \mathbb{N}$. This becomes $\dfrac{1}{a^{-n} a^{-m}} = \dfrac{1}{a^{-(m+n)}} = a^{m+n}$, and we are finished at last.

Parts 2, 3, and 4 are left as exercises for the masochistic reader. □

Example 4.6.6. If x and y are non-zero field elements, simplify the given expression and write the result using only positive exponents.

1. $\dfrac{3x^{-2}y^4}{x^3y^2}$ 2. $\left(\dfrac{3x^{-2}y^4}{x^3y^2}\right)^{-3}$ 3. $\dfrac{x^{-3}+y^{-2}}{x^{-1}+y^{-3}}$

Solution:

1.

$$\begin{aligned} \frac{3x^{-2}y^4}{x^3y^2} &= 3x^{(-2-3)}y^{(4-2)} && \text{Division of like bases} \\ &= 3x^{-5}y^2 && \text{\_\_\_\_\_\_\_\_\_\_\_\_\_\_\_\_} \\ &= \frac{3y^2}{x^5} && \text{\_\_\_\_\_\_\_\_\_\_\_\_\_\_\_\_} \end{aligned}$$

2.

$$\begin{aligned}
\left(\frac{3x^{-2}y^4}{x^3y^2}\right)^{-3} &= \left(\frac{3y^2}{x^5}\right)^{-3} && \text{Previous problem} \\
&= \left(\frac{x^5}{3y^2}\right)^{3} && \text{Defn. negative exponent} \\
&= \frac{(x^5)^3}{(3y^2)^3} && ________________ \\
&= \frac{x^{15}}{27y^6} && ________________
\end{aligned}$$

3.

$$\begin{aligned}
\frac{x^{-3}+y^{-2}}{x^{-1}+y^{-3}} &= \frac{\frac{1}{x^3}+\frac{1}{y^2}}{\frac{1}{x}+\frac{1}{y^3}} && ________________ \\
&= \left(\frac{\frac{1}{x^3}+\frac{1}{y^2}}{\frac{1}{x}+\frac{1}{y^3}}\right)\cdot\frac{x^3y^3}{x^3y^3} && ________________ \\
&= \frac{\left(\frac{1}{x^3}+\frac{1}{y^2}\right)\frac{x^3y^3}{1}}{\left(\frac{1}{x}+\frac{1}{y^3}\right)\frac{x^3y^3}{1}} && ________________ \\
&= \frac{y^3+x^3y}{x^2y^3+x^3} && ________________
\end{aligned}$$

□

Section 4.6 Exercises

Write each expression with a single positive exponent.

1. 5^6x^6
2. $[(x+y)^3]^{-3}$
3. z^5z^{-3}
4. $\dfrac{(x-3y)^7}{(x-3y)^{-2}}$
5. $\dfrac{x^{-3}x^7}{x^{-4}}$
6. $[(2x-5y)^{-2}]^5$
7. $(x+2)^{-3}$
8. $(2.45\times10^4)(1.42\times10^{-6})$
9. $\dfrac{(3.14\times10^{-6})(-7.26\times10^{-4})}{(5.11\times10^{-8})(4.43\times10^{-2})}$
10. Evaluate the expression $4\cdot10^5+7\cdot10^4+2\cdot10^3+0\cdot10^2+1\cdot10^1+9\cdot10^0$.
11. Evaluate the expression $2\cdot10^3-2\cdot10^2+8\cdot10^1-6\cdot10^0+1\cdot10^{-1}+7\cdot10^{-2}+3\cdot10^{-3}$.

Perform the indicated operation and simplify each expression. (Remember, to simplify a fraction, the first thing you need to do is **factor the numerator and denominator**.)

12. $\dfrac{27y^3}{225y^5}$

13. $(4x^5)(5x^3)$

14. $\dfrac{2^{-3}x^2y^{-3}}{2^{-2}x^{-4}y^2}$

15. $\dfrac{3^{-4}x^{-3}y^3}{3^2x^{-5}y^2}$

16. $\dfrac{2^4 3^5 x^{-4} y^{-2}}{2^{-2} 3^2 x^6 y^{-5}}$

17. $\dfrac{2x}{x+y} - \dfrac{2y}{x-y}$

18. $\dfrac{8x^3y^2}{15xz^4} \div \dfrac{12xyz}{25x^2y^2z}$

19. $(2t^3)^{-2}$

20. $\dfrac{(3^{-2}x^{-4}y^2)^{-4}}{(3^3x^{-2}y^{-3})^3}$

21. $\dfrac{x^5}{x^3+x^2}$

22. $\dfrac{x^4+y^{-2}}{x^3-y^{-3}}$

23. $\dfrac{x^{-2}-y^2}{x^3-y^{-2}}$

24. $\dfrac{x^{-1}+y^{-1}}{x^{-2}-y^{-2}}$

25. $\dfrac{3x^{-3}-3y^{-2}}{3x^{-2}+y^{-2}}$

26. Show that raising to powers is not an associative operation; that is, find a, b, and c such that $(a^b)^c \neq a^{(b^c)}$.

27. Use mathematical induction to prove that $(a^m)^n = a^{mn}$.

28. Use mathematical induction to prove that if $a_1 \cdot a_2 \cdot a_3 \cdots a_n = 0$, then $a_i = 0$ for some $i, 1 \leq i \leq n$.

29. Prove parts 2, 3, and 4 of Theorem 35.

Chapter 5

Complex Numbers and Order Axioms

In this unit, we introduce the set of complex numbers as a two-dimensional number system, define addition, multiplication and scalar multiplication on this set, and investigate which field properties are satisfied by these operations. We discover a copy of the real numbers embedded in the complex numbers and introduce (and investigate) the order axioms to contrast the field of real numbers with the field of complex numbers.

Objectives

- To introduce the set of complex numbers as a set of ordered pairs
- To verify that the field properties hold in the set of complex numbers
- To exhibit a copy of the real numbers as a subset of the complex numbers
- To demonstrate the correspondence between the field of complex numbers as ordered pairs and the standard representation of complex numbers
- To introduce and investigate the order axioms
- To contrast the field of complex numbers with the field of real numbers

Terms

- complex numbers
- scalar multiplication
- imaginary unit
- standard form
- conjugate
- modulus
- inequality
- positive
- negative
- absolute value
- distance
- circle

5.1 Complex Numbers

It is often the case that a one-dimensional number system is insufficient and a two-dimensional number system is necessary. For example, in applications involving both length and direction, two dimensions are necessary. To locate one's car in a parking lot or to find a point on a map, two coordinates are needed. One type of two-dimensional numbers is defined below.

Definition 5.1.1. Let $\mathbb{C}$ be the set of ordered pairs of real numbers; that is

$$\mathbb{C} = \mathbb{R} \times \mathbb{R} = \{(a,b) | a,b \text{ are real numbers }\}.$$

Definition 5.1.2. Two ordered pairs are **equal** provided their first coordinates are equal and their second coordinates are equal; *i.e.,* $(a,b) = (c,d)$ if and only if $a = c$ and $b = d$.

Example 5.1.1. If $(2x+1, y-4) = (x+4, 3y-8)$, what are x and y?

Solution: Equality of these ordered pairs means that $2x+1 = x+4$ and $y-4 = 3y-8$. Therefore $x = 3$ and $y = 2$.

□

Definition 5.1.3. Define the operations **addition**, $+$, and **multiplication**, $\cdot$, on $\mathbb{C}$ as follows: For any pairs (r,s) and (t,u) in $\mathbb{C}$,

$$(r,s) + (t,u) = (r+t, s+u) \text{ and}$$
$$(r,s) \cdot (t,u) = (rt - su, ru + st)$$

Note that $\mathbb{C}$ is closed under addition and multiplication since $(r+t, s+u) \in \mathbb{C}$ and $(rt - su, ru + st) \in \mathbb{C}$ whenever $(r,s) \in \mathbb{C}$ and $(t,u) \in \mathbb{C}$.

Example 5.1.2. Using the above definitions we have the following calculations:

- $(3, -2.4) + (-1.6, 4) = (3 + (-1.6), -2.4 + 4) = (1.4, 1.6)$
- $(-2,3) \cdot (4,-5) = ((-2)(4) - (3)(-5), (-2)(-5) + (3)(4)) = (7,27)$

□

Theorem 36. $(0,0)$ *is an additive identity for* $\mathbb{C}$.

Proof. Let $(a,b) \in \mathbb{C}$. Then

$$\begin{aligned} (0,0) + (a,b) &= (0+a, 0+b) && \text{Definition of complex number addition} \\ &= (a,b) && 0 \text{ is the additive identity in } \mathbb{R} \end{aligned}$$

Therefore, $(0,0) + (a,b) = (a,b)$, by transitivity of equality. Similarly, $(a,b) + (0,0) = (a,b)$. Thus, by definition of an additive identity, $(0,0)$ is an additive identity for $\mathbb{C}$. □

Theorem 37. $(1,0)$ *is a multiplicative identity for* $\mathbb{C}$.

Proof. Let $(a,b) \in \mathbb{C}$. Then

$$\begin{array}{rcll} (1,0)\cdot(a,b) & = & (1\cdot a - 0\cdot b, 1\cdot b + 0\cdot a) & \text{Definition of complex multiplication} \\ & = & (a-0, b+0) & 1 \text{ is the multiplicative identity in } \mathbb{R}, x\cdot 0 = 0 \\ & = & (a,b) & 0 \text{ is the additive identity in } \mathbb{R}. \end{array}$$

Therefore, $(1,0)\cdot(a,b) = (a,b)$, by transitivity of equality. Similarly, $(a,b)\cdot(1,0) = (a,b)$. Thus, by definition of a multiplicative identity, $(1,0)$ is an multiplicative identity for $\mathbb{C}$. □

The proofs of the previous theorems give the "flavor" of many of the proofs of theorems for complex numbers: because complex addition and multiplication are defined *in terms of* real number addition and multiplication, the strategy will be to reduce one computation in $\mathbb{C}$ to two computations in $\mathbb{R}$, where we already understand the properties at work.

Theorem 38. *Addition on* $\mathbb{C}$ *is associative and commutative.*

Proof. Let $(r,s), (t,u)$, and $(v,w) \in \mathbb{C}$. Then

$$\begin{array}{rcll} [(r,s)+(t,u)]+(v,w) & = & (r+t, s+u)+(v,w) & \text{Definition of complex addition} \\ & = & ((r+t)+v, (s+u)+w) & \text{Definition of complex addition} \\ & = & (r+(t+v), s+(u+w)) & \textbf{Real} \text{ addition is associative} \\ & = & (r,s)+(t+v, u+w) & \text{Definition of complex addition} \\ & = & (r,s)+[(t,u)+(v,w)] & \text{Definition of complex addition.} \end{array}$$

Therefore, complex addition is associative. That complex addition is commutative is similar and left as an exercise. □

Theorem 39. *Multiplication on* $\mathbb{C}$ *is associative and commutative.*

Proof. Let $(r,s), (t,u)$, and $(v,w) \in \mathbb{C}$. Then

$$\begin{array}{rcll} [(r,s)\cdot(t,u)]\cdot(v,w) & = & (rt-su, ru+st)\cdot(v,w) & \_\_\_\_\_\_\_\_ \\ & = & ((rt-su)v-(ru+st)w, (rt-su)w+(ru+st)v) & \_\_\_\_\_\_\_\_ \\ & = & (rtv-suv-ruw-stw, rtw-suw+ruv+stv) & \_\_\_\_\_\_\_\_ \\ & = & (rtv-ruw-stw-suv, rtw+ruv+stv-suw) & \_\_\_\_\_\_\_\_ \\ & = & (r(tv-uw)-s(tw+uv), r(tw+uv)+s(tv-uw)) & \_\_\_\_\_\_\_\_ \\ & = & (r,s)\cdot(tv-uw, tw+uv) & \_\_\_\_\_\_\_\_ \\ & = & (r,s)\cdot[(t,u)\cdot(v,w)] & \_\_\_\_\_\_\_\_ \end{array}$$

Therefore, complex multiplication is associative. That complex multiplication is commutative is similar and left as an exercise. □

Theorem 40. $(-a,-b)$ *is an additive inverse of* (a,b).

Proof. Let $(a,b) \in \mathbb{C}$. Then since $a,b \in \mathbb{R}$, $-a,-b \in \mathbb{R}$. Thus, $(-a,-b) \in \mathbb{C}$ and

$$\begin{array}{rcll} (a,b)+(-a,-b) & = & (a+(-a), b+(-b)) & \text{Definition of complex number addition} \\ & = & (0,0) & \text{additive inverse in } \mathbb{R}. \end{array}$$

Therefore, $(a,b)+(-a,-b) = (0,0)$, by transitivity of equality. By commutativity of complex addition, $(-a,-b)+(a,b) = (0,0)$. Thus, by definition of an additive inverse, $(-a,-b)$ is an additive inverse for (a,b). □

Using our standard notation for additive inverses, this means that $-(a,b) = (-a,-b)$.

Theorem 41. *If* $(a,b) \neq (0,0)$, *then*

$$\left(\frac{a}{a^2+b^2}, \frac{-b}{a^2+b^2}\right)$$

is a multiplicative inverse of (a,b).

Proof. Let $(a,b) \neq (0,0) \in \mathbb{C}$. Then $a \neq 0$, $b \neq 0$, and $a^2+b^2 \neq 0$. We calculate the following product:

$$\begin{aligned}
&(a,b)\cdot\left(\frac{a}{a^2+b^2}, \frac{-b}{a^2+b^2}\right) \\
&= \left(a\cdot\frac{a}{a^2+b^2} - b\cdot\frac{-b}{a^2+b^2}, a\cdot\frac{-b}{a^2+b^2} + (-b)\cdot\frac{a}{a^2+b^2}\right) && \underline{\hspace{4cm}} \\
&= \left(\frac{a^2}{a^2+b^2} + \frac{b^2}{a^2+b^2}, \frac{-ab}{a^2+b^2} + \frac{ba}{a^2+b^2}\right) && \text{Fraction multiplication} \\
&= \left(\frac{a^2+b^2}{a^2+b^2}, \frac{-ab+ba}{a^2+b^2}\right) && \text{Fraction Addition I} \\
&= \left(\frac{a^2+b^2}{a^2+b^2}, \frac{-ab+ab}{a^2+b^2}\right) && \underline{\hspace{4cm}} \\
&= (1,0) && \text{Fraction simplification.}
\end{aligned}$$

Therefore, $(a,b)\cdot\left(\frac{a}{a^2+b^2}, \frac{-b}{a^2+b^2}\right) = (1,0)$, by transitivity of equality. By commutativity of complex multiplication, $\left(\frac{a}{a^2+b^2}, \frac{-b}{a^2+b^2}\right)\cdot(a,b) = (1,0)$. Thus, by definition of a multiplicative inverse, $\left(\frac{a}{a^2+b^2}, \frac{-b}{a^2+b^2}\right)$ is a multiplicative inverse for (a,b). □

Example 5.1.3. The additive inverse of $(-2,5)$ is $-(-2,5) = (-(-2),-5) = (2,-5)$. The multiplicative inverse of $(-2,5)$ is $\left(\frac{-2}{(-2)^2+5^2}, \frac{-5}{(-2)^2+5^2}\right) = \left(-\frac{2}{29}, -\frac{5}{29}\right)$.

□

Theorem 42. *Complex multiplication distributes over complex addition; that is,*

$$(r,s)\cdot((t,u)+(v,w)) = (r,s)\cdot(t,u)+(r,s)\cdot(v,w).$$

Proof. The proof of this theorem is calculational in nature and left to the reader. □

To this point, we have seen that $\mathbb{C}$ is closed under addition and multiplication, addition and multiplication in $\mathbb{C}$ are associative and commutative, $\mathbb{C}$ contains an additive identity and a multiplicative identity, every element of $\mathbb{C}$ has an additive inverse, every nonzero element of $\mathbb{C}$ has a multiplicative inverse, and complex multiplication distributes over complex addition. That is, $\mathbb{C}$ satisfies all of the field properties!

Theorem 43. *$\mathbb{C}$ is a field.*

This marvelous fact means that we may use any theorem we have proved about fields at any time, and all of the techniques we have developed for solving equations are still applicable.

Example 5.1.4. Solve $(2,3)z+(4,-1)=(-1,5)$. [Note: z traditionally represents a complex variable.]

Solution: This equation has the form $ax+b=c$; we solve it just as we would any other such equation.

$$\begin{array}{rcll}
(2,3)z+(4,-1) & = & (-1,5) & \text{Given} \\
[(2,3)z+(4,-1)]+(-4,1) & = & (-1,5)+(-4,1) & \text{Additive Inverse} \\
(2,3)z+[(4,-1)+(-4,1)] & = & (-1+(-4),5+1) & \text{Associativity} \\
(2,3)z+(0,0) & = & (-5,6) & \text{Simplification} \\
(2,3)z & = & (-5,6) & \text{Additive Identity in } \mathbb{C} \\
\left(\frac{2}{2^2+3^2},\frac{-3}{2^2+3^2}\right)\cdot(2,3) & = & \left(\frac{2}{2^2+3^2},\frac{-3}{2^2+3^2}\right)\cdot(-5,6) & \text{Multiplicative Inverse} \\
(1,0)z & = & \left(\frac{8}{13},\frac{27}{13}\right) & \text{Multiplication in } \mathbb{C}.
\end{array}$$

Therefore, $z=(8/13,27/13)$ since $(1,0)$ is the multiplicative identity in $\mathbb{C}$.

□

The procedure is the same; only the *forms* of the numbers have changed.

Theorem 44. *The set $R=\{(r,0): r$ is a real number$\}$ is closed under addition and multiplication.*

Proof. Let $(a,0),(b,0)\in R$. Then

$$\begin{array}{rcll}
(a,0)+(b,0) & = & (a+b,0+0) & \text{Definition of complex number addition} \\
& = & (a+b,0) & \text{Real number additive identity.}
\end{array}$$

Therefore, $(a,0)+(b,0)\in R$, and R is closed under complex addition.
Also, we have

$$\begin{array}{rcll}
(a,0)\cdot(b,0) & = & (ab-0\cdot 0,a\cdot 0+0\cdot b) & \text{Definition of complex multiplication} \\
& = & (ab,0) & \text{Real additive identity, } a\cdot 0=0
\end{array}$$

Therefore, $(a,0)\cdot(b,0)\in R$, and R is closed under complex multiplication. □

Note that

1. the set described in the theorem above "acts" just like the set of real numbers,

2. any real number determines exactly one element of this set, and

3. any element of this set uniquely determines a real number.

Consequently, when referring to any pair of the form $(a,0)$, we frequently just use the real number a and write $(a,0)=a$.

Definition 5.1.4. For any real numbers a, b, and c, define $a \cdot (b,c) = (ab, ac)$. This operation is called **scalar multiplication**. (Notice that we get the same result for this as if we use $(a,0)$; see the exercises.)

Theorem 45. *For any real numbers a and b, $(a,b) = a \cdot (1,0) + b \cdot (0,1)$.*

Proof. Let a and b be real numbers. Then

$$\begin{aligned} a(1,0)+b(0,1) &= (a\cdot 1, a\cdot 0)+(b\cdot 0, b\cdot 1) && \text{Definition of scalar multiplication} \\ &= (a,0)+(0,b) && \text{Real additive identity, } a\cdot 0 = 0 \\ &= (a+0, 0+b) && \text{Definition of complex addition} \\ &= (a,b) && \text{Real additive identity.} \end{aligned}$$

Therefore, $a \cdot (1,0) + b \cdot (0,1) = (a,b)$ by transitivity of equality. □

This means that $(1,0)$ and $(0,1)$ are some kind of a "basis" for the complex numbers. We have already identified $(1,0) = 1$ as something special (the multiplicative identity); we now consider $(0,1)$.

Definition 5.1.5. Let $i = (0,1)$. The number i is called the **imaginary unit**.

Theorem 46. $i^2 = (-1,0) = -1$.

Proof. We will use definitions to calculate i^2 :

$$\begin{aligned} i^2 &= i\cdot i && \text{Definition of squaring} \\ &= (0,1)\cdot(0,1) && \text{Definition of } i \\ &= (0\cdot 0 - 1\cdot 1, 0\cdot 1 + 1\cdot 0) && \text{Definition of complex multiplication} \\ &= (0-1, 0+0) && \text{Multiplication in } \mathbb{R} \\ &= (-1,0) && \text{Additive Identity in } \mathbb{R} \\ &= -1 && \text{Theorem 44.} \end{aligned}$$

□

Corollary 5.1.1. $i = \sqrt{-1}$.

Example 5.1.5. Solve $z^2 + 1 = 0$ for all complex solutions z.

Solution: If we rewrite $z^2 + 1 = z^2 - (-1)$, then we have the difference of two perfect squares.

$$\begin{array}{rcll} z^2+1 & = & 0 & \text{Given} \\ z^2-(-1) & = & 0 & \text{Subtraction} \\ (z-i)(z+i) & =0 & \text{Factor} & \\ z-i=0 & \text{or} & z+i=0 & \text{Zero Product} \\ z-i & \text{or} & z=-i & \text{Addition} \end{array}$$

□

Theorem 47. *For any real numbers a and b, $(a,b) = a + bi$. This is called the* **standard form** *for a complex number. The real number a is called the* **real part** *of $a + bi$, and the real number b is called the* **imaginary part** *of $a + bi$.*

Notice that the imaginary part does not include the i.

Proof. Let a and b denote real numbers. Then

$$\begin{array}{rcll} a+bi & = & (a,0)+b(0,1) & \text{Theorem 44, Definition of } i \\ & = & (a,0)+(b\cdot 0, b\cdot 1) & \text{Definition of scalar multiplication} \\ & = & (a,0)+(0,b) & \text{Real number arithmetic} \\ & = & (a+0,0+b) & \text{Complex addition} \\ & = & (a,b) & \text{Real number additive identity.} \end{array}$$

□

We may use either form in computations.

Example 5.1.6. $(2+5i)+(1-3i)=(2+1)+(5-3)i=3+2i$ may also be represented as $(2,5)+(1,-3)=(3,2)$.

□

Definition 5.1.6. The **complex conjugate** of the complex number (a,b) is the complex number

$$\overline{(a,b)}=(a,-b).$$

Example 5.1.7. The complex conjugate of $(-3,5)$ is $\overline{(-3,5)}=(-3,-5)$. This can also be written as $\overline{-3+5i}=-3-5i$.

□

Example 5.1.8. The complex conjugate of $\left(2,-\frac{\sqrt{3}}{2}\right)$ is $\overline{\left(2,-\frac{\sqrt{3}}{2}\right)}=\left(2,\frac{\sqrt{3}}{2}\right)$.

□

Definition 5.1.7. The **modulus** of (a,b) is the real number

$$|(a,b)|=\sqrt{a^2+b^2}.$$

Example 5.1.9. The modulus of $(-3,5)$ is $|(-3,5)|=\sqrt{(-3)^2+5^2}=\sqrt{34}$. This can also be written as $|-3+5i|=\sqrt{34}$.

□

Example 5.1.10. The modulus of $\left(2,\frac{\sqrt{3}}{2}\right)$ is $\left|\left(2,\frac{\sqrt{3}}{2}\right)\right|=\sqrt{(2)^2+\left(\frac{\sqrt{3}}{2}\right)^2}=\frac{\sqrt{19}}{2}$.

□

Notice that $(a,b)\cdot\overline{(a,b)}=a^2+b^2=|(a,b)|^2$. That is, the product of a complex number and its conjugate is always a real number.

Example 5.1.11. Compute the following:

- $(3,2)\overline{(3,2)} = (3,2)(3,-2) = ((3)(3)-(2)(-2),(3)(2)+(3)(-2)) = (9+4,0) = 13$
- $|(3,2)|^2 = (\sqrt{3^2+2^2})^2 = 13.$

□

Section 5.1 Exercises

Compute. Subtraction and division are defined as they are for any field.

1. $(-\sqrt{2}+1,4)-(\sqrt{2},6)$
2. $(x+2,3)+(3x-7,y)$
3. $(2-4i)-(12-i)$
4. $(-1,1)\cdot(4,4)$
5. $(2,-0.5)\cdot(8,-2)$
6. $(3-2i)(5+4i)$
7. $(3.1,-7)^{-1}$
8. $(x,2)^{-1}$
9. $-1(4,2)$
10. $-2(5,-1)$
11. $(2.2,5)\div(7,13)$
12. $(4,-8)\div(-12,5)$
13. $|(-6,10)|$
14. $|2-6i|$
15. $\overline{(2,2)}$
16. $\overline{12-9i}$
17. $\dfrac{3-4i}{2+i}$
18. $\dfrac{(2,2)}{(-1,1)}$

Compute each power of i.

19. i^3
20. i^4
21. i^5
22. i^{12}
23. i^{48}
24. $(i^5)^4$
25. $(i^{17})^9$
26. i^{95}
27. i^{753}
28. i^{-1}
29. i^0
30. $i^{208,334}$

Solve each equation for the complex number z.

31. $(3,1)z-(1,-1)=(4,2)$
32. $(2,-5)z+(2,6)=(-3,8)$
33. $z^2+4=0$
34. $z^2+9=0$
35. $2z+(2,-1)z=(4,3)$
36. $5z-(2,9)z+(2,5)=(-2,8)$

37. $3z+2z=(4,-12)$

38. $7z-3z=(-8,9)$

39. $(2,1)z+(13,-3)=(-4,2)z+(7,2)$

40. $(1,7)z-(2,19)=(5,19)z=(1,2)$

41. Verify that $a(b,c)=(a,0)\cdot(b,c)$.

42. Prove that addition in $\mathbb{C}$ is commutative.

43. Prove that multiplication in $\mathbb{C}$ is commutative.

44. Prove that complex multiplication distributes over complex addition.

45. Show that $(a+bi)\cdot(c+di)=(ac-bd)+(ad+bc)i$. Does this match our original definition of multiplication of complex numbers?

5.2 Order Axioms

So far, the axioms we have could describe lots of different fields, but we want to consider the real numbers. We are going to impose another axiom, called the **order axiom**.

We define an ordering $>$ ("greater than") on a field F by way of the following axiom, if possible.

Order Axiom: The field F contains a subset F^+ that has the following properties:

1. (Closure) F^+ is closed under addition and multiplication.
2. (Trichotomy) For each $x \in F$, one and only one of the following holds:
   (a) $x \in F^+$
   (b) $x = 0$
   (c) $-x \in F^+$

Definition 5.2.1. A field that satisfies the order axiom is called an **ordered field**, and the set F^+ is the set of positive elements of F.

Not all fields are ordered, as we will see at the end of the section. Throughout this section F refers to an ordered field.

Example 5.2.1. We define $\mathbb{Q}^+$ by $\mathbb{Q}^+ = \left\{ \frac{a}{b} \middle| a, b \in \mathbb{N}, ab > 0 \right\}$. It is not hard to show that $\mathbb{Q}^+$ is closed under addition and multiplication and that the trichotomy property holds. Therefore, $\mathbb{Q}$ is an ordered field.

□

Definition 5.2.2. We say that x is **greater than** y, written $x > y$, if $x - y \in F^+$. Alternatively, we write $y < x$ for "y is less than x" if $x > y$. We also say that x is greater than or equal to y, written $x \geq y$, if $x > y$ or $x = y$; similarly, x is less than or equal to y, written $x \leq y$, if $x < y$ or $x = y$. These relations are all referred to as **inequalities**.

Example 5.2.2. In $\mathbb{Q}$, we have $\frac{2}{3} > \frac{3}{5}$ since $\frac{2}{3} - \frac{3}{5} = \frac{1}{15} \in \mathbb{Q}^+$. In alternate notation we have that $\frac{3}{5} < \frac{2}{3}$ and $\frac{2}{3} \geq \frac{3}{5}$.

□

Theorem 48 (Transitivity of Order). *Let $a, b, c \in F$, where F is an ordered field. If $a > b$ and $b > c$, then $a > c$.*

Proof. If $a > b$ and $b > c$, then $a - b \in F^+$ and $b - c \in F^+$ by definition. By the closure of F^+ under addition, we must have $(a - b) + (b - c) = a - c \in F^+$ as well. Thus $a > c$ by definition. □

Definition 5.2.3. A number a is **negative** if $a < 0$, and **positive** if $a > 0$; 0 itself is neither positive nor negative.

Note first that $a > 0$ if and only if $a \in F^+$, since $a > 0$ if and only if $a - 0 \in F^+$ by definition. Please note: $-a$ **need not be a negative number**. For example: consider $a = -2$; what is $-a$? The symbol $-a$ refers to the additive inverse of a, not whether it is positive or negative.

Definition 5.2.4. The **compound inequality** $a < b < c$ means $a < b$ and $b < c$. The compound inequalities $a < b \leq c, a \leq b < c$, and $a \leq b \leq c$ are defined similarly, as are $a > b > c, a \geq b > c, a > b \geq c$, and $a \geq b \geq c$.

Example 5.2.3. The inequalities $x \leq 2$ and $x > -4$ can be combined into the single compound inequality $-4 < x \leq 2$.

□

Note that by transitivity, if $a < b < c$, then $a < c$ as well. Also, we won't mix opposite inequalities in a compound inequality; that is, it is **incorrect** to write $-2 < 5 > 0$.

Definition 5.2.5. The following symbols represent the indicated sets in **interval notation**:

1. $(a,b) = \{x | a < x < b\}$
2. $[a,b] = \{x | a \leq x \leq b\}$
3. $(a,b] = \{x | a < x \leq b\}$
4. $[a,b) = \{x | a \leq x < b\}$
5. $(-\infty, b) = \{x | x < b\}$
6. $(-\infty, b] = \{x | x \leq b\}$
7. $(a, \infty) = \{x | x > a\}$
8. $[a, \infty) = \{x | x \geq a\}$

We see that parentheses ("(" and ")") indicate that the given element is **not** a part of the set, and square brackets ("[" and "]") indicate the the given element **is** a part of the set. The "infinity" symbol ∞ is never a part of our set and always gets a parenthesis.

Example 5.2.4. $\{x | -2 < x \leq 12\} = (-2, 12]$.
$\{x | x < 4 \text{ or } x > 10\} = (-\infty, 4) \cup (10, \infty)$.

□

Interval notation is a convenient way to express the set of all numbers between two given numbers.

Theorem 49. *Let $a,b,c \in F$, where F is an ordered field. If $a < b$, then $a + c < b + c$.*

Proof. Let $a,b,c \in F$, where F is an ordered field. Then F contains a subset F^+ that is additively and multiplicatively closed and trichotomy holds.

$$\begin{array}{rcll} a & < & b & \text{Hypothesis} \\ b-a & \in & F^+ & \text{Definition of } < \\ b+0-a & \in & F^+ & \text{Additive Identity} \\ b+c-c-a & \in & F^+ & \text{Additive Inverses} \\ (b+c)+(-a-c) & \in & F^+ & \text{Summand Permutation} \\ (b+c)-(a+c) & \in & F^+ & \text{Gen. Distributive Law} \\ a+c & < & b+c & \text{Definition of } < \end{array}$$

□

Note that since subtraction in a field is essentially addition, we have the following corollary, the proof of which is an application of the previous theorem.

Corollary 5.2.1. *Let $a,b,c \in F$, where F is an ordered field. If $a+c < b+c$, then $a < b$.*

Proof.

$$\begin{array}{rcll} a+c & < & b+c & \text{Hypothesis} \\ (a+c)+(-c) & < & (b+c)+(-c) & \text{Previous theorem} \\ a+(c+(-c)) & < & b+(c+(-c)) & \text{Associativity of Addition} \\ a+0 & < & b+0 & \text{Additive Inverses} \\ a & < & b & \text{Additive Identity} \end{array}$$

□

Theorem 50. *Let $a,b,c \in F$, where F is an ordered field. If $a < b$ and if $c > 0$, then $ac < bc$.*

Proof.

$$\begin{array}{rl} a < b,\ c > 0 & \text{Hypotheses} \\ b-a \in F^+,\ c \in F^+ & \text{Definition of } < \\ (b-a)c \in F^+ & F^+ \text{ Multiplicatively Closed} \\ bc-ac \in F^+ & \text{Gen. Distributive Law} \\ ac < bc & \text{Definition of } < \end{array}$$

□

The above theorems can also be interpreted as "if $b > a$, then $b+c > a+c$. If also $c > 0$, then $bc > ac$." This follows from the relationship between "less than" and greater than. Notice also the extra hypothesis required for multiplication: *if* $c > 0$, then multiplication by c maintains the inequality.

Theorem 51 (Multiplication by a Negative). *Let $a,b \in F$, where F is an ordered field. If $a < b$ and $c < 0$, then $ac > bc$.*

Proof. Since $a < b$, $b-a \in F^+$. Since $c < 0$, $-c \in F^+$. Therefore, since F^+ is closed under multiplication, we get $(b-a)(-c) \in F^+$. But

$$\begin{array}{rcll} (b-a)(-c) & = & b(-c)-a(-c) & \underline{\hspace{5cm}} \\ & = & -(bc)-(-ac) & x(-y) = -(xy) \\ & = & -(bc)+ac & x-(-y) = x+y \\ & = & bc-ac & \text{Commutativity of Addition; Definition of Subtraction.} \end{array}$$

That is, $(a-b)(-c) = bc - ac \in F^+$. Therefore we get $bc > ac$, which is the same as $ac < bc$. □

Theorem 52 (Additive Inverses and Inequality). *Let $a, b \in F$, where F is an ordered field. Then $a > b$ if and only if $-a < -b$, and $a < b$ if and only if $-a > -b$. In particular, $a > 0$ if and only if $-a < 0$.*

Proof. First, we'll assume that $a > b$ and logically conclude that $-a < -b$. Then we'll "reverse" the argument to obtain that if $-a < -b$, then $a > b$.

a	$<$	b	Hypothesis
$(-a)+a$	$<$	$(-a)+b$	Add $(-a)$
$0+(-b)$	$<$	$-a+b+(-b)$	Additive Identity; add $(-b)$
$-b$	$<$	$-a$	________________

Reversing this argument, we have that

$-b$	$<$	$-a$	Hypothesis
$(-b)+0$	$<$	$-a+(b+(-b))$	________________
0	$<$	$(-a)+b$	________________
$a+0$	$<$	$a+(-a)+b$	________________
a	$<$	b	________________

Using the symbol $\iff$ to stand for "if and only if," we can have the above two arguments combined into the following:

$$a < b \iff (-a)+a < (-a)+b \iff (-b)+0 < -a+b+(-b) \iff -b < -a.$$

The second part follows from the first part and the definition of "less than." □

Example 5.2.5. This theorem says that **if** $2 > 0$, then $-2 < 0$, and that **if** $-\pi < 0$, then $-(-\pi) = \pi > 0$. We will show shortly that $2 > 0$.

□

Theorem 53. *Let $a, b, c, d \in F$, where F is an ordered field. If $a > b$ and $c > d$, then $a+c > b+d$. If $a, b, c, d > 0$, then $ac > bd$.*

Proof. Since $a > b$, $a+c > b+c$. Also, since $c > d$, $b+c > b+d$. Therefore, by transitivity, $a+c > b+d$. The proof of the multiplicative portion is left as an exercise. □

These order relations aren't free. Again, many of the properties we take for granted really do need to be proved from the axioms — even things like $1 > 0$.

Theorem 54. *In an ordered field, $1 > 0$.*

Proof. Remember that 1 represent the multiplicative identity and 0 represents the additive identity in the ordered field F. By the trichotomy property, we know that **exactly one** of these holds: $1 = 0$, $1 < 0$, or $1 > 0$. We will eliminate each of the first two possibilities leaving the third as correct.

We already know $1 \neq 0$ because for $a \neq 0$, $0 \cdot a \neq a$ and $1 \cdot a = a$.

If $1 < 0$, then $-1 > 0$ by the trichotomy property. In this case, we have $(-1)(-1) > 0$ by the closure of F^+ under multiplication. But this says $1 > 0$, a contradiction to our original assumption! Thus, $1 \not< 0$.

Therefore $1 > 0$. □

We have the following corollary:

Corollary 5.2.2. *In an ordered field,* $-1 < 0$.

We can now prove that, for example, $2 > 1$ using the theorems above. In fact, all of the natural numbers are positive.

Theorem 55. $2 > 1$.

Proof. Since $1 > 0$, we have that $1 + 1 > 0 + 1$. Thus, $2 > 1$. □

Theorem 56. *If* $n \in \mathbb{N}$, *then* $n > 0$. *(That is,* $\mathbb{N} \subseteq \mathbb{R}^+$.*)*

Proof. The proof is by induction on n; we have already proved the base step $1 > 0$. Assume that for some $k \in \mathbb{N}$, $k > 0$. Since $k > 0$ and $1 > 0$, by the above theorem, we know that $k + 1 > 0 + 0$. Thus, $k + 1 > 0$. Therefore, by the principle of mathematical induction, if $n \in \mathbb{N}$, then $n > 0$. □

How do we compare fractions?

Theorem 57. *If* $a \in F$, *where* F *is an ordered field, then* $a > 0$ *if and only if* $a^{-1} > 0$.

Proof. Assume that $a > 0$ in the ordered field F. By the trichotomy property, we have that either $a^{-1} = 0, a^{-1} < 0$, or $a^{-1} > 0$. We will eliminate the first two possibilities by contraction. Since $aa^{-1} = 1$, $a^{-1} \neq 0$. Also, if $a^{-1} < 0$, then the Multiplication by a Negative Theorem would imply that since $a > 0$ and $a^{-1} < 0$, $aa^{-1} < 0 \cdot a^{-1}$. Thus we get the contradiction that $1 < 0$. Therefore, by the trichotomy property, $a^{-1} > 0$. Similarly, if $a^{-1} > 0$, then $a > 0$. □

Theorem 58. *If* $a > b > 0$, *then* $\frac{1}{a} < \frac{1}{b}$.

Proof. If $a > b > 0$, then $a^{-1} > 0$ and $b^{-1} > 0$ from the previous theorem. Then

$$\begin{array}{rcll}
a & > & b & \text{Hypothesis} \\
aa^{-1} & > & ba^{-1} & \underline{\hspace{4cm}} \\
1 & > & ba^{-1} & \underline{\hspace{4cm}} \\
b^{-1} \cdot 1 & > & b^{-1}(ba^{-1}) & \underline{\hspace{4cm}} \\
b^{-1} & > & (b^{-1}b)a^{-1} & \underline{\hspace{4cm}} \\
b^{-1} & > & 1 \cdot a^{-1} & \underline{\hspace{4cm}} \\
b^{-1} & > & a^{-1} & \underline{\hspace{4cm}} \\
\frac{1}{b} & > & \frac{1}{a} & \underline{\hspace{4cm}}
\end{array}$$

□

These ideas allow us to *solve* inequalities.

Example 5.2.6. Solve $2x+5<7x-2$ for x.

Solution:

$$\begin{array}{rcll} 2x+5 & < & 7x-2 & \text{Given} \\ 2x+5+(-5) & < & 7x-2+(-5) & \text{\_\_\_\_\_\_\_\_\_\_\_\_\_\_\_\_} \\ 2x & < & 7x-7 & \text{Simplification} \\ -7x+2x & < & -7x+7x-7 & \text{\_\_\_\_\_\_\_\_\_\_\_\_\_\_\_\_} \\ -5x & < & -7 & \text{Simplification.} \end{array}$$

Here we pause to take note of what is happening. Because we know that $5>0$, the trichotomy property implies that $-5<0$. Now Theorem 57 implies that $-\frac{1}{5}<0$. Therefore, when we multiply both sides of our inequality by $-\frac{1}{5}$, we reverse the sense of the inequality:

$$\left(-\frac{1}{5}\right)(-5x) > \left(-\frac{1}{5}\right)(-7).$$

Therefore, $x>7/5$; the solution set is $(7/5,\infty)$. (Recall that $(7/5,\infty)$ is **interval notation** for $\{x|x>7/5\}$.)

□

Example 5.2.7. Gerard knows that the revenue from his store must be at least \$10,000 more than twice his expenses if he is going to expand his business. If his store's revenue is fixed at \$38,000, what is the most his expenses can be?

□

Solution: Let x denote the expenses. The revenue must be *at least* $10,000+2x$. That is, his revenue must be greater than or equal to $10,000+2x$. This means $10,000+2x \leq 38,000$. Thus $2x \leq 28,000$, so $x \leq \$14,000$. Gerard can realize his dream only if he can keep his expenses under \$14,000.

Theorem 59 (Sign of a Product)**.** *Let $a,b \in F$. Then $ab>0$ if and only if a and b are both positive or both negative.*

Proof. If either a or b is zero, then $ab=0$, so we may assume that $a,b \neq 0$.
Remember that "if and only if" indicates that we have two theorems to prove:

(1) If $ab>0$, then a and b are both positive or both negative;

(2) if a,b are both positive or both negative, then $ab>0$.

We will begin with (2). If a,b are both positive, then $a>0$ and $b>0$, so $ab>0$ by closure of F^+ under multiplication. If a and b are both negative, then $a,b<0$, so that $-a,-b>0$. Thus $(-a)(-b)=ab>0$, again by closure.

We will prove (1) using its contrapositive: If one of a,b is positive and the other negative (remember, neither is zero), then $ab<0$. We may assume that $a<0$ and $b>0$. Then $-a>0$, so $-(ab)=(-a)b>0$ by closure of F^+ under multiplication. Therefore, $ab<0$. Since the contrapositive of a conditional statement is logically equivalent to the original statement, we have proved the theorem. □

The above theorem allows us to solve quadratic inequalities.

Example 5.2.8. Solve $x^2+5x+3>-3$.

Solution:

$$\begin{array}{rcll} x^2+5x+3 & > & -3 & \text{Given} \\ x^2+5x+6 & > & 0 & \text{Add 3 to both sides} \\ (x+2)(x+3) & > & 0 & \text{Factor} \\ (x+2)>0 & \text{and} & x+3>0 & \\ & \text{or} & & \\ (x+2)<0 & \text{and} & x+3<0 & \text{Sign of a Product Theorem} \end{array}$$

- If $x+2>0$ and $x+3>0$, then $x>-2$ and $x>-3$. Since $2<3$, we have that $-2>-3$. Thus, if $x>-2$, then $x>-3$ automatically by transitivity.
- If $x+2<0$ and $x+3<0$, then $x<-2$ and $x<-3$. Since $2<3$, we again have that $-2>-3$. Thus, if $x<-3$, then $x<-2$ automatically by transitivity.

Therefore, if $x^2+5x+3>-3$, then either $x>-2$ or $x<-3$; the solution set is $(-\infty,-3)\cup(-2,\infty)$.

□

Corollary 5.2.3 (Squares are Nonnegative). *Let $a \in F$. Then $a^2 \geq 0$.*

Proof. Let $a \in F$. Then, either $a=0$ or $a \neq 0$.
If $a=0$, then $a^2=0$, so $a^2 \geq 0$.
If $a \neq 0$, then a and a are both positive or both negative, so $a^2>0$ by the Sign of a product theorem.
Therefore, in either case, $a^2 \geq 0$. □

The ordered fields we are most familiar with are $\mathbb{R}$, the field of real numbers, and $\mathbb{Q}$, the field of rational numbers. We are taught very early how to compare two numbers (using a number line model, for example). However, not all fields can be ordered.

Theorem 60. *$\mathbb{C}$ cannot be ordered (in the sense of the Order Axiom).*

Proof. If $\mathbb{C}$ could be ordered, then for any $a \in \mathbb{C}$, $a^2 \geq 0$. Also, we have shown that in an ordered field $-1<0$. Consider $a=i$. If $\mathbb{C}$ is ordered, then $i^2 \geq 0$, while also $i^2=-1<0$. Since these are not both possible, we have arrived at a contradiction. The cause of our contradiction was the original assumption that $\mathbb{C}$ can be ordered; therefore, $\mathbb{C}$ cannot be ordered. □

This means that there is no sense of one complex number being "larger" than another.

Section 5.2 Exercises

Determine whether each quantity is positive or negative, if possible.

1. $-(-3)$
2. $-\pi$
3. $(-2)^{-1}$
4. -2^{-1}
5. $-a$
6. $(-3)^2$
7. -3^2
8. a^2 (for $a \neq 0$)
9. $(-x) \cdot x$ (for $x \neq 0$)

Solve each inequality. Express your answer in interval notation.

10. $\frac{2}{3}x - 2 \geq \frac{4}{5} - 5x$
11. $3x - 7 < 5x + 4$
12. $2 - x \geq x + 5$
13. $(x+1)(x-5) > 0$
14. $x - \frac{1}{2} \leq \frac{3}{7} + 3$
15. $2x^2 - 3x > x^2 - 2$
16. $(3x-2)(x+4)(x-1) \leq 0$
17. $x^2 - 8x + 2 \leq x^2 - 4$
18. $x^2 + 1 \geq 0$

19. One long distance plan costs $0.99 for the first 20 minutes, and then $0.07 per minute thereafter. Another plan costs a flat $0.10 per minute. After how many minutes does the $0.99 plan become less expensive?

20. One brand of car costs $10,000 and gets 28 miles per gallon. Another brand of car costs only $8,000, but only gets 25 miles per gallon. After how many miles will the $10,000 car be "cheaper" than the $8,000 car? Assume that the fuel is $1.50 per gallon. [Hint: first determine the cost to drive each car *d* miles.]

21. Prove that for $a, b, c \in F$, where F is an ordered field, if $a + c < b + c$, then $a < b$.

22. Prove that if $a > b, c > d$ and $a, b, c, d > 0$, then $ac > bd$.

23. Prove that $\mathbb{Q}^+ = \left\{\frac{a}{b} | a, b \in \mathbb{N}, ab > 0\right\}$ is closed under addition and multiplication and that the trichotomy property holds. Conclude that $\mathbb{Q}$ is an ordered field, as claimed.

24. Let $\mathbb{R}^+$ denote the positive real numbers, and let $q \in \mathbb{R}^+$. Define $a \bigoplus b = a \cdot b$ and $a \bigotimes b = a^{\log_q b}$. Show that $\mathbb{R}^+$ is a field with respect to the addition $\bigoplus$ and multiplication $\bigotimes$.

5.3 Coordinate systems and distance formulas

In any ordered field, there is a notion of inequality. Therefore, in an ordered field, we can compare any two quantities. We are taught very early how to compare two (real) numbers using a number line model. In this section, we extend this idea to any ordered field.

Consider any straight line. To each point on the line, we associate a unique real number as follows: Choose a point O on the line, and label it with a 0 (zero). We will call O the **origin**. Now choose any other point to establish the size of one unit. Label the second point 1. With these markings, we can lay out all of the integers on the line by marking divisions of this size. Note that the negatives are on one side of 0, and the positives are on the other side of 0.

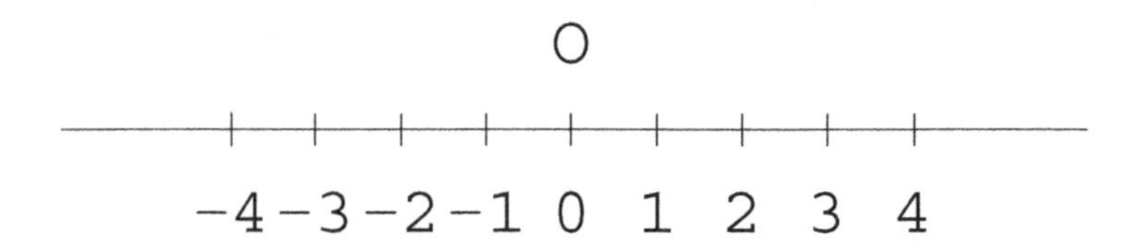

To obtain rational numbers, just divide the unit segments into appropriate lengths. The irrational numbers are harder; we will not deal with the construction of the rest of the real numbers, but assume that there is a one-to-one correspondence between real numbers and points on the line. That is, to every point on the line, there corresponds a unique real number, and to every real number, there corresponds a unique point on the line. The number corresponding to a given point is its **coordinate**.

We usually adopt the convention that numbers to the right of 0 are positive, if the line has been laid out horizontally. The distance between two points on the line is the (positive) difference in their coordinates. Thus the distance from 7 to 4 is 3, and so is the distance from 4 to 7.

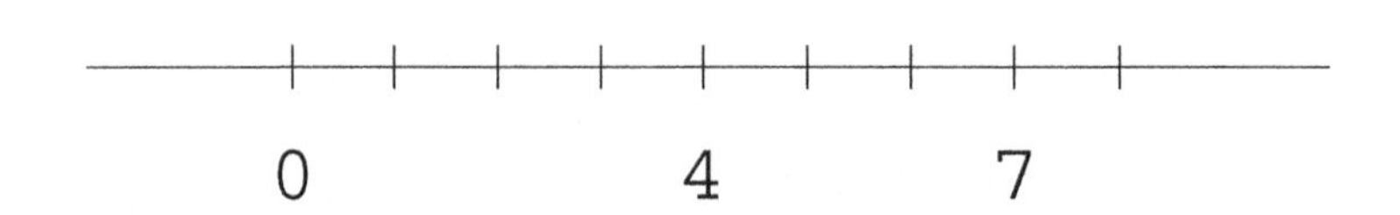

Definition 5.3.1. We introduce the **absolute value** of a number: $|x| = \begin{cases} x & \text{if } x \geq 0 \\ -x & \text{if } x < 0 \end{cases}$

Notice that the absolute value of a number is always nonnegative; that is, for any real number x, $|x| \geq 0$.

Definition 5.3.2. The **distance** between points x_1 and x_2 on a number line is $|x_2 - x_1|$.

Example 5.3.1. $|-3| = -(-3) = 3$ since $-3 < 0$. The distance between the points 5 and 2 is $|5-2| = |3| = 3$ since $3 > 0$. Notice that the absolute value sign is another grouping symbol.

□

Recall that $\sqrt{a}$ means the positive square root of a, provided that $a \geq 0$. Thus $\sqrt{x^2} = x$ if $x \geq 0$. If $x < 0$, then $\sqrt{x^2} = \sqrt{(-x)^2} = -x$ (since we want the POSITIVE square root). This means that $\sqrt{x^2} = |x|$. That is, we have a formula for computing $|x|$. Notice that it is **not** always true that $\sqrt{x^2} = x$; see the example below.

Example 5.3.2. $\sqrt{(-3)^2} = 3 = |-3|$. (Remember, $\sqrt{a}$ means the **positive** square root.)

□

Theorem 61. *If $a \in \mathbb{R}$, then $-|a| \leq a \leq |a|$.*

Proof. If $a \geq 0$, then $-|a| = -a \leq 0 \leq a = |a|$. By transitivity of inequality, we get that $-|a| \leq a \leq |a|$.

If $a < 0$, then $-|a| = -(-a) = a < 0 < -a = |a|$, and we again get $-|a| \leq a \leq |a|$. □

We have another theorem that is useful in interpreting absolute values.

Theorem 62. *Let $c, d \in \mathbb{R}$. Then $|c| \leq d$ if and only if $-d \leq c \leq d$.*

The utility of this theorem lies in the fact that we can take an inequality involving absolute values and change it into two inequalities which do not involve absolute values.

Proof. Since this theorem is an "if and only if" theorem, we need to prove two things:
First, we will show that if $|c| \leq d$, then $-d \leq c \leq d$. Assume $|c| \leq d$. Since $0 \leq |c|$ and $|c| \leq d$, we have $0 \leq d$, by transitivity. Notice that there are two cases.

Case I: $c \geq 0$. In this case, $|c| = c$. Also, $d \geq 0$ implies $-d \leq 0$. Combining these statements, we have that $-d \leq 0 \leq c = |c| \leq d$, so that $-d \leq c \leq d$.

Case II: $c < 0$. In this case, $|c| = -c$. Since $|c| \leq d$, we have $-c \leq d$, so $c \geq -d$. Since $c < 0$, $-c > 0$, and we arrive at $-d \leq c < 0 < -c \leq d$. Therefore, $-d \leq c \leq d$.

Next, we must show that if $-d \leq c \leq d$, then $|c| \leq d$. If $c \geq 0$, then we already know $|c| \leq d$ since $|c| = c$. If $c < 0$, then $-d \leq c \implies d \geq -c$, or $-c \leq d$, and again, $|c| \leq d$ since $|c| = -c$.
This completes the proof. □

The same ideas would also show that $|c| < d$ if and only if $-d < c < d$; we will use this in the example below.

Example 5.3.3. Solve $|3x+5| < 2$ for the possible values of x.

Using the above theorem, we have $-2 < 3x+5 < 2$. Remember that $a < b < c$ is a compound inequality which says that $a < b$ and $b < c$ (and we get $a < c$ by transitivity). Therefore, using theorems from the previous section, we may add -5 to all three pieces of this inequality as follows:

$$\begin{array}{rcccl} -2 & < & 3x+5 & < & 2 \\ -2+(-5) & < & 3x+5+(-5) & < & 2+(-5) \\ -7 & < & 3x & < & -3 \end{array}$$

By the same reasoning, we may multiply each member by $\frac{1}{3}$ without changing the sense of the inequality (since $3 > 0 \implies \frac{1}{3} > 0$). Thus we have

$$\begin{array}{rcccl} -7 & < & 3x & < & -3 \\ \frac{1}{3}(7) & < & \frac{1}{3}(3x) & < & \frac{1}{3}(-3) \\ -\frac{7}{3} & < & x & < & -1 \end{array}$$

That is, x may be any real number strictly between $-\frac{7}{3}$ and -1; the solution set is $(-\frac{7}{3},-1)$.

□

Similarly, we have the following theorem:

Theorem 63. *Let $c,d \in \mathbb{R}$. Then $|c| \geq d$ if and only if either $c \geq d$ or $c \leq -d$.*

Example 5.3.4. Solve $|3x+5| > 7$ for the possible values of x.

Using the preceding theorem we have two inequalities to solve:

$$\begin{array}{rclcrcl} 3x+5 & < & -7 & \text{or} & 3x+5 & > & 7 \\ 3x & < & -12 & \text{or} & 3x & > & 2 \\ x & < & -4 & \text{or} & x & > & \frac{2}{3} \end{array}$$

Therefore, the solution set is $(-\infty,-4) \cup (\frac{2}{3},\infty)$.

□

We also have the following theorem:

Theorem 64. *For $a,b \in \mathbb{R}$, $|ab| = |a||b|$. If $b \neq 0$, then $\left|\frac{a}{b}\right| = \frac{|a|}{|b|}$.*

Proof. Using what we know of square roots,

$$|ab| = \sqrt{(ab)^2} = \sqrt{a^2b^2} = \sqrt{a^2}\sqrt{b^2} = |a||b|.$$

Similarly, $\left|\frac{a}{b}\right| = \sqrt{\left(\frac{a}{b}\right)^2} = \sqrt{\frac{a^2}{b^2}} = \frac{\sqrt{a^2}}{\sqrt{b^2}} = \frac{|a|}{|b|}$. □

Sometimes it is easier to deal with factors separately.

Example 5.3.5. Thus $|(-2)(3)| = |-2||3| = 2(3) = 6$.

□

Theorem 65 (Triangle inequality). *If $a, b \in \mathbb{R}$, then $|a+b| \leq |a| + |b|$.*

Proof. Since $x^2 \geq 0$ for all $x \in \mathbb{R}$, $|x|^2 = |x^2| = x^2$ for all $x \in \mathbb{R}$. Consider

$$\begin{aligned}
(|a|+|b|)^2 - |a+b|^2 &= |a|^2 + 2|a||b| + |b|^2 - (a+b)^2 \\
&= a^2 + 2|ab| + b^2 - (a^2 + 2ab + b^2) \\
&= a^2 + 2|ab| + b^2 - a^2 - 2ab - b^2 \\
&= 2(|ab| - ab) \\
&\geq 0 \qquad \text{since } |x| \geq x \text{ for all } x \in \mathbb{R}
\end{aligned}$$

Observe also that if x and y are positive, then $x > y \iff x^2 > y^2$. (Consider $x^2 - y^2$). Therefore, the above calculation shows that $|a| + |b| \geq |a+b|$, which is what we wanted to show. □

Example 5.3.6. $2 = |2| = |(-2) + 4| \leq |(-2)| + |4| = 2 + 4 = 6$.

□

We have seen a system of two-dimensional numbers in the complex numbers. We wish to examine another interpretation of two-dimensional numbers generalizing the ideas from one-dimensional coordinate systems in the previous section.

Suppose that we are given a line (horizontal, for convenience); we know from the previous section that we may assign 0 to any point on that line; say O. Having done this, we now take another line that is **perpendicular to the first** and that also passes through O. We will assign the 0 coordinate of this line to O as well, as in the diagram. These lines are known as **axes**, and each one is an **axis**. From geometry, we know that they lie in a single plane.

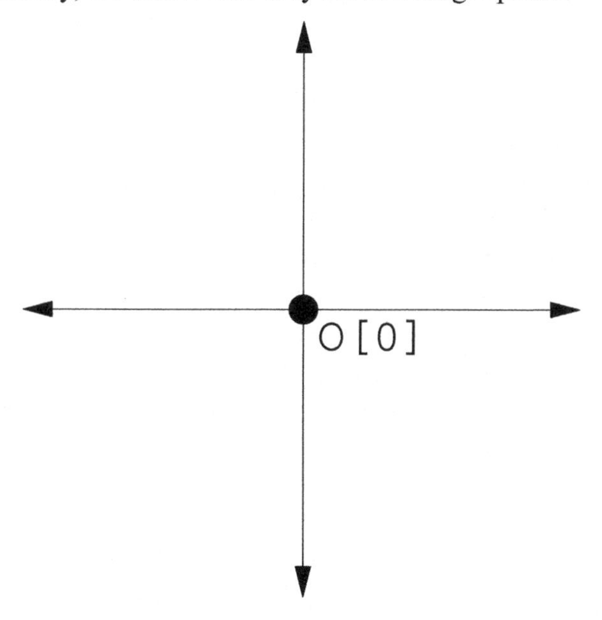

Now consider again the set $\mathbb{C} = \mathbb{R} \times \mathbb{R} = \{(x,y) | x,y \in \mathbb{R}\}$. We will assign to any ordered pair of the form $(x,0)$ the point on our horizontal axis that has coordinate x. Similarly, we will assign to an ordered pair of the form $(0,y)$ the point on the other (necessarily vertical) axis that has coordinate y.

Suppose we are given some ordered pair (x,y), with neither x nor y equal to zero (so that the point lies on neither axis). Our horizontal line has a point P with coordinate x and our vertical axis has a point Q with coordinate y. We now construct a line through P perpendicular to the horizontal axis, and a line through Q perpendicular to the vertical axis. (There are theorems from geometry that allow us to do this.) These two new lines will meet at a unique point R, which we will associate with the ordered pair (x,y). See the diagram below.

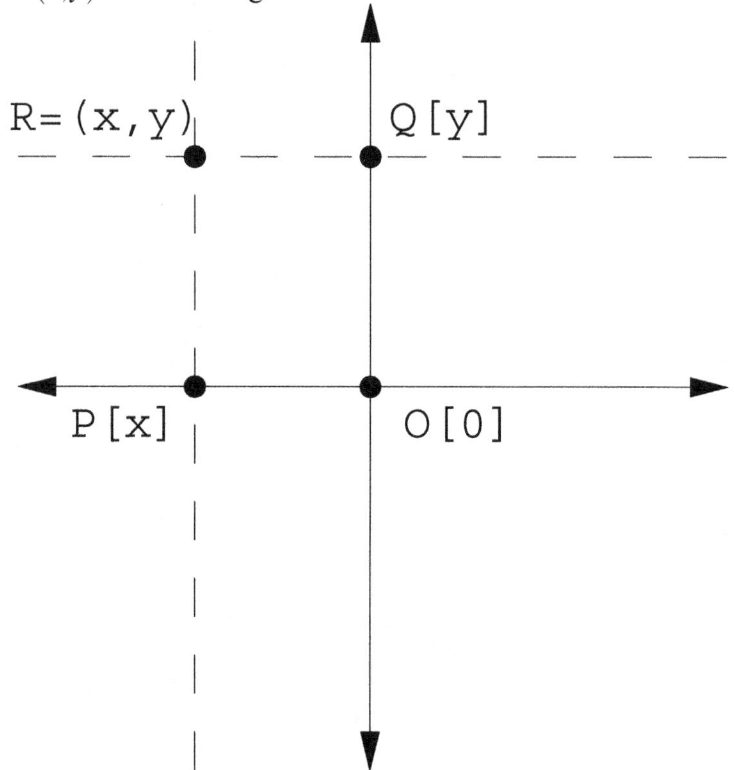

In this way, every ordered pair in $\mathbb{R} \times \mathbb{R}$ is associated with a unique point in the plane, just as previously we associated every real number in $\mathbb{R}$ with a unique point on a given line.

Now consider any point R in the plane. Again, using theorems from geometry, we know that we can construct a line through R that is perpendicular to the horizontal axis; these lines intersect in a unique point P with some coordinate x. We can also construct a line through R that is perpendicular to the vertical axis; these lines intersect in a unique point Q with coordinate y. (See the figure above.) We will associate the point R with the coordinate (x,y). In this way, every point in the plane is associated with a unique ordered pair in $\mathbb{R} \times \mathbb{R}$.

That is, we have shown that there is a one-to-one correspondence between points in the plane and ordered pairs in $\mathbb{R} \times \mathbb{R}$. We will use the ordered pairs as coordinates of points in the plane. When the plane has been coordinatized in this way, we refer to it as the **Cartesian plane**, after René Descartes.

One very important point is that *where* we decide to place the lines is irrelevant; it only gives us a frame of reference. On a map, for example, it might be convenient to put the origin in the lower left corner, or in the very center of the map. Cities will have different coordinates depending on which choice you make, but the actual *positions* of the cities will not change.

Example 5.3.7. Locate the points $(-1,3),(2,5),(-1,-1)$, and $(4,-2)$ in the plane.

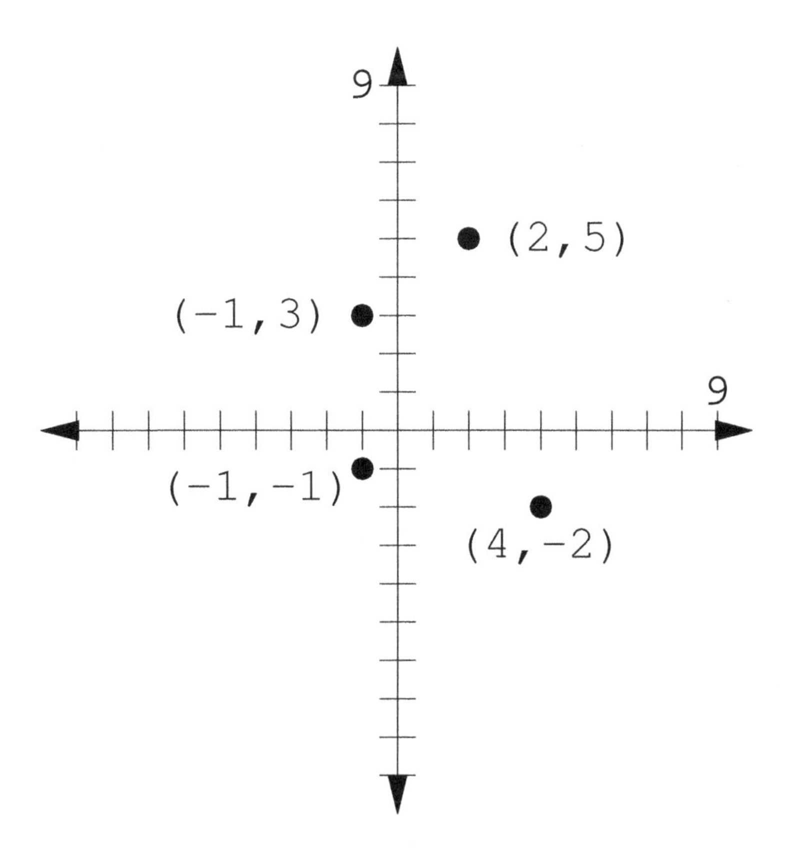

Notice that $(-1,3)$ and $(-1,-1)$ have the same x-coordinate, and they lie on a vertical line (parallel to the vertical axis).

□

We can "see" the complex numbers in the plane by identifying each complex number (x,y) with the appropriate point. In the figure, we have shown i, 1, and $(a,b) = a+bi$. Note that the imaginary part moves the point "vertically," while the real part moves the point "horizontally." We can represent the complex numbers with arrows (vectors) drawn from the origin to the point associated with the complex number. In this context, we refer to the vertical axis as the **imaginary axis** and the horizontal axis as the **real axis**.

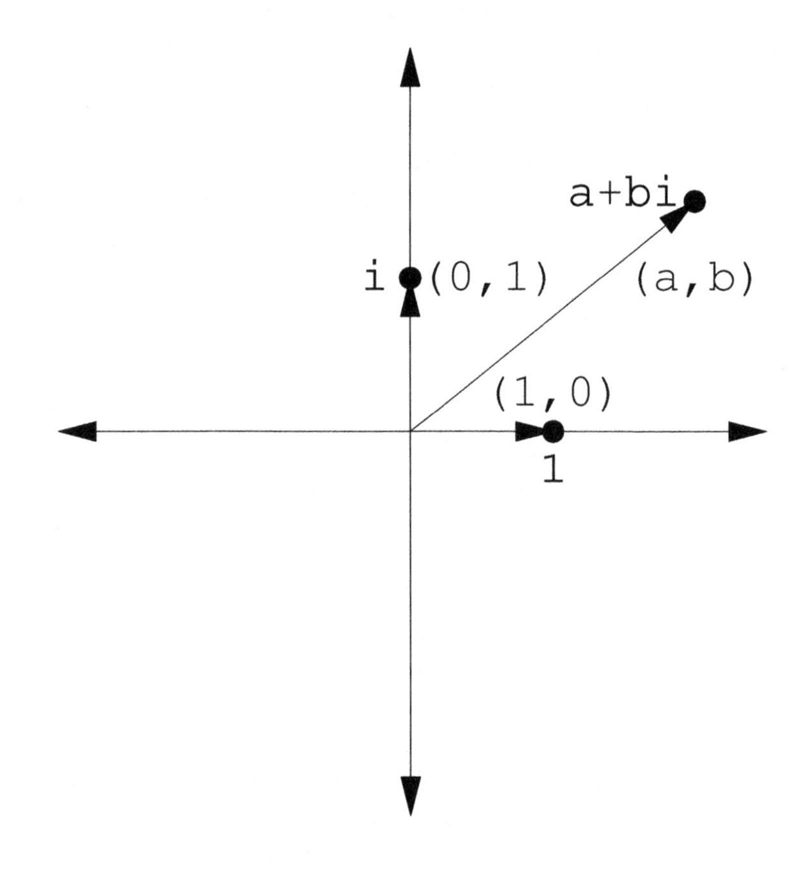

Furthermore, addition in $\mathbb{C}$ has a very pleasant geometric interpretation known as the "parallelogram law." Since $(a,b)+(c,d)=(a+c,b+d)$, we can locate the sum $(a+c,b+d)$ by beginning at the point (a,b) and then moving horizontally by c and vertically by d. This corresponds to placing the arrow for (c,d) at the end of the arrow for (a,b), or vice-versa. The sum is then just the diagonal of the parallelogram so formed!

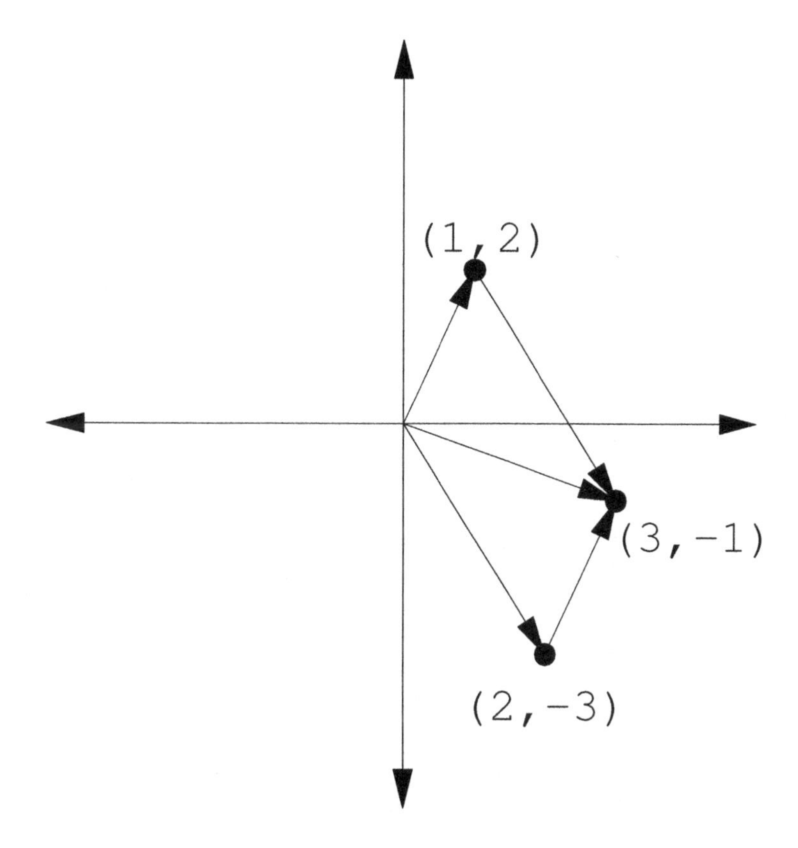

There is also a beautiful interpretation of complex multiplication, but we are not prepared to deal with that at present.

We've already considered the **distance** between two points on a *coordinate line* with coordinates x_1 and x_2; this distance is given by $d(x_1, x_2) = |x_2 - x_1|$. We'll now consider the distance between any two points in the *coordinate plane*; the distance formula is a direct consequence of the Pythagorean theorem and is a generalization of the distance formula for points on a one-dimensional coordinate line.

Consider the following diagram.

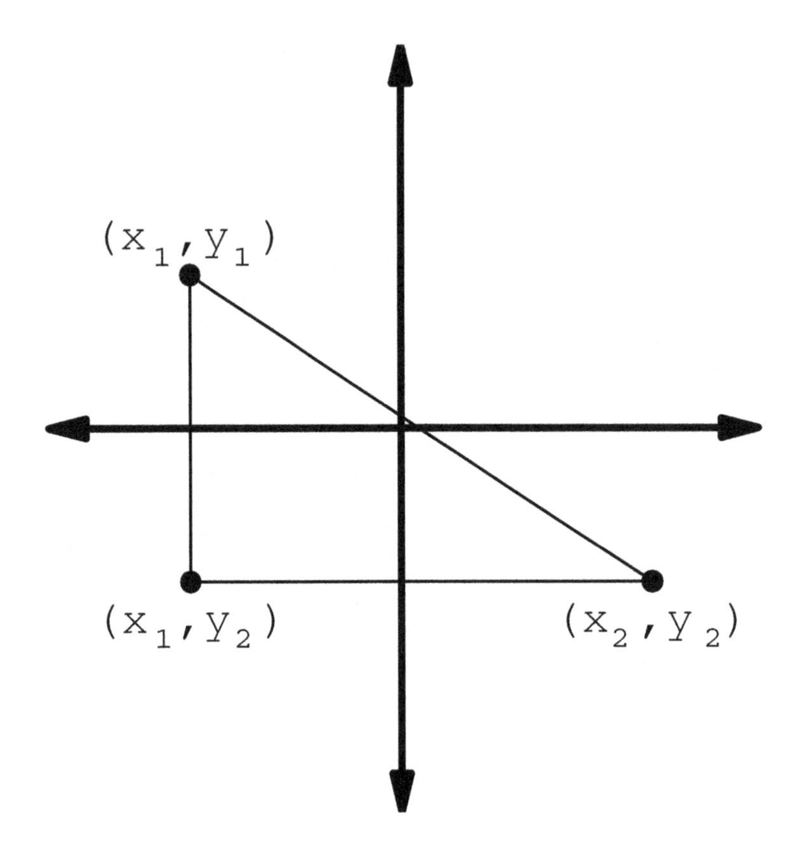

Notice that the legs of the triangle are perpendicular to each other; this is a consequence of the way we set up our coordinate system. (The axes are perpendicular.) Therefore, the distance from (x_1, y_1) to (x_2, y_2) is just the length of the hypotenuse of the right triangle we've formed! The lengths of the legs are easily determined since they are parallel to the axes: the horizontal leg is $|x_2 - x_1|$ units long, and the vertical leg is $|y_2 - y_1|$ units long. Therefore, by the Pythagorean Theorem, the hypotenuse is $\sqrt{|x_2 - x_1|^2 + |y_2 - y_1|^2} = \sqrt{(x_2 - x_1)^2 + (y_2 - y_1)^2}$ units long.

Definition 5.3.3. The **distance** between two points $P_1 = (x_1, y_1)$ and $P_2 = (x_2, y_2)$ **in the Cartesian plane** is given by

$$d(P_1, P_2) = \sqrt{(x_2 - x_1)^2 + (y_2 - y_1)^2}.$$

Example 5.3.8. The distance between $P_1 = \left(-1, \frac{1}{3}\right)$ and $P_2 = (2, -3)$ is

$$\begin{aligned} d(P_1, P_2) &= \sqrt{(-1-2)^2 + \left(\frac{1}{3} - (-3)\right)^2} \\ &= \sqrt{(-3)^2 + \left(\frac{10}{3}\right)^2} \\ &= \sqrt{9 + \frac{100}{9}} \\ &= \sqrt{\frac{181}{9}} \\ &= \frac{\sqrt{181}}{3} \\ &\approx 4.484541. \end{aligned}$$

□

Example 5.3.9. The distance between two points $P_1 = (1, 2)$ and $P_2 = (5, 2)$ is given by

$$d(P_1, P_2) = \sqrt{(5-1)^2 + (2-2)^2} = \sqrt{16 - 0} = 4.$$

Note that this is a generalization of the distance between two points on a vertical coordinate line with coordinates 1 and 5; the distance is $|1 - 5| = 4$.

□

Example 5.3.10. Prove that the distance formula in two-dimensions is a generalization of the distance formula in the one-dimensional case; that is, prove that if two points in the plane lie on a single horizontal or vertical line, then the two-dimensional distance formula yields the one-dimensional distance formula.

Solution: Let $P_1 = (x_1, y_1)$ and $P_2 = (x_2, y_1)$ denote two points (lying on the same horizontal line) in the plane. Then

$$\begin{aligned} d(P_1, P_2) &= \sqrt{(x_2 - x_1)^2 + (y_1 - y_1)^2} \\ &= \sqrt{(x_2 - x_1)^2 + (0)^2} \\ &= \sqrt{(x_2 - x_1)^2} \\ &= |x_2 - x_1| \\ &= d(x_1, x_2) \end{aligned}$$

□

Example 5.3.11. Prove that the four points $A = (-2, 3)$, $B = (3, 5)$, $C = (4, 1)$, and $D = (-1, -1)$ form the vertices of a parallelogram.

Solution: A **parallelogram** is a quadrilateral (four-sided polygon) with exactly two pairs of congruent and parallel sides. Therefore, we need to compute distances between adjacent points and slopes of the respective sides. Calculations yield the desired result.

$$\begin{aligned}
d(A,B) &= \sqrt{(3+2)^2+(5-3)^2} &= \sqrt{29} \\
d(C,D) &= \sqrt{(4+1)^2+(1+1)^2} &= \sqrt{29} \\
d(A,D) &= \sqrt{(-2+1)^2+(3+1)^2} &= \sqrt{17} \\
d(B,C) &= \sqrt{(3-4)^2+(5-1)^2} &= \sqrt{17}
\end{aligned}$$

$$m_{AB} = \frac{(5-3)}{(3+2)} = \frac{2}{5}; \quad m_{CD} = \frac{(1+1)}{(4+1)} = \frac{2}{5}$$

$$m_{AD} = \frac{(3+1)}{(-2+1)} = -4; \quad m_{BC} = \frac{(5-1)}{(3-4)} = -4$$

□

Definition 5.3.4. A **circle** is the set of all points (x,y) in the Cartesian plane that are a fixed distance r from a fixed point (h,k). The fixed distance r is called the **radius** and the fixed point (h,k) is called the **center** of the circle.

Example 5.3.12. An equation for the circle with center $(1,2)$ and radius 3 may be found as follows. Let (x,y) denote an arbitrary point on the circle. Then the distance from (x,y) to $(1,2)$ is 3. Hence,

$$\begin{aligned}
\sqrt{(x-1)^2+(y-2)^2} &= 3 \\
(x-1)^2+(y-2)^2 &= 9 \\
x^2+y^2-2x-4x-4 &= 0
\end{aligned}$$

□

Example 5.3.13. In general, we have that a circle with center (h,k) and radius r has equation $(x-h)^2+(y-k)^2=r^2$. This is known as the **standard form** of the equation of a circle.

□

Example 5.3.14. A special-case circle, known as the **unit circle**, is centered at the origin with radius 1 and equation $x^2+y^2=1$.

□

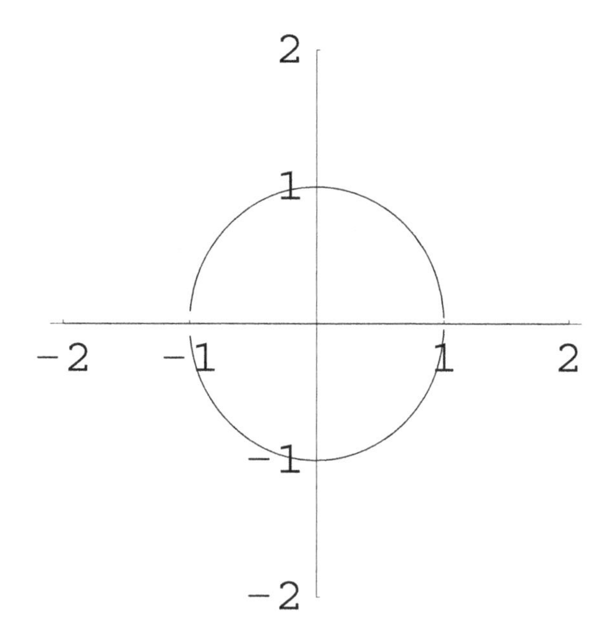

Example 5.3.15. Expanding the standard form equation of an arbitrary circle produces the **general form** of the equation of a circle $x^2+y^2+Ax+By+C=0$, where A, B, and C are constants. Thus, three parameters, A,B,C, characterize a circle. Notice that three points (x,y) will uniquely determine a circle and allow us to "solve" for the unknown constants A, B, and C. A solution technique will be taken up in the matrix unit.

□

Section 5.3 Exercises

Determine each absolute value, if possible.

1. $|3|$
2. $|5-9|$
3. $\left|\frac{-2}{3}\right|$
4. $|x|$
5. $|-a|$
6. $|x^2|$

Determine the exact distance between the two given points on the number line.

7. -2 and -8
8. 6 and 9
9. 8 and -1.7
10. 3 and π
11. 1.41 and $\sqrt{2}$
12. x and $x-2$

Solve each of the following.

13. $|x-3|=5$
14. $|x+2|=\frac{2}{3}$
15. $|3x+7|=\frac{1}{8}$
16. $|2x+4|\leq 3$
17. $|-3x-2|\leq 5$
18. $|\frac{1}{2}x+5|<3$

19. $|3x+2| \geq 7$

20. $|-5x-4| \leq 2$

21. $|\frac{2}{3}x+8| < 9$

22. $|x^2| \geq 0$

23. $|x^2| < 16$

24. $|x^2| \geq 16$

25. Your car runs out of gas one night, and you call for help on your cellular phone. When you give your location, you are told that a gas station is one mile away on the same road. If your position on the road is x, and the position of the gas station is y, what absolute value equation models this situation? What are the possibilities for y? Does this fit with the ambiguity of the directions you received?

26. Show that $|x^2| = |x|^2 = x^2$.

27. Prove that if $x,y > 0$, then $x > y$ if and only if $x^2 > y^2$.

28. If $x > 1$ and $y < -3$, what can $|x-y|$ be? (Your answer should be a range of values.)

Plot the given points.

29. $(-3,4)$

30. $(5,-2)$

31. $(0,7)$

32. $(2,3)$

33. $(3,0)$

34. $(\frac{1}{2},6)$

35. $(\sqrt{2},3)$

36. $(\sqrt{3},-5)$

37. $(\sqrt{7},\sqrt{7})$

Use the parallelogram law to find the sum of the given complex numbers, and compare to the definition of complex addition. We suggest you use graph paper.

38. $(1,2)+(3,4)$

39. $(3,\frac{1}{2})+(1,-\frac{3}{2})$

40. $(2,3)+(-4,-5)$

41. $(0,3)+(-5,-2)$

Plot the given complex numbers and their product. Try to find a pattern connecting the numbers to their product.

42. $(1,2)$ and $(2,3)$

43. $(1,1)$ and $(1,3)$

44. $(1,-1)$ and $(2,1)$

45. $(3,1)$ and $(3,-1)$

46. Recall the Pythagorean theorem from geometry: if $\triangle ABC$ is a right triangle with hypotenuse of length c and legs of length a and b, then $a^2+b^2=c^2$. Use the Pythagorean theorem to show that $|(a,b)|$ is the length of the vector (a,b) in the plane.

47. Show geometrically that if $(a,b) \in \mathbb{C}$, then $(a,b)+\overline{(a,b)} \in \mathbb{R}$. Then show it from the definitions.

48. Determine as accurately as possible the coordinates of each point shown.

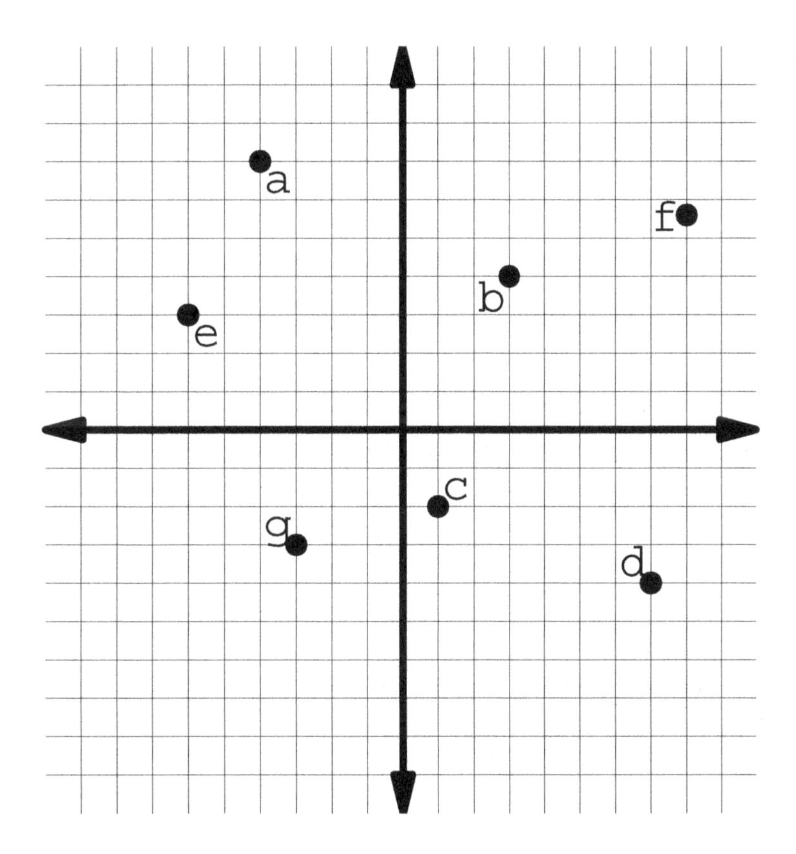

Compute the distance between the given points.

49. $(15,7)$ and $(-2,5)$
50. $(1,4)$ and $(1,7)$
51. $(2,3)$ and $(1,-5)$
52. $(-1,4)$ and $(-3,-7)$
53. $\left(\frac{2}{5},\frac{3}{5}\right)$ and $(9,-2)$
54. $\left(\frac{1}{2},3\right)$ and $\left(2,\frac{3}{5}\right)$

55. Write an equation of the circle centered at $(2,-0.8)$ with radius 2.

56. Write an equation of the circle with center $(-2,3)$ and radius 0.3.

57. If $(5,3)$ and $(2,-4)$ are on opposite ends of a diameter of a circle, determine an equation of the circle.

58. A **rhombus** is a parallelogram with all four sides of equal length. Show that the four points $(1,2),(3,4),(5,2)$, and $(3,0)$ are the vertices of a rhombus.

59. An **ellipse** is defined as follows: given two points (a,b) and (c,d), the ellipse is the set of all points (x,y) in the plane such that the sum of the distances from (x,y) to (a,b) and (x,y) to (c,d) is a constant. Find an equation describing the set of points (x,y) on an ellipse.

60. Suppose you need to drive from $(2,5)$ to $(-12,16)$, but you must pass through $(-4,2)$. (The coordinates are in miles from some central location). If the cities at these points are connected by straight roads and you can average 50 mph, how long will your trip take?

61. While driving west to Houston, you see at one point that Houston is 143 miles away. Some time later, a sign tells you that Houston is 78 miles away.

   (a) Determine reasonable two-dimensional coordinates for the given points.

   (b) Use the two-dimensional distance formula to compute the distance between the two points.

   (c) Explain why you could have just used the one-dimensional distance formula to find the distance.

62. The vertices of an arbitrary rectangle can be represented by the points $(0,0),(a,0),(a,b)$, and $(0,b)$. Show that the diagonals of a rectangle are of equal length.

63. Show that if two points lie on the same vertical line, then the two-dimensional distance formula becomes the one-dimensional distance formula.

64. Prove that for $c,d \in \mathbb{R}$, $|c| \geq d$ if and only if either $c \geq d$ or $c \leq -d$.

65. Show that any circle represented in the form $(x-h)^2+(y-k)^2=r^2$ can also be represented in the form $x^2+y^2+Ax+By+C=0$. Identify A, B, and C.

Chapter 6

Matrices

This unit introduces the notion of a matrix as an algebraic object and defines operations on sets of matrices. Students will investigate which field properties are satisfied by the matrix operations and explore some applications of matrices. In particular, we use an inverse matrix multiplication technique to solve square, nonsingular, linear systems of equations.

Objectives

- To introduce the set of matrices and define operations on the set
- To investigate which of the field properties hold in the set of matrices
- To explore solving systems of equations as an application of matrices
- To introduce a matrix multiplication technique for solving nonsingular square linear systems
- To review the techniques of substitution and elimination (specifically for solving singular or nonlinear systems)

Terms

- matrix
- scalar multiplication
- square matrix
- identity matrix
- inverse matrix
- nonsingular
- invertible
- determinant
- system of equations
- linear system
- substitution
- elimination

6.1 Matrices

Definition 6.1.1. An $m \times n$ **matrix** is a rectangular array of numbers with m rows and n columns. (Note that we specify the number of rows *first*.) An $m \times 1$ matrix is called a **column matrix**; a $1 \times n$ matrix is a **row matrix**. An $n \times n$ matrix is a **square matrix**. The plural of matrix is matrices.

We usually enclose matrices in brackets; as the following example illustrates. Also, we will name matrices with capital letters. In our case, the numbers appearing as entries in matrices will be real numbers, but be aware that they could also be complex numbers.

Example 6.1.1. Let A, B, and C be the matrices defined below.

$$A = \begin{bmatrix} 1 & -3 \\ \pi & x \end{bmatrix}, B = \begin{bmatrix} 2 & -1 & 5 \\ 4 & 2 & 0 \end{bmatrix}, C = \begin{bmatrix} c_{11} & c_{12} & c_{13} & \cdots & c_{1n} \\ c_{21} & c_{22} & c_{23} & \cdots & c_{2n} \\ c_{31} & c_{32} & c_{33} & \cdots & c_{3n} \\ \vdots & \vdots & \vdots & \ddots & \vdots \\ c_{m1} & c_{m2} & c_{m3} & \cdots & c_{mn} \end{bmatrix}.$$

The matrix A is a 2×2 matrix; it is square. We will sometimes write $A_{2\times 2}$.

The matrix B is 2×3, which we could denote by $B_{2\times 3}$.

The matrix $C_{m\times n}$ is a generic matrix; it is $m \times n$, and the symbol c_{ij} represents the entry in row i and column j. This is a useful notation, and we will use it regularly. We can also write C in the more compact form $C = [c_{ij}]$, where the square brackets indicate that we are referring to a matrix, and the symbol c_{ij} is the generic name for the entry in row i and column j.

In this notation, if we put $A = [a_{ij}]$, then $a_{21} = \pi$ since π is the entry in row 2 and column 1 of A. Similarly, if $B = [b_{ij}]$, then $b_{22} = 2$ and $b_{23} = 0$.

□

Definition 6.1.2. Two matrices are **equal** if and only if they have the same dimension and the elements in corresponding positions are equal.

Example 6.1.2. Let B be as above, and let $D = \begin{bmatrix} x & y & z \\ r & s & t \end{bmatrix}$. Then B and D are both 2×3, and they are equal if and only if $x = 2, y = -1, z = 5, r = 4, s = 2$, and $t = 0$.

□

Example 6.1.3. Let $A = \begin{bmatrix} x+2 & 3 \\ y-1 & 2z+4 \\ 3 & w \end{bmatrix}$ and let $B = \begin{bmatrix} 4-x & 3 \\ 2 & z-1 \\ a & 4 \end{bmatrix}$. Then $A = B$ if and only if matrices A and B have equal corresponding entries. Thus we need the following to hold:

$$\begin{aligned} x+2 &= 4-x \\ 3 &= 3 \\ y-1 &= 2 \\ 2z+4 &= z-1 \\ 3 &= a \\ w &= 4 \end{aligned}$$

The first equation becomes $2x = 2$, or $x = 1$; the second is already true; the third requires $y = 3$; the fourth requires $z = -5$, and the last two will be satisfied if $a = 3$ and $w = 4$.

□

Matrices are extremely useful in the sciences. They have wide applications, including computer graphics, solving systems of linear equations, optimization (of business plans, plane routes, etc.), fitting a curve to a set of data points, and many others. We will explore their properties, treating them as a new kind of "number."

With this in mind, we need to define addition and multiplication of matrices.

Definition 6.1.3. Let $A_{m\times n}$ and $B_{m\times n}$ be matrices of the same dimensions. Then define the **sum** $A+B$ by

$$A+B = [a_{ij}] + [b_{ij}] = [a_{ij} + b_{ij}].$$

Addition of matrices is said to be performed **componentwise**. Thus, to find the entry in the second row and third column of the sum (the 2,3 entry), add the entry in the second row and third column of A to the entry in the second row and third column of B.

Example 6.1.4. Let $A = \begin{bmatrix} 1 & -3 & x \\ -4 & 2 & 1 \end{bmatrix}$ and $B = \begin{bmatrix} 2 & 2 & -3 \\ 1 & -5 & 0 \end{bmatrix}$. Then

$$A+B = \begin{bmatrix} 1+2 & -3+2 & x+(-3) \\ -4+1 & 2+(-5) & 1+0 \end{bmatrix} = \begin{bmatrix} 3 & -1 & x-3 \\ -3 & -3 & 1 \end{bmatrix}.$$

□

As with complex numbers, we also have a **scalar multiplication**.

Definition 6.1.4. Let A be an $m \times n$ matrix, and let $r \in \mathbb{R}$. The matrix rA is the matrix obtained by multiplying each entry of A by r.

Example 6.1.5. If $A = \begin{bmatrix} 1 & 2 \\ 3 & 4 \end{bmatrix}$ and $r = -2$, we get

$$-2A = \begin{bmatrix} -2(1) & -2(2) \\ -2(3) & -2(4) \end{bmatrix} = \begin{bmatrix} -2 & -4 \\ -6 & -8 \end{bmatrix}.$$

□

Finally, we define multiplication of matrices as follows:

Definition 6.1.5. Let $A = [a_{ij}]$ be an $m \times n$ matrix and let $B = [b_{ij}]$ be an $n \times r$ matrix. The **product** AB of A and B is the $m \times r$ matrix $C = [c_{ij}]$, where

$$c_{ij} = a_{i1}b_{1j} + a_{i2}b_{2j} + a_{i3}b_{3j} + \ldots + a_{in}b_{nj}.$$

The product is only defined if the number of columns in A is the same as the number of rows in B. One way to think of this is as "multiplying" row i of A by column j of B componentwise and adding the products together.

This is considerably more complicated than addition. It also should convince you that you need to be able to quickly interpret a_{ij} as the entry in row i and column j of matrix A. This definition may seem odd, but we will see a little later why it is a reasonable way to approach the idea of multiplying matrices.

Example 6.1.6. Let $A = \begin{bmatrix} 1 & 2 \\ 5 & -2 \end{bmatrix}$ and $B = \begin{bmatrix} 2 & -3 & 1 \\ -1 & 2 & 4 \end{bmatrix}$. Note that A is 2×2 and B is 2×3, so the product AB is defined. Also, observe that AB will be 2×3. We must find each entry in the product separately.

To find the 1,1 entry, use entries from row 1 of A and column 1 of B : $1(2) + 2(-1) = 0$.
To find the 1,2 entry, use entries from row 1 of A and column 2 of B : $1(-3) + 2(2) = 1$.
To find the 1,3 entry, use entries from row 1 of A and column 3 of B : $1(1) + 2(4) = 9$.
To find the 2,1 entry, use entries from row 2 of A and column 1 of B : $5(2) + (-2)(-1) = 12$.
To find the 2,2 entry, entries from row 2 of A and column 2 of B : $5(-3) + (-2)(2) = -19$.
To find the 2,3 entry, use entries from row 2 of A and column 3 of B : $5(1) + (-2)(4) = -3$.

Therefore, the product $AB = \begin{bmatrix} 0 & 1 & 9 \\ 12 & -19 & -3 \end{bmatrix}$.

□

Section 6.1 Exercises

(These are not in the back of the book.)

1. Using matrices A, B, C, and D find the following as indicated.

$$A = \begin{bmatrix} 1 & 0 & 2 \\ 1 & -1 & 3 \end{bmatrix} \quad B = \begin{bmatrix} 2 & 3 & 4 \\ 1 & 0 & -1 \\ -1 & 1 & 1 \end{bmatrix} \quad C = \begin{bmatrix} 1 & 0 & 1 \\ 1 & 2 & 3 \\ -2 & 1 & 2 \end{bmatrix} \quad D = \begin{bmatrix} 1 & 1 \\ 0 & 2 \\ 1 & 3 \end{bmatrix}$$

   (a) $B + C$ and $C + B$
   (b) BC and CB
   (c) AD and DA
   (d) $(B + C) + DA$ and $B + (C + DA)$
   (e) $A(BC)$ and $(AB)C$
   (f) $A(B + C)$ and $AB + AC$
   (g) $A + \begin{bmatrix} 0 & 0 & 0 \\ 0 & 0 & 0 \end{bmatrix}$ and $\begin{bmatrix} 0 & 0 & 0 \\ 0 & 0 & 0 \\ 0 & 0 & 0 \end{bmatrix} + B$

(h) $-A$ and $-B$

(i) $(2+3)A$ and $2A+3A$

(j) $4(B+C)$ and $4B+4C$

(k) $(2\cdot 7)D$ and $2(7D)$ and $7(2D)$

2. Based on the explorations above, make conjectures concerning the properties of matrix addition, scalar multiplication, and matrix multiplication and discuss whether or not matrix subtraction and matrix division can be defined in a manner analogous to the definitions of real number subtraction and division.

Now: we have a set of "numbers" and operations of addition and scalar multiplication. We want to know which field properties hold; do we have a field? We know from the definitions that the set of $m \times n$ matrices is closed under addition. The set of $n \times n$ matrices is closed under multiplication. However, if m and n are different, then we cannot even multiply two $m \times n$ matrices. We need to modify our question to the following: if we are given n, is the set of $n \times n$ matrices a field?

We begin by listing several properties, in some cases without proof. The proofs would be of the same nature as we had for complex numbers; that is, we would be using properties known for real numbers (commutativity, associativity, etc.) to prove the corresponding properties for matrices.

Theorem 66. *Addition of matrices is associative and commutative. That is, assuming that the necessary sums are defined,* $(A+B)+C = A+(B+C)$ *and* $A+B = B+A$.

Proof. Let $A_{m\times n} = [a_{ij}], B_{m\times n} = [b_{ij}]$, and $C_{m\times n} = [c_{ij}]$. We must show that

$$(A+B)+C = A+(B+C).$$

$$\begin{aligned}
(A+B)+C &= ([a_{ij}]+[b_{ij}])+[c_{ij}] && \text{Notation} \\
&= [a_{ij}+b_{ij}]+[c_{ij}] && \text{Definition of Matrix Addition} \\
&= [(a_{ij}+b_{ij})+c_{ij}] && \text{Definition of Matrix Addition} \\
&= [a_{ij}+(b_{ij}+c_{ij})] && \text{Associativity of Addition in } \mathbb{R} \\
&= [a_{ij}]+[b_{ij}+c_{ij}] && \text{Definition of Matrix Addition} \\
&= [a_{ij}]+([b_{ij}]+[c_{ij}]) && \text{Definition of Matrix Addition} \\
&= A+(B+C) && \text{Notation}
\end{aligned}$$

Therefore, $(A+B)+C = A+(B+C)$ by transitivity of equality. □

Theorem 67. *Multiplication of matrices is associative. That is, assuming the necessary products are defined,* $(AB)C = A(BC)$.

The proof of this is beyond what we intend for this course.

Theorem 68. *Multiplication of matrices is NOT commutative.*

We saw this above when we defined matrix multiplication. For example, if A is 2×2 and B is 2×3, then AB is a 2×3 matrix, but BA is not even defined! This already lets us down — the set of $n \times n$ matrices cannot be a field.

Although matrix multiplication fails to be commutative, we do get some of the other field properties.

Theorem 69. *Matrix multiplication distributes over matrix addition. That is, assuming the necessary sums and products are defined,* $A(B+C) = AB+AC$, *and* $(A+B)C = AC+BC$.

Theorem 70. *The matrix* $0_{m\times n} = \begin{bmatrix} 0 & 0 & \cdots & 0 \\ 0 & 0 & \cdots & 0 \\ \vdots & \vdots & \ddots & \vdots \\ 0 & 0 & \cdots & 0 \end{bmatrix}$ *is an additive identity for the set of* $m\times n$ *matrices. The matrix* $-A = (-1)A$ *is an additive inverse for the* $m\times n$ *matrix* A.

It is not too hard to see that the matrix 0 is an additive identity since addition is performed componentwise; for the same reason, $-A$ will be an additive inverse for A.

Example 6.1.7. Let

$$A = \begin{bmatrix} 2 & 1 & -3.5 \\ -2.71 & -3/8 & 2 \\ 1 & 9 & -4 \\ 0 & -\pi & 6 \end{bmatrix}.$$

Then

$$A + 0_{4\times 3} = \begin{bmatrix} 2 & 1 & -3.5 \\ -2.71 & -3/8 & 2 \\ 1 & 9 & -4 \\ 0 & -\pi & 6 \end{bmatrix} + \begin{bmatrix} 0 & 0 & 0 \\ 0 & 0 & 0 \\ 0 & 0 & 0 \\ 0 & 0 & 0 \end{bmatrix} = \begin{bmatrix} 2 & 1 & -3.5 \\ -2.71 & -3/8 & 2 \\ 1 & 9 & -4 \\ 0 & -\pi & 6 \end{bmatrix} = A.$$

Also,

$$A + (-A) = \begin{bmatrix} 2 & 1 & -3.5 \\ -2.71 & -3/8 & 2 \\ 1 & 9 & -4 \\ 0 & -\pi & 6 \end{bmatrix} + \begin{bmatrix} -2 & -1 & -(-3.5) \\ -(-2.71) & -(-3/8) & -2 \\ -1 & -9 & -(-4) \\ -0 & -(-\pi) & -6 \end{bmatrix}$$

$$= \begin{bmatrix} 2-2 & 1-1 & -3.5-(-3.5) \\ -2.71-(-2.71) & -3/8-(-3/8) & 2-2 \\ 1-1 & 9-9 & -4-(-4) \\ 0-0 & -\pi-(-\pi) & 6-6 \end{bmatrix} = \begin{bmatrix} 0 & 0 & 0 \\ 0 & 0 & 0 \\ 0 & 0 & 0 \\ 0 & 0 & 0 \end{bmatrix} = 0_{4\times 3}.$$

□

Recall the definition of a multiplicative identity: A matrix I is a multiplicative identity on the set of $m \times n$ matrices if and only if $I \cdot A = A \cdot I = A$ for all $m \times n$ matrices A. This means that we must be able to multiply on both sides of A by I. In order to multiply an $m \times n$ matrix by an $m \times n$ matrix, however, we must have $n = m$. That is, the only possibility for a multiplicative identity is a square matrix.

Theorem 71. *The matrix* $I_n = \begin{bmatrix} 1 & 0 & 0 & \cdots & 0 & 0 \\ 0 & 1 & 0 & \cdots & 0 & 0 \\ 0 & 0 & 1 & \cdots & 0 & 0 \\ \vdots & \vdots & \vdots & \ddots & \vdots & \vdots \\ 0 & 0 & 0 & \cdots & 1 & 0 \\ 0 & 0 & 0 & \cdots & 0 & 1 \end{bmatrix}$ *is a multiplicative identity for the set of* $n \times n$ *matrices.*

Example 6.1.8.

$$\begin{bmatrix} 2 & -1 & 4 \\ 8 & -12 & 3 \\ 7 & -2 & -5 \end{bmatrix} \begin{bmatrix} 1 & 0 & 0 \\ 0 & 1 & 0 \\ 0 & 0 & 1 \end{bmatrix} = \begin{bmatrix} 2+0+0 & 0+(-1)+0 & 0+0+4 \\ 8+0+0 & 0+(-12)+0 & 0+0+3 \\ 7+0+0 & 0+(-2)+0 & 0+0+(-5) \end{bmatrix}$$

$$= \begin{bmatrix} 2 & -1 & 4 \\ 8 & -12 & 3 \\ 7 & -2 & -5 \end{bmatrix}$$

□

You should verify that the product above has been computed correctly.

Now that we have a multiplicative identity, we can discuss multiplicative inverses. Consider the following example.

Example 6.1.9. The matrix $\begin{bmatrix} 3 & 1 \\ 2 & 2 \end{bmatrix}$ has a multiplicative inverse; in fact,

$$\begin{bmatrix} 3 & 1 \\ 2 & 2 \end{bmatrix}^{-1} = \begin{bmatrix} 2/4 & -1/4 \\ -2/4 & 3/4 \end{bmatrix}.$$

Indeed,

$$\begin{bmatrix} 3 & 1 \\ 2 & 2 \end{bmatrix}\begin{bmatrix} 2/4 & -1/4 \\ -2/4 & 3/4 \end{bmatrix} = \begin{bmatrix} 1 & 0 \\ 0 & 1 \end{bmatrix} \text{ and } \begin{bmatrix} 2/4 & -1/4 \\ -2/4 & 3/4 \end{bmatrix}\begin{bmatrix} 3 & 1 \\ 2 & 2 \end{bmatrix} = \begin{bmatrix} 1 & 0 \\ 0 & 1 \end{bmatrix}.$$

□

We know that not all matrices have inverses; for example, the 0 matrix cannot. Do all nonzero matrices have multiplicative inverses?

Example 6.1.10. Consider $\begin{bmatrix} 1 & 1 \\ 0 & 0 \end{bmatrix}$. This matrix is not zero. However, if $\begin{bmatrix} a & b \\ c & d \end{bmatrix}$ is any other 2×2 matrix, then $\begin{bmatrix} 1 & 1 \\ 0 & 0 \end{bmatrix}\begin{bmatrix} a & b \\ c & d \end{bmatrix} = \begin{bmatrix} a+c & b+d \\ 0 & 0 \end{bmatrix}$, which cannot equal I_2 since the 2,2 entry is 0, not 1. Therefore, our given matrix **cannot** have a multiplicative inverse even though it is not the zero matrix.

□

This means that having a multiplicative inverse is a fairly special property, perhaps worthy of its own definition!

Definition 6.1.6. A square matrix that has a multiplicative inverse is called **invertible** or **nonsingular**. If A is nonsingular, we denote its multiplicative inverse A^{-1}; we will generally refer to a multiplicative inverse simply as an inverse.

We will work with matrix inverses more explicitly in the next section.

Section 6.1 Exercises

Let $A = \begin{bmatrix} 2 & -1.4 & \sqrt{5} \\ 1 & 0 & -3 \\ -\frac{5}{7} & 3 & 2.718 \end{bmatrix}$.

1. What are the dimensions of A?

2. Determine the 3,2 entry of A.

3. Determine the 1,3 entry of A.

4. Determine the 2,2 entry of A.

5. Determine the 3,1 entry of A.

6. If $\begin{bmatrix} 4-x & 2 \\ y+1 & 3z \end{bmatrix} = \begin{bmatrix} 2x+5 & w+4 \\ 1-y & z-6 \end{bmatrix}$, find x, y, z, and w.

Compute.

7. $\begin{bmatrix} 2 & 7 & -3 \\ 4 & -1 & 6.2 \end{bmatrix} + \begin{bmatrix} 5 & -2 & 2 \\ -4 & 2.5 & 1 \end{bmatrix}$

8. $\begin{bmatrix} -1 & 0 & 2.4+a & 6 \\ -3.5 & 0 & 12 & 7.142 \\ x+2 & 4 & -15 & 2 \end{bmatrix} + \begin{bmatrix} 0 & -4 & 1.09 & 6 \\ 2.43 & x+5 & -4 & -13 \\ 9 & -4 & \sqrt{2} & 12-5x \end{bmatrix}$

9. $\begin{bmatrix} 1 & 4 \\ -2 & 0 \\ 3 & 1 \end{bmatrix} \begin{bmatrix} 4 & 8 \\ 2 & -3 \end{bmatrix}$

10. $\begin{bmatrix} 5 & 3 & -2 \\ 1 & -1 & 2 \end{bmatrix} \begin{bmatrix} 2 & 4 \\ -1 & -6 \\ 2 & 0 \end{bmatrix}$

11. $4\begin{bmatrix} 2 & -3 \\ 5 & 3.4 \end{bmatrix}$

12. $-A$, where $A = \begin{bmatrix} -2 & 3 & -1 \\ 0 & 14 & -9 \end{bmatrix}$

Determine which products are defined (AB, BA, CBA, etc.), if any, for matrices of the given dimensions. If a product is defined, determine the dimensions of the product.

13. $A_{2\times 2}, B_{2\times 5}$

14. $A_{3\times 4}, B_{4\times 3}$

15. $A_{4\times 7}, B_{4\times 7}$

16. $A_{3\times 5}, B_{3\times 4}, C_{5\times 3}$.

Let $A = \begin{bmatrix} 2 & 4 \\ -1 & 0 \end{bmatrix}$, $B = \begin{bmatrix} -2 & 3 \\ -3 & 2 \end{bmatrix}$, and $X = \begin{bmatrix} x & y \\ z & w \end{bmatrix}$. Solve each of the following equations for the matrix X.

17. $2A + 3B = X$
18. $3A - X = B$
19. $A + 2X = B$
20. $7A + 5X = 4B$
21. $\frac{2}{3}B + X = \frac{1}{2}A$
22. $A + \frac{2}{5}X = \frac{2}{5}B$
23. $AB + X = 2A$
24. $BA + X = 2A$
25. $AB + X = BA$

26. Find a multiplicative inverse for the matrix $\begin{bmatrix} 2 & -1 \\ 0 & 1 \end{bmatrix}$.

    [Hint: find x, y, z, w such that $\begin{bmatrix} 2 & -1 \\ 0 & 1 \end{bmatrix} \begin{bmatrix} x & y \\ z & w \end{bmatrix} = \begin{bmatrix} 1 & 0 \\ 0 & 1 \end{bmatrix}$.]

27. We can think of a point (x, y) in the plane as a column matrix $\begin{bmatrix} x \\ y \end{bmatrix}$. Let $A = \begin{bmatrix} 1 & 2 \\ -2 & 1 \end{bmatrix}$.

    For each point, plot the given point, and the point given by the product $A \cdot \begin{bmatrix} x \\ y \end{bmatrix}$. Record any patterns you observe. (Plot all 12 points on the same set of coordinate axes; we suggest you use one color for the given points and another color for the new points.)

    (a) $(2, 3)$
    (b) $(-2, -3)$
    (c) $(4, 6)$
    (d) $(-4, 6)$
    (e) $(1, 5)$
    (f) $(5, -1)$

28. Prove that matrix multiplication distributes over matrix addition.

29. Prove that matrix addition is commutative.

6.2 Matrix Inverses

Let's try to find a multiplicative inverse for a generic 2×2 matrix, $A = \begin{bmatrix} a & b \\ c & d \end{bmatrix}$. If $B = \begin{bmatrix} x & y \\ z & w \end{bmatrix}$ is to be a multiplicative inverse for A, then we need $AB = I_2$. We have

$$AB = \begin{bmatrix} a & b \\ c & d \end{bmatrix} \begin{bmatrix} x & y \\ z & w \end{bmatrix} = \begin{bmatrix} ax+bz & ay+bw \\ cx+dz & cy+dw \end{bmatrix} = \begin{bmatrix} 1 & 0 \\ 0 & 1 \end{bmatrix}.$$

Our job is to determine x, y, z, and w so that this equation is satisfied.

That is, we need the following four equations to be satisfied:

$$\begin{aligned} ax+bz &= 1 \\ ay+bw &= 0 \\ cx+dz &= 0 \\ cy+dw &= 1 \end{aligned}$$

Multiplying both sides of the first equation by d and of the second equation by b changes them to

$$adx+bdz = d \text{ and } bcx+bdz = 0.$$

Subtraction yields

$$adx+bdz-(bcx+bdz) = d-0,$$

so that $adx-bcx = d$. Factoring out an x gives $(ad-bc)x = d$, so that

$$x = \frac{d}{ad-bc},$$

provided $ad-bc \neq 0$.

Multiplying both sides of the first equation by c and the third by a gives

$$acx+bcz = c \text{ and } acx+adz = 0,$$

so that subtraction again yields

$$acx+adz-(acx-bcz) = 0-c.$$

Therefore, $adz-bcz = -c$, and $(ad-bc)z = -c$. That is,

$$z = \frac{-c}{ad-bc},$$

provided that $ad-bc \neq 0$.

Next, multiply the second equation by c and the fourth equation by a to get

$$acy+bcw = 0 \text{ and } acy+adw = a,$$

giving

$$acy+adw-(acy+bcw) = a-0,$$

or $adw - bcw = a$. This time we get

$$w = \frac{a}{ad - bc},$$

provided that $ad - bc \neq 0$.

Finally, multiply the second equation by d and the fourth equation by b to obtain that

$$ady + bdw = 0 \text{ and } bcy + bdw = b.$$

Then

$$ady + bdw - (bcy + bdw) = 0 - b \implies ady - bcy = -b,$$

so that

$$y = \frac{-b}{ad - bc},$$

provided $ad - bc \neq 0$.

Consider now the matrix we have built:

$$\begin{bmatrix} x & y \\ z & w \end{bmatrix} = \begin{bmatrix} \dfrac{d}{ad-bc} & \dfrac{-b}{ad-bc} \\ \dfrac{-c}{ad-bc} & \dfrac{a}{ad-bc} \end{bmatrix} = \frac{1}{ad-bc}\begin{bmatrix} d & -b \\ -c & a \end{bmatrix}.$$

We need to verify that this is in fact a multiplicative inverse for our original matrix.

$$\begin{aligned} \frac{1}{ad-bc}\begin{bmatrix} d & -b \\ -c & a \end{bmatrix}\begin{bmatrix} a & b \\ c & d \end{bmatrix} &= \frac{1}{ad-bc}\begin{bmatrix} da-bc & db-bd \\ -ca+ac & -cb+ad \end{bmatrix} \\ &= \begin{bmatrix} \dfrac{ad-bc}{ad-bc} & 0 \\ 0 & \dfrac{ad-bc}{ad-bc} \end{bmatrix} \\ &= \begin{bmatrix} 1 & 0 \\ 0 & 1 \end{bmatrix}. \end{aligned}$$

You should verify that taking the product in the reverse order also gives I_2.

We have found that a matrix $\begin{bmatrix} a & b \\ c & d \end{bmatrix}$ will have an inverse provided that $ad - bc \neq 0$; it is interesting that the same quantity arose in all four equations above. This suggests that perhaps the quantity $ad - bc$ is special! In fact, it is, and we give it its own name.

Definition 6.2.1. The **determinant** of a 2×2 matrix $A = \begin{bmatrix} a & b \\ c & d \end{bmatrix}$ is $|A| = ad - bc$. (Although the symbol $|A|$ looks like an absolute value, it is not. The notation $\det(A)$ is also used.)

Determinants can also be defined for any $n \times n$ matrix. Your calculator can handle the details of computing them, and we will not learn here how to compute them by hand. (Your calculator should also be able to find inverses, sums, and products.)

The significance of determinants is given by the following theorem.

Theorem 72. *The matrix A has a multiplicative inverse if and only if $|A| \neq 0$. For a 2×2 matrix $A = \begin{bmatrix} a & b \\ c & d \end{bmatrix}$, $A^{-1} = \frac{1}{|A|}\begin{bmatrix} d & -b \\ -c & a \end{bmatrix}$.*

The proof of this theorem is far beyond the scope of this course. If you would like to know more of the details about how matrices work, be patient and take a linear algebra course later, or talk to your teacher about it.

Example 6.2.1. Let $A = \begin{bmatrix} 3 & -1 \\ 4 & 2 \end{bmatrix}$. Notice that $|A| = 3 \cdot 2 - (-1) \cdot 4 = 10$, so the theorem above implies that A has a multiplicative inverse. This inverse is

$$A^{-1} = \frac{1}{10}\begin{bmatrix} 2 & 1 \\ -4 & 3 \end{bmatrix}.$$

□

Having matrix inverses allows us to solve some matrix equations.

Theorem 73. *If A is a square matrix with $|A| \neq 0$, then the matrix equation $AX = B$ has a unique solution $X = A^{-1}B$.*

Proof. Suppose the A is a square matrix with $|A| \neq 0$. Then

$$\begin{array}{rcll} AX & = & B & \text{Given equation} \\ A^{-1} \cdot (AX) & = & A^{-1} \cdot B & A^{-1} \text{ exists, Multiplication by equals} \\ (A^{-1}A) \cdot X & = & A^{-1} \cdot B & \text{Associativity of matrix multiplication} \\ I \cdot X & = & A^{-1} \cdot B & \text{Multiplicative inverses} \\ X & = & A^{-1} \cdot B & \text{Multiplicative identity.} \end{array}$$

□

Example 6.2.2. Solve $AX = B$ for X, where $A = \begin{bmatrix} 2 & -3 \\ 1 & 1 \end{bmatrix}$ and $B = \begin{bmatrix} 1 & 2 \\ -2 & 0 \end{bmatrix}$.

Solution: Notice that $|A| = 2 \cdot 1 - (-3) \cdot 1 = 5$, so A has a multiplicative inverse. Therefore, the matrix equation $AX = B$ has a unique solution $X = A^{-1}B$, so $X = \frac{1}{5}\begin{bmatrix} 1 & 3 \\ -1 & 2 \end{bmatrix}\begin{bmatrix} 1 & 2 \\ -2 & 0 \end{bmatrix} = \frac{1}{5}\begin{bmatrix} -5 & 2 \\ -5 & -2 \end{bmatrix}$.

□

Note that when confronted with an equation such as $AX = B$, many students will write

$$A^{-1}(AX) = BA^{-1}$$

for the next step. However, the property of equality we know is that if $x = y$, then $cx = cy$, not $cx = yc$. In a field it will not matter since multiplication in a field is commutative, but for matrices, *this is an error*. You should proceed to $A^{-1}(AX) = A^{-1}B$.

Section 6.2 Exercises

Compute the determinant of each 2×2 matrix. If the determinant is not 0, find the inverse of the given matrix.

1. $\begin{bmatrix} 4 & 3 \\ 3 & 2 \end{bmatrix}$

2. $\begin{bmatrix} 2 & -6 \\ -3 & 9 \end{bmatrix}$

3. $\begin{bmatrix} 0 & -4 \\ -2 & 3 \end{bmatrix}$

4. $\begin{bmatrix} 1 & 2 \\ -5 & -9 \end{bmatrix}$

5. $\begin{bmatrix} 1/2 & 2/3 \\ 5/7 & -1/8 \end{bmatrix}$

6. $\begin{bmatrix} 1 & a \\ b & 2 \end{bmatrix}$

Solve each equation. Do not use your calculator unless instructed to do so.

7. $\begin{bmatrix} 2 & -3 \\ 4 & 1 \end{bmatrix} X = \begin{bmatrix} 2 & -1 \\ 6 & -9 \end{bmatrix}$

8. $\begin{bmatrix} 1 & 2 \\ -3 & 5 \end{bmatrix} X + \begin{bmatrix} 1 & 3 \\ 5 & -7 \end{bmatrix} = \begin{bmatrix} 7 & 9 \\ -2 & 4 \end{bmatrix}$

9. $\begin{bmatrix} 4 & -2 & 16 \\ 2 & 1 & 1 \end{bmatrix} + X = \begin{bmatrix} -6 & 12 & 2 \\ 17 & 53 & -5 \end{bmatrix}$

10. $\begin{bmatrix} 4 & -2 & 16 \\ 2 & 1 & 1 \end{bmatrix} - 3X = \begin{bmatrix} -6 & 12 & 2 \\ 17 & 53 & -5 \end{bmatrix}$

11. $3\begin{bmatrix} 28 & -46 \\ 51 & 13 \end{bmatrix} + \begin{bmatrix} 15 & 12 \\ -10 & 7 \end{bmatrix} X = \begin{bmatrix} -6 & 14 \\ 20 & 48 \end{bmatrix}$

12. $\begin{bmatrix} 5 & -2 & 4 \\ 2 & 3 & -1 \\ 5 & 2 & 2 \end{bmatrix} X = \begin{bmatrix} 3 & 7 & -1 \\ 2 & 0 & -4.6 \\ 1 & 4 & 7 \end{bmatrix}$ [You will need to do this one on your calculator.]

13. Show that if $A = \begin{bmatrix} a & b \\ ka & kb \end{bmatrix}$, then $\det(A) = 0$.

14. Show that if $\begin{bmatrix} a & b \\ c & d \end{bmatrix}$ has determinant 0, then $c = ka$ and $d = kb$ for some $k \in \mathbb{R}$, or $a = b = 0$.

15. Using the definition of matrix equality, determine what conditions must be satisfied in order to have $\begin{bmatrix} 2 & 5 \\ -1 & 3 \end{bmatrix} \begin{bmatrix} x \\ y \end{bmatrix} = \begin{bmatrix} -4 \\ 2 \end{bmatrix}$. Do not solve the equation, just list which entries must be equal; you will need to multiply the matrices on the left-hand side of the equation first.

6.3 Systems of Equations

Many problems involve "solving" for a set of variables subject to several conditions. The resulting algebraic representation is a system of equations or inequalities. There exist multiple (algebraic and graphical) solution techniques for systems of equations. We will review the techniques of **substitution** and **elimination** and introduce a **matrix multiplication** technique for square, linear systems. We will also discuss the graphical analysis of both linear and non-linear systems of equations.

Definition 6.3.1. A **system of equations** is a collection of two or more equations in one or more variables. A **solution** to a system of equations is a set of values for the variables that makes all of the equations true.

Definition 6.3.2. A **linear** system of equations is a system of equations consisting entirely of linear equations.

Example 6.3.1. The linear system given by

$$\begin{array}{rcrcrcl} x & + & y & - & z & = & -1 \\ 4x & - & 3y & + & 2z & = & 16 \\ 2x & - & 2y & - & 3z & = & 5 \end{array}$$

has solution $x = 2, y = -2$, and $z = 1$, as may be verified by direct substitution:

$$\begin{array}{rcrcl} 2+(-2)-1 & = & 2-2-1 & = & -1 \\ 4(2)-3(-2)+2(1) & = & 8+6+2 & = & 16 \\ 2(2)-2(-2)-3(1) & = & 4+4-3 & = & 5. \end{array}$$

Such a solution is typically denoted as an **ordered triple** $(x,y,z) = (2,-2,1)$.

□

Example 6.3.2. At a local cinema, regular tickets sell for \$6.25 while senior citizen discounted tickets sell for \$4.25. Over a particular weekend in which 600 tickets were sold, ticket sales totaled \$2966.00. Write a system of equations to model this situation.

Solution: Let x denote the number of regular tickets sold, and let y denote the number of senior citizen tickets sold. Then we have two conditions by which we may relate the variables x and y. First, we know that there were a total of 600 tickets sold; that is, $x+y = 600$. Also, we know total ticket sales came to \$2966.00; thus, $6.25x+4.25y = 2966$. Therefore, we have the following *linear system of two equations in two unknowns*:

$$\begin{array}{rcrcl} x & + & y & = & 600 \\ 6.25x & + & 4.25y & = & 2966 \end{array}$$

□

If we desire the number of each type of ticket sold in the previous example, then we need to "solve" the system; that is, find all values of the variables that make the system true. This solution will consist of ordered pairs (x,y).

One method of solution for a square linear system consists of converting the system to an appropriate matrix equation $AX = B$ and, if possible, using the inverse matrix A^{-1} to obtain a solution.

Example 6.3.3. Express the system

$$\begin{array}{rcrcr} 2x & + & 3y & = & 7 \\ -4x & - & y & = & 3 \end{array}$$

as an equivalent matrix equation.

Solution: Let $A = \begin{bmatrix} 2 & 3 \\ -4 & -1 \end{bmatrix}$ denote the **coefficient matrix** (that is, the matrix of coefficients), let $X = \begin{bmatrix} x \\ y \end{bmatrix}$ denote the **variable matrix**, and let $B = \begin{bmatrix} 7 \\ 3 \end{bmatrix}$ denote the **constant matrix**. Then

$$AX = \begin{bmatrix} 2 & 3 \\ -4 & -1 \end{bmatrix} \cdot \begin{bmatrix} x \\ y \end{bmatrix} = \begin{bmatrix} 2x+3y \\ -4x-y \end{bmatrix} = \begin{bmatrix} 7 \\ 3 \end{bmatrix} = B.$$

This is equivalent to the original system of equations by the *definition* of matrix equality.

□

Example 6.3.4. Solve the system

$$\begin{array}{rcrcr} 2x & + & 3y & = & 7 \\ -4x & - & y & = & 3 \end{array}$$

by solving an equivalent matrix equation.

Solution: Let $A = \begin{bmatrix} 2 & 3 \\ -4 & -1 \end{bmatrix}$ denote the coefficient matrix, let $X = \begin{bmatrix} x \\ y \end{bmatrix}$ denote the variable matrix, and let $B = \begin{bmatrix} 7 \\ 3 \end{bmatrix}$ denote the constant matrix, as above. Then $AX = B$, as noted above. Since the determinant of the coefficient matrix is $|A| = -2-(-12) = 10 \neq 0$, we know that A^{-1} exists and that $X = A^{-1}B$ is the unique solution to the system. Hence,

$$X = A^{-1}B = \begin{bmatrix} -0.1 & -0.3 \\ 0.4 & 0.2 \end{bmatrix} \cdot \begin{bmatrix} 7 \\ 3 \end{bmatrix} = \begin{bmatrix} -1.6 \\ 3.4 \end{bmatrix}$$

Thus, we see that $\begin{bmatrix} x \\ y \end{bmatrix} = \begin{bmatrix} -1.6 \\ 3.4 \end{bmatrix}$. Therefore, the solution to the original system is $x = -1.6$ and $y = 3.4$, as may be verified by direct substitution.

□

Example 6.3.5. Solve the following 3×3 linear system.

$$\begin{array}{rcrcrcr} 3x & - & 3y & + & 6z & = & 20 \\ x & - & 3y & + & 10z & = & 40 \\ -x & + & 3y & - & 5z & = & 30 \end{array}$$

Solution: This linear system can be translated into the matrix equation $AX = B$, where $A = \begin{bmatrix} 3 & -3 & 6 \\ 1 & -3 & 10 \\ -1 & 3 & -5 \end{bmatrix}$, $X = \begin{bmatrix} x \\ y \\ z \end{bmatrix}$, and $B = \begin{bmatrix} 20 \\ 40 \\ 30 \end{bmatrix}$. Since $|A| = -30 \neq 0$, the matrix A^{-1} exists and

$$X = A^{-1}B = \begin{bmatrix} 0.5 & -0.1 & 0.4 \\ 0.1\overline{6} & 0.3 & 0.8 \\ 0 & 0.2 & 0.2 \end{bmatrix} \cdot \begin{bmatrix} 20 \\ 40 \\ 30 \end{bmatrix} = \begin{bmatrix} 18 \\ 39.\overline{3} \\ 14 \end{bmatrix}.$$

Hence, the original system has solution $x = 18$, $y = 39.\overline{3}$, and $z = 14$. (Here, we used a calculator to find A^{-1}.)

□

Example 6.3.6. Find an equation of the circle ($x^2 + y^2 + Ax + By + C = 0$) that passes through the points $(1,1)$, $(-2,5)$ and $(2,-3)$.

Solution: Specifying the three unknown constants A, B, and C will solve this problem. We create a 3×3 linear system in the unknowns A, B, and C by substituting each of the three given points into the equation $x^2 + y^2 + Ax + By + C = 0$.

- For the point $(1,1)$ we get $1^2 + 1^2 + A \cdot 1 + B \cdot 1 + C = 0$.
- For the point $(-2,5)$ we get $(-2)^2 + 5^2 + A \cdot (-2) + B \cdot 5 + C = 0$.
- For the point $(2,-3)$ we get $2^2 + (-3)^2 + A \cdot 2 + B \cdot (-3) + C = 0$.

Thus, we obtain the following 3×3 linear system:

$$\begin{array}{rcrcrcr} A & + & B & + & C & = & -2 \\ -2A & + & 5B & + & C & = & -29 \\ 2A & - & 3B & + & C & = & -13 \end{array}$$

This linear system can be translated into the matrix equation

$$\begin{bmatrix} 1 & 1 & 1 \\ -2 & 5 & 1 \\ 2 & -3 & 1 \end{bmatrix} \cdot \begin{bmatrix} A \\ B \\ C \end{bmatrix} = \begin{bmatrix} -2 \\ -29 \\ -13 \end{bmatrix}.$$

Since $\left| \begin{bmatrix} 1 & 1 & 1 \\ -2 & 5 & 1 \\ 4 & -3 & 1 \end{bmatrix} \right| = 8 \neq 0$, the matrix $\begin{bmatrix} 1 & 1 & 1 \\ -2 & 5 & 1 \\ 4 & -3 & 1 \end{bmatrix}^{-1}$ exists and

$$\begin{bmatrix} A \\ B \\ C \end{bmatrix} = \begin{bmatrix} 1 & 1 & 1 \\ -2 & 5 & 1 \\ 4 & -3 & 1 \end{bmatrix}^{-1} \begin{bmatrix} -2 \\ -29 \\ -13 \end{bmatrix} = \begin{bmatrix} 19 \\ 7.5 \\ -28.5 \end{bmatrix}.$$

Hence, an equation of the desired circle is given by $x^2 + y^2 + 19x + 7.5y - 28.5 = 0$.

□

As mentioned earlier, *if* $|A| \neq 0$, then $AX = B$ has a unique solution. This says nothing about the case where $|A| = 0$. For our purposes, we will revert to the elementary techniques of substitution and/or elimination in these cases. It should be noted, however, that there do exist matrix techniques to find solutions to linear systems in the cases with $|A| = 0$, but these techniques are more appropriately left to a linear algebra course.

Section 6.3 Exercises

1. Verify that $x = 8, y = 1$ is a solution to the system

$$\begin{aligned} 2x - 12y &= 8 \\ 3x - y &= 23 \end{aligned}$$

2. Verify that $x = 2, y = -1, z = 4$ is a solution to the system

$$\begin{aligned} 3x - 5y + 8z &= 43 \\ -x + 4y + z &= -2 \\ 4x + y &= 7 \end{aligned}$$

Express each system of equations as a single matrix equation. Then solve each system.

3. $$\begin{aligned} 3x - 2y &= 4 \\ -7x + 3y &= 5 \end{aligned}$$

4. $$\begin{aligned} 4x + 5y &= -6 \\ 2y &= 11 \end{aligned}$$

5. $$\begin{aligned} -2x + 4y - 2z &= -3 \\ 3x - 2y + 2z &= 1 \\ x + 5y - z &= 0 \end{aligned}$$

6. $$\begin{aligned} 2y + 3z &= 0 \\ x - y + 3z &= 0 \\ x + 2y + 7z &= 0 \end{aligned}$$

Write a system of equations to model each situation. Solve each system.

7. Papa Joe is twice as old as his daughter. Ten years ago, Papa Joe was three times as old as his daughter. How old are they now?

8. The University of Kentucky Wildcats made 40 field goals in a recent basketball game, some 2-pointers and some 3-pointers. Altogether the 40 baskets accounted for 89 points. How many field goals of each type did they make?

9. Mountainside Fleece sold 40 neckwarmers. Solid color neckwarmers sold for \$9.90 each and print ones sold for \$12.75 each. In all, \$421.65 was taken in for the neckwarmers. How many of each type were sold?

10. W&G, Inc. makes three kinds of toys: balls, (toy) cars, and pogo sticks, and they are able to sell all that they make. The balls cost \$0.75 cents to make, and sell for \$1.65; the cars cost \$0.52 to make, and sell for \$0.99; the pogo sticks cost \$2.31 to make, and sell for \$5.95. Last month, W&G, Inc. sold 361 toys. Their production costs were \$313.05, and their income was \$708.03. How many of each kind of toy did they make and sell?

11. What happens if you attempt to find an equation of the circle through the three points $(-2,5),(1,3)$, and $(7,-1)$? Why?

12. Find an equation of the circle that passes through the points $(-2,4),(5,-1)$, and $(3,1)$.

13. Find an equation of the circle that passes through the points $(1,2),(3,-1)$, and $(4,4)$.

14. The set of all points (x,y) satisfying $y=Ax^2+Bx+C$ is called a **parabola**. Find an equation of a parabola passing through the three points $(2,0),(-1,3)$, and $(5,2)$.

15. Find an equation of a parabola passing through the points $(-4,1),(2,-3)$, and $(0,5)$.

16. Explain how to solve the system

$$\begin{array}{rcrl} 4x & + & 5y & = -6 \\ & & 2y & = 11 \end{array}$$

without using matrix methods. What makes it "easy"?

6.4 Substitution and Elimination

We will briefly review the techniques of **substitution** and **elimination** and provide examples of each.

Substitution refers to the technique of solving one equation in a system for a particular variable and substituting the resulting quantity for that variable in all other equations. In a system of two equations with two unknowns, this reduces the system to a single equation in one variable, which we can solve using our techniques derived from the field properties.

Example 6.4.1. To solve the system

$$\begin{array}{rcrcl} 2x & + & y & = & 5 \\ -4x & + & 6y & = & 12 \end{array}$$

by substitution, we will solve the first equation for $y = 5 - 2x$ and substitute into the second equation; we get $-4x + 6(5 - 2x) = 12$. This equation can be solved in a straightforward manner to obtain $x = \frac{9}{8}$. Substituting this value for x into the first equation yields $2\left(\frac{9}{8}\right) + y = 5$ which can be further solved to obtain $y = \frac{11}{4}$. The resultant solution to the original system is the ordered pair $(x, y) = \left(\frac{9}{8}, \frac{11}{4}\right)$.

□

Example 6.4.2. To solve the system

$$\begin{array}{rcrcl} x & + & y & = & 600 \\ 6.25x & + & 4.25y & = & 2966 \end{array}$$

that resulted from the cinema scenario, we may solve the first equation for $y = 600 - x$ and substitute this into the second equation; we get $6.25x + 4.25(600 - x) = 2966$. Solving this equation by applying the appropriate field properties, we get $x = 208$. Upon substituting this into the first equation, we get $y = 392$. Therefore, the cinema sold 208 regular tickets and 392 senior citizen tickets over the weekend in question.

□

Elimination refers to a technique for solving systems of equations whereby one multiples one equation by an appropriate real number so that when this equation is added to another equation from the system, one or more variables is eliminated. Note that the legitimacy of "multiplying an equation on both sides by the same amount" has already been established. However, we've not yet discussed the legitimacy of "adding two equations." Or have we? In fact, we *have* established that "the same amount can be added to both sides of an equation." This allows for the addition of "equal amounts" to both sides of an equation, which is precisely what occurs in the "addition of two equations."

Example 6.4.3. To solve the system

$$\begin{array}{rcrcl} 2x & + & y & = & 5 \\ -4x & + & 6y & = & 12 \end{array}$$

by elimination, we may multiply the first equation by 2. This produces the equivalent system

$$\begin{array}{rcrcl} 4x & + & 2y & = & 10 \\ -4x & + & 6y & = & 12. \end{array}$$

Together, these look like $a = b$ and $c = d$, so $a+c = b+d$. The equation $a+c = b+d$ is what we mean by "adding" the two equations $a = b$ and $c = d$. Upon adding the two equations in our problem, the variable x is "eliminated" and we are left with $8y = 22$. This reduces to the equivalent equation $y = \frac{11}{4}$. We substitute this y-value into either equation to find the corresponding x-value: $x = \frac{5-y}{2} = \frac{(9/4)}{2} = \frac{9}{8}$. The solution is therefore $\left(\frac{9}{8}, \frac{11}{4}\right)$.

□

Example 6.4.4. Solve the following linear system.

$$\begin{array}{rcrcl} 2.5x & + & 6.1y & = & 5.3 \\ -0.4x & + & 6.8y & = & 12.7 \end{array}$$

Solution: To solve a system of equations with decimal coefficients using elimination, we may first multiply *both* equations by an appropriate power of 10 to produce an equivalent system with integer coefficients. In this example, we multiply both equations by 10 to obtain the equivalent system

$$\begin{array}{rcrcl} 25x & + & 61y & = & 53 \\ -4x & + & 68y & = & 127, \end{array}$$

which we solve using elimination. To eliminate x, we may multiply the first equation by 4 and the second by 25. We get

$$\begin{array}{rcrcl} 100x & + & 244y & = & 212 \\ -100x & + & 1700y & = & 3175, \end{array}$$

which reduces to $1944y = 3387$. We see that $y = \frac{1129}{648}$. Substituting this y-value into the first equation yields $25x + 61\left(\frac{1129}{648}\right) = 53$, so that $x = -\frac{1381}{648}$. Therefore, the solution to the original system is $\left(-\frac{1381}{648}, \frac{1129}{648}\right)$.

□

Example 6.4.5. Solve the given general linear system

$$\begin{array}{rcrcl} ax & + & by & = & c \\ dx & + & ey & = & f. \end{array}$$

Solution: We'll first attempt to eliminate x by multiplying the first equation by $-d$ and the second by a.

$$\begin{array}{rcrcl} -dax & - & dby & = & -dc \\ adx & + & aey & = & af \end{array}$$

Adding these equations eliminates x, and we obtain $(ae-bd)y = af-cd$ using commutativity of real multiplication. Provided that $ae-bd \neq 0$, we get that $y = \frac{af-cd}{ae-bd}$. Substituting this for y in the original first equation produces $ax+b\left(\frac{af-cd}{ae-bd}\right) = c$, and solving this yields

$$x = \frac{c-b\left(\frac{af-cd}{ae-bd}\right)}{a} = \frac{cae-cdb-baf+bcd}{a(ae-bd)} = \frac{ce-bf}{ae-bd}.$$

Thus, provided $ae-bd \neq 0$, we obtain the ordered pair $\left(\frac{ce-bf}{ae-bd}, \frac{af-cd}{ae-bd}\right)$ as the solution to the original system. (Notice that $ae-bd$ would be the determinant of the coefficient matrix if we used matrix techniques!)

We had a division by a when we solved for x; if $a=0$, this is a problem. However, you can verify for yourself that even if $a=0$, the solution will be the same anyway.

□

Thus far, we've mostly considered what are known as 2×2 **linear systems**; that is, a linear system involving exactly two equations and two unknowns. Let us consider a graphical interpretation in this case. It is well known that the graph of a linear equation in two variables is a line in the plane. Therefore, the graph of two linear equations in two variables (or unknowns) consists of two lines in the plane. (We will discuss graphs more formally later.)

Example 6.4.6. Consider a previous example of a 2×2 linear system,

$$\begin{array}{rcrcl} 2x & + & y & = & 5 \\ -4x & + & 6y & = & 12. \end{array}$$

We see that the first equation can be rewritten in slope-intercept form as $y=-2x+5$; the graph is a line with slope $m=-2$ and y-intercept $(0,5)$. The second equation can be rewritten in slope-intercept form as $y=\frac{2}{3}x+2$; the graph is a line with slope $m=\frac{2}{3}$ and y-intercept $(0,2)$.

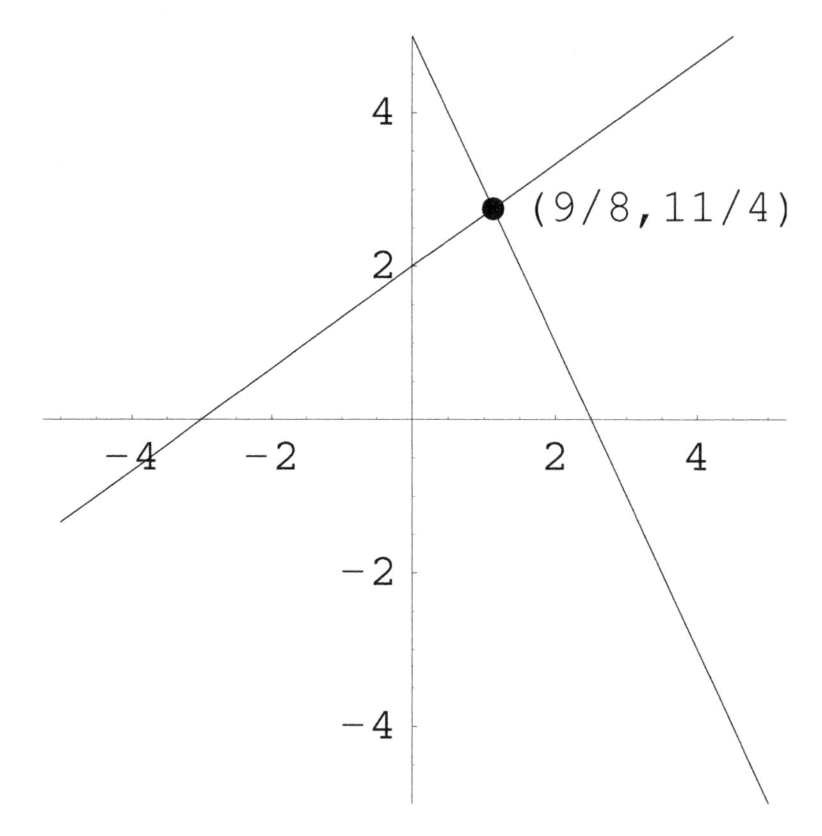

These two distinct, non-parallel lines intersect in the point $\left(\frac{9}{8}, \frac{11}{4}\right)$. This point of intersection is the (unique) solution to the original system; the point satisfying both equations.

□

When considering the possible solutions to a given 2×2 linear system, it is helpful to keep the graphical representation in mind. Two lines in a plane are related in exactly one of three ways:

1. The lines intersect in exactly one point. *(intersecting lines)*
2. The lines do not intersect at all. *(parallel lines)*
3. The lines intersect in infinitely many points. *(coincident lines)*

Therefore, there are exactly three possibilities for the number of solutions to a particular 2×2 linear system.

Number of Solutions	Graphical Interpretation
1	Intersecting Lines
0	Parallel Lines
∞	Coincident Lines

We've seen the algebraic manifestation in the first case; various solution techniques produce the unique solution to the system. We are now concerned with the algebraic manifestations in the remaining cases. Let's look at the following examples.

Example 6.4.7. Consider the following 2×2 linear system.

$$\begin{array}{rcrcr} 2x & + & y & = & 5 \\ 4x & + & 2y & = & 8 \end{array}$$

To solve this system using elimination, one might multiply the first equation by -2 to produce the equivalent system

$$\begin{array}{rcrcr} -4x & - & 2y & = & -10 \\ 4x & + & 2y & = & 8 \end{array}$$

Addition of the two equations results in the *contradiction* that $0 = 2$. This contradiction means that the understood *assumption* that the original system had a solution was incorrect; the original system has no solutions. This is due to the fact that the original system consisted of two equations of parallel lines. (How can this be verified?)

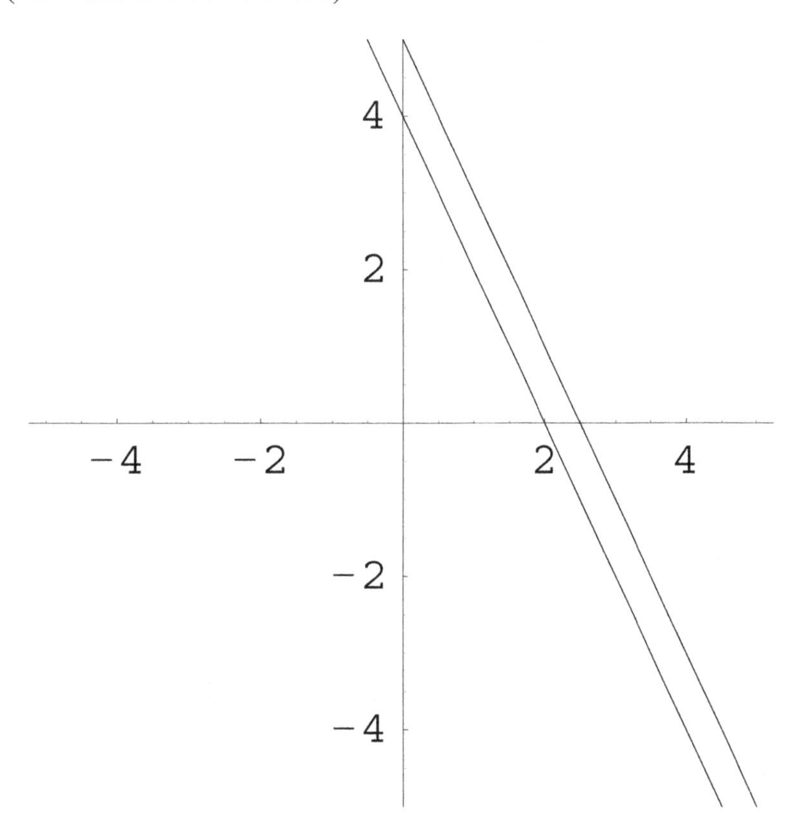

□

Example 6.4.8. Consider the following 2×2 linear system.

$$\begin{array}{rcrcr} 2x & + & y & = & 5 \\ 4x & + & 2y & = & 10 \end{array}$$

To solve this system using elimination, one might multiply the first equation by -2 to produce the equivalent system

$$\begin{array}{rcrcr} -4x & - & 2y & = & -10 \\ 4x & + & 2y & = & 10 \end{array}$$

Addition of the two equations results in the *identity* that $0 = 0$. This means that the original system consisted of two equations of the same line! Hence, there are infinitely many solutions to the original system. Five of them are $(0,5)$, $(1,3)$, $(2,1)$, $(-1,7)$, and $\left(\frac{1}{2},4\right)$.

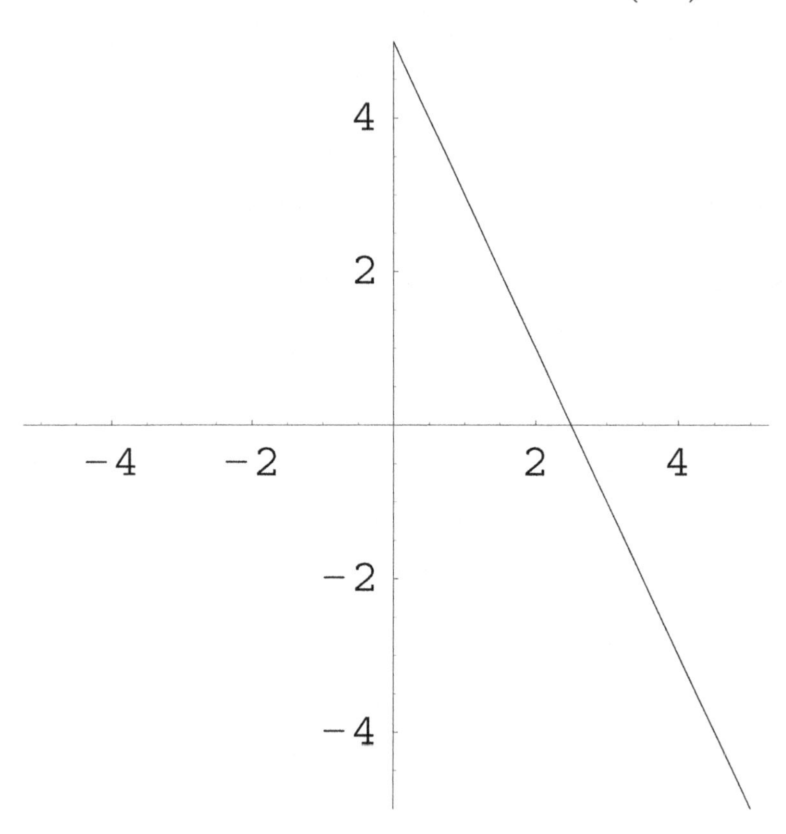

□

Section 6.4 Exercises

Solve each system by substitution.

1. $x+y=12, 3x-y=4$

2. $4x+2y=6, 3x+5y=8$

Solve each system by elimination.

3. $2x+5y=10, 4x+3y=-3$

4. $5x+7y=13, 6x-4y=9$

Solve each system by either substitution or elimination. (Decide which will be more efficient.)

5. $12x+5y=2, 6x+9y=12$

6. $2x - y = -7, 4x = 20$

7. $5x + y = 17, -15x + 13y = 22$

8. $2x - y = 5, 7x = 21$

9. **Kirchhoff's laws** are equations that describe current flow in a circuit. The laws state that (a) the current entering any point in the circuit equals the current leaving that point, and (b) the sum of the voltage drops around any loop is zero. In a certain circuit, these equations are $I_1 = I_2 + I_3, 5I_1 + 2I_3 = 12$, and $5I_1 + 3I_2 = 12$. Find the three currents I_1, I_2, and I_3 using the techniques of this section.

10. Solve the general system $ax + by = c, dx + ey = f$ by matrix methods, and compare your result with the results of this section.

11. Ashley grows apple trees and pear trees. She has room to add 150 new trees to her stock, and she has $885 to spend. If apple trees cost $8 each and pear trees cost $5 each, how many of each kind can she get?

Determine each solution as accurately as possible by graphing the given lines. Then use a method from an earlier section to find the exact solution.

12. $x + 2y = 6, 2x - y = 4$

13. $2x - 4y = 3, 3x + y = 8$

14. $6x + 8y = 9, 9x + 12y = 12$

15. $-4x + 5y = 9, 2x + 7y = 5$

16. Show that a 2×2 system of equations has no solution or infinitely many solutions if and only if the determinant of the coefficient matrix is 0.

17. Solve the following *nonlinear* system.

$$\begin{array}{rcrcrcl} x^2 & - & x & - & y & = & 1 \\ & - & x & + & y & = & 1 \end{array}$$

18. Show that the nonlinear system given below has no real solutions.

$$\begin{array}{rcrcl} -x & + & y & = & 4 \\ x^2 & + & y & = & 3 \end{array}$$

What is the graphical interpretation of this system?

19. Show that the nonlinear system given below has four solutions.

$$\begin{array}{rcrcl} x^2 & + & y^2 & = & 13 \\ x^2 & - & y & = & 7 \end{array}$$

What is the graphical interpretation of this system?

Chapter 7

Functions

This unit defines and investigates functions as algebraic objects. First, we define functions and discuss various means of representing them. Then we introduce operations on functions and examine which field properties hold for the operations, with emphasis on composition of functions.

Objectives

- To define function and introduce operations on the set of functions
- To investigate which of the field properties hold in the set of functions
- To explore various ways of representing functions
- To examine graphs of "basic" functions and simple transformations of them
- To study inverse functions in the context of the field properties and introduce algebraic methods for determining whether a function has an inverse and, if so, to find it
- To introduce special functions (polynomial, exponential, logarithmic, and sequences)

Terms

- function
- domain
- range
- sequence
  - term
  - recursively defined sequence
  - arithmetic sequence
  - geometric sequence
- zero (of a function)
- intercept
- vertical line test
- equality of functions
- operations on functions
  - sum (difference)
  - product (quotient)
  - composition

- one-to-one functions
- horizontal line test
- inverse functions

7.1 Functions

Most students are already familiar with the idea of a function. Here we formalize this idea and examine which field properties hold for a collection of functions.

Definition 7.1.1. A **function** f from a set X to a set Y is a pairing of elements from X with elements from Y in such a way that every element of X is associated with exactly one element of Y; we write $f : X \to Y$.

If $y \in Y$ is paired with $x \in X$ in the function f, we write $f(x) = y$ and say that y is the **image** of x under f. We sometimes refer to x as the **input** to f and y as the **output** from f.

Cautionary note: The symbol $f(x)$ does NOT signify multiplication, even though the notation is similar. You will need to be alert for context to decide which is intended. Generally, we will reserve the letters f, g, and h for functions.

The set X is called the **domain** of f, and the set of all images

$$R = \{y | y = f(x) \text{ for some } x \in X\}$$

is called the **range** of f. The domain is usually considered as the largest subset of the real numbers for which the function is defined, unless otherwise stated. The range is usually somewhat hard to find, but a graph of the function can help, as we will see in another section.

Most of the functions we are concerned with involve performing operations on real numbers. To determine the domain, we must determine which operations will *yield* real numbers. Are we trying to find a multiplicative inverse for 0? Are we trying to find a square root of a negative number? Or are we just adding and multiplying real numbers?

Example 7.1.1. Determine the domain and range of each function, and compute $f(1)$ and $f(-4)$, if possible.

1. $f(x) = \frac{1}{x}$.

   Remember that $\frac{1}{x}$ means the multiplicative inverse of x. This means that given the input x, the function f returns its multiplicative inverse as the output. This can be done as long as $x \neq 0$; therefore, the domain of f is $\{x \in \mathbb{R} | x \neq 0\}$.

   For the range, we need to find out what values of $f(x)$ can be obtained. Set $y = f(x)$ for convenience. Then we need to know what values of y can be obtained from $y = \frac{1}{x}$. If $y = \frac{1}{x}$, then also $x = \frac{1}{y}$, so it is clear that $y \neq 0$ for the same reason that $x \neq 0$. However, any other value of y is obtainable since if $y \neq 0$, then we can set $x = \frac{1}{y}$, and then $f(x) = \frac{1}{1/y} = y$. Therefore, the range of f is $\{y \in \mathbb{R} | y \neq 0\}$.

   $f(1) = \frac{1}{1} = 1$, and $f(-4) = \frac{1}{-4} = -\frac{1}{4}$.

2. $f(x) = \sqrt{x}$.

Recall that $\sqrt{x}$ means a positive number a such that $a^2 = x$. However, we know from earlier work that if $a \in \mathbb{R}$, then $a^2 \geq 0$. Therefore, we must have $x \geq 0$. It is a fact (that we have not discussed) that every positive real number has a positive square root. This means that the domain of f is $\{x \in \mathbb{R} | x \geq 0\} = [0, \infty)$.

If y is in the range of f, then $f(x) = \sqrt{x} = y$ for some real number x, so $y \geq 0$. (Remember, $\sqrt{x} \geq 0$ for all $x \geq 0$.) Conversely, if $y \geq 0$, then let $x = y^2$. We get $f(x) = \sqrt{x} = \sqrt{y^2} = |y| = y$ since $y \geq 0$.

That is, y is in the range of f if and only if $y \geq 0$, so the range of f is $\{y \in \mathbb{R} | y \geq 0\} = [0, \infty)$.

$f(1) = \sqrt{1} = 1$, but $f(-4)$ is not defined in the real numbers.

3. $f(x) = \dfrac{x+2}{x^2 - 4}$

   We have $f(x) = \dfrac{x+2}{(x+2)(x-2)}$. Remember that $\dfrac{a}{a} = 1$, *provided that* $a \neq 0$; when we are finding the domain, we may not simplify. We can also represent f by

$$f(x) = (x+2) \cdot \frac{1}{x+2} \cdot \frac{1}{x-2}.$$

   We are now in a position to determine the domain of f. This function requires finding multiplicative inverses for $x+2$ and $x-2$; if $x = -2$, then $x+2$ does not have a multiplicative inverse (so f is not defined), and if $x = 2$ then $x-2$ does not have a multiplicative inverse (so f is not defined). Since $x = 2$ and $x = -2$ are the only values of x that will not yield a real number for y, they are the only real numbers not in the domain of f. Therefore, the domain of f is
   $\{x \in \mathbb{R} | x \neq 2 \text{ and } x \neq -2\}$.

   $f(1) = \dfrac{1+2}{1^2 - 4} = -1$. $f(-4) = \dfrac{-4+2}{(-4)^2 - 4} = -\dfrac{1}{6}$.

   We are not equipped to determine the range right now.

4. $f(x) = 2x^3 - 5$.

   When we are given x, we must first find $x^3 = x \cdot x \cdot x$, then multiply it by 2, and then subtract 5 (recalling the order of operations). Since all of these operations may be performed on any real number, the domain of f is $\mathbb{R}$. The range is also $\mathbb{R}$ since all real numbers have a real cube root: given y, let $x = \sqrt[3]{(y+5)/2}$. Then $f(x) = 2(\sqrt[3]{(y+5)/2})^3 - 5 = 2[(y+5)/2] - 5 = y + 5 - 5 = y$.

   $f(1) = 2(1)^3 - 5 = -3$. $f(-4) = 2(-4)^3 - 5 = -133$.

□

The examples above illustrate two ways that a number x can fail to be in the domain of a function: x can cause a division by 0 (which is not defined), or x can cause us to try to find a (real) square root for a negative number, which is also not defined. There are other ways that x can fail to be in the domain of a function, as well, and we will discuss some of these later.

We have a variety ways to represent a function:

1. Tabular representation: We can arrange the information in a table, as follows:

x	$f(x)$
-2	-10
-1	-7
0	-4
1	-1
2	2

   This works fine for simpler functions or for data gathered experimentally, but it can be awkward to extrapolate from.

2. Ordered pair representation: We can think of functions as ordered pairs in $X \times Y$, where every element of X appears in exactly one ordered pair:

$$f = \{(x,y) | f(x) = y \text{ and } x \in X\}.$$

   For example, the function above is $f = \{(x, 3x-4) | x \in \mathbb{R}\}$ if we extrapolate, or

$$f = \{(-2,-10), (-1,-7), (0,-4), (1,-1), (2,2)\}$$

   if we do not. Notice that $(x,y) \in f$ if and only if $y = f(x)$.

   This is perhaps the most natural representation from the point of view of the definition; functions are *defined* as pairings, and here we are just being explicit about what the pairings are.

3. Graphical representation: Because we can think of functions as ordered pairs, we can graph them, too! We plot points on a Cartesian coordinate system until we have enough to figure out what the graph looks like. (The points we plot are just those corresponding to ordered pairs that belong to the function.) We can be fooled, though, so we must be careful. We can also determine whether some relations are functions by using the **vertical line test** (see below), but we have to be able to graph the relation to begin with. Your calculator can help you with some. We will look at graphing in more detail later. The graph of the function in parts 1 and 2 is below, with a few specific points plotted, and the rest filled in by extrapolation.

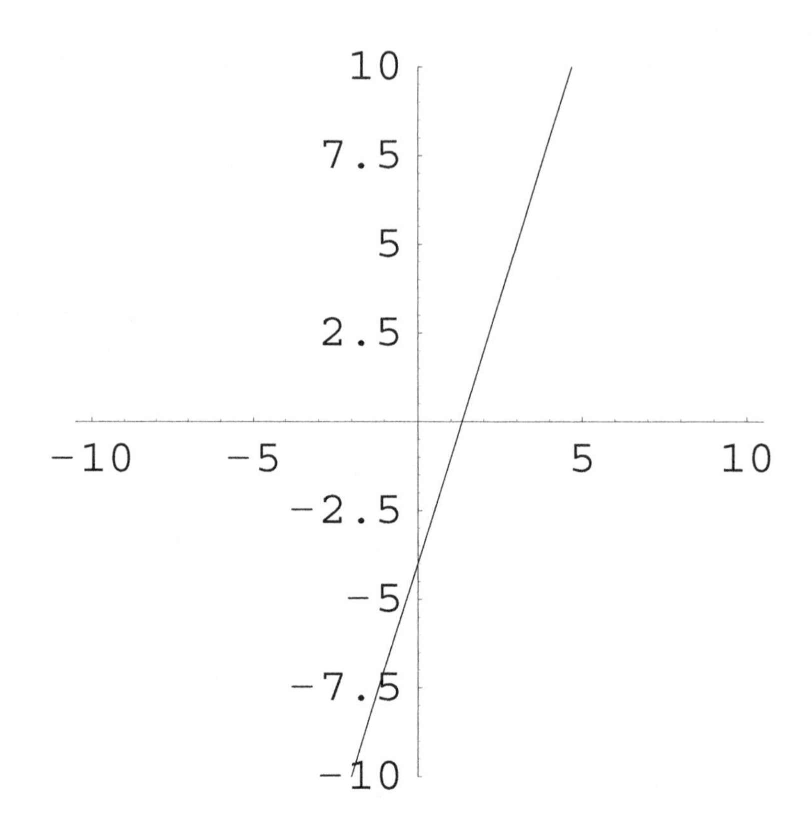

4. Algebraic representation: The examples above can be interpreted in terms of some formula or equation, namely, $f(x) = 3x - 4$.

5. Verbal representation: A function can also be given by a description, such as "Each person is paired with the table at which that person is sitting."

Example 7.1.2. Consider $f : \{1,2,3,4\} \to \{0,1\}$ given by $f(1) = 1, f(2) = 0, f(3) = 1$, and $f(4) = 0$. Then f is a function with domain $\{1,2,3,4\}$ and range $\{0,1\}$. The function f can also be represented as a set of ordered pairs $f = \{(1,1),(2,0),(3,1),(4,0)\}$; note that each element of the domain appears in exactly one ordered pair.

□

Example 7.1.3. Let $g : \{1,2,3,4\} \to \{0,1\}$ satisfy $g(1) = 0, g(2) = 1, g(1) = 1, g(3) = 0$, and $g(4) = 1$. Then g is not a function since 1 is paired with both 0 and 1. As a set of ordered pairs, $g = \{(1,0),(2,1),(1,1),(3,0),(4,1)\}$. Since the element 1 is used twice as a first-coordinate, g is not a function.

□

Example 7.1.4. Let $h : \mathbb{R} \to \mathbb{R}$ be given by $h(x) = x^2 - 3x + 1$. Then h is a function with domain $\mathbb{R}$ and a range that we are not equipped to determine at present. We have

$$h(3) = 3^2 - 3(3) + 1 = 1 \text{ and } h(-2) = (-2)^2 - 3(-2) + 1 = 11.$$

□

Example 7.1.5. Function evaluation is to be taken quite literally. A formula for $f(x)$ means that *no matter what x is*, its value is to be substituted unchanged into the formula. Consider $f(x) = \frac{x}{x+1}$.

$f(w-3) = \frac{w-3}{(w-3)+1}$.

□

In the following sections we will consider some special functions that are noteworthy; in particular, we'll explore polynomial functions, exponential functions, logarithmic functions, and sequences.

Section 7.1 Exercises

Determine the domain of each function and express your answer in interval notation.

1. $f(x) = \frac{1}{2x+5}$
2. $f(x) = \frac{x-1}{x+1}$
3. $f(x) = \frac{x^2-5x-3}{x^2-5x-6}$
4. $f(x) = \frac{x^2}{\sqrt{x-3}}$
5. $f(x) = \sqrt{x^2-9}$
6. $f(x) = x^2 - 5x$

Evaluate each function as indicated and simplify.

7. For $f(x) = \sqrt{x^2-2x+2}$, find $f(0), f(-2)$, and $f(3)$.
8. For $p(t) = \frac{t}{t+1}$, find $p(4), p(-5.2), p(100)$, and $p(0)$.
9. For $x(t) = t^t$, find $x(1), x(-3)$, and $x(2)$.
10. For $f(x) = x^2$, find $f(2)$, and $\frac{f(x+h)-f(x)}{h}$.
11. Given the function $f(x) = \frac{1}{1+x^2}$, create a tabular representation with 6 values for x (of your choosing). Then plot the 6 points from your table and connect them with a smooth curve.
12. In professional football games, the analysts often discuss the "hang time" of a punt. If the hang time is T seconds, then it can be shown that a reasonable estimate for the height of the football after t seconds have elapsed is $h(t) = 16t(T-t)$ feet. Also, since the time the ball spends as much time going up as coming down (roughly), the maximum height occurs at $t = \frac{1}{2}T$. If a punt has a hang time of 5 seconds, how high does it go?
13. The same equation as in the previous problem can be used to model jumps of basketball players. If a basketball player has a hang time of 1 second, how high did the player jump? (Compare to Clyde Drexler's 44 inch vertical leap.)

14. Let $f(x) = 3x^3 - 5x^2 + 7$, and let $g(x) = 3x^2 - 2x + 5$.

    (a) Compute the new function $h(x) = f(x) + g(x)$ and simplify.

    (b) Compute $h(2)$.

    (c) Compute $f(2)$ and $g(2)$ and add them.

    (d) What do you notice about your answers for parts (b) and (c)?

7.2 Sequences

We consider special type of function known as a sequence; sequences are functions with a special (simple) domain.

Definition 7.2.1. A **sequence** is a function whose domain is the set of positive integers.

We usually represent a sequence by listing its values in order: $a(1), a(2), \ldots$. The numbers are the **terms** of the sequence.

Example 7.2.1. The formula $a(n) = 1/n$, with $n \in \mathbb{N}$, gives the sequence $1, \frac{1}{2}, \frac{1}{3}, \frac{1}{4}, \ldots$ since $a(1) = 1, a(2) = \frac{1}{2}$, etc.

□

Example 7.2.2. The formula $b(n) = 3n+1$ gives the sequence $4, 7, 10, 13, 16, \ldots$.

□

Instead of $a(n)$, we often use subscripts and write a_n to represent a general term (nth term) and list $a_1, a_2, \ldots, a_n, \ldots$ to represent a listing of the terms of the sequence in order. In the examples above, we had the general terms $a_n = 1/n$ and $b_n = 3n+1$. (The letter we choose corresponds to the name of the function.)

Definition 7.2.2. The notation $\{a_n\}$ stands for the sequence whose nth term is a_n.

Example 7.2.3. List the first five terms of the sequence $\{a_n\} = \{(n-1)/n\}$.

Solution: We have $a_1 = \frac{1-1}{1} = 0$, $a_2 = \frac{2-1}{2} = \frac{1}{2}$, $a_3 = \frac{3-1}{3} = \frac{2}{3}$, $a_4 = \frac{3}{4}$, and $a_5 = \frac{4}{5}$.

□

Example 7.2.4. List the first seven terms of $\{c_n\}$, where $c_n = n$ if n is even, and $c_n = 1/n$ if n is odd.

Solution: This function is defined piecewise: $c_n = \begin{cases} n & n \text{ even} \\ 1/n & n \text{ odd} \end{cases}$.

Thus we have $c_1 = \frac{1}{1} = 1, c_2 = 2, c_3 = \frac{1}{3}, c_4 = 4, c_5 = \frac{1}{5}, c_6 = 6$, and $c_7 = \frac{1}{7}$.

□

Definition 7.2.3. For a positive integer n, we define $n!$ by $n! = n(n-1)(n-2)\cdots(3)(2)(1)$. In other words, $n!$ is the product of the first n natural numbers. We also adopt the convention that $0! = 1$. The symbol $n!$ is pronounced "n factorial" or "factorial n."

Notice that $(n+1)! = (n+1)n!$; this observation will prove useful later on.

Example 7.2.5. Simple calculations yield the following.

- $5! = (5)(4)(3)(2)(1) = 120$

- $13! = (13)(12)(11)\cdots(3)(2)(1) = 6,227,020,800$

□

Definition 7.2.4. A **recursively defined sequence** is a sequence whose terms are specified by a formula involving the preceding terms.

Example 7.2.6. The Fibonacci sequence is given by

$$a_1 = 1, a_2 = 1, \ldots, a_{n+2} = a_{n+1} + a_n.$$

List the first eight terms.

Solution: We have $a_1 = 1$ and $a_2 = 1$ given. We compute $a_3 = a_2 + a_1 = 1 + 1 = 2$, $a_4 = a_3 + a_2 = 2 + 1 = 3$, $a_5 = a_4 + a_3 = 3 + 2 = 5$, $a_6 = a_5 + a_4 = 5 + 3 = 8$, $a_7 = 8 + 5 = 13$, and $a_8 = 13 + 8 = 21$. Thus the Fibonacci sequence is $1, 1, 2, 3, 5, 8, 13, 21, \ldots$.

□

Example 7.2.7. Find the first five terms of the sequence $f_1 = 1, f_{n+1} = (n+1)f_n$.

Solution: We have that $f_2 = 2(1) = 2$, $f_3 = 3(2) = 6$, $f_4 = 4(6) = 24$, and $f_5 = 5(24) = 120$. Looking at this pattern, we see that we have $f_n = n!$, the factorial sequence.

□

Frequently a sequence is indicated by a partial listing of its terms and we desire a formula for the general term. This problem is technically unsolvable *without more information.* In fact, for any finite listing of sequence terms, there are infinitely many (correct) completing patterns and general terms.

Example 7.2.8. Find the next six terms and explain the pattern based on the partial sequence $1, 2, 4,$ ____, ____, ____ .

Solution 1: Using the pattern of doubling, we have $1, 2, 4, 8, 16, 32, 64, 128, 256\ldots, 2^{n-1}, \ldots$. In this case, a closed formula is $a_n = 2^{n-1}$.

Solution 2: Using the pattern of adding one more to each successive difference, we have $1, 2, 4, 7, 11, 16, 22, 29, 37\ldots$. A *recursive pattern* is given by $a_1 = 1$ and $a_n = a_{n-1} + (n-1)$.

Solution 3: Using an "odd, even, even" pattern, we have $1, 2, 4, 5, 6, 8, 9, 10, 12, \ldots$.

Solution 4: Skipping all multiples of three, we have $1, 2, 4, 5, 7, 8, 10, 11, 13\ldots$.

$\vdots$

□

We will temporarily restrict ourselves to **arithmetic sequences** and investigate their special properties.

Definition 7.2.5. An **arithmetic sequence** is a sequence in which successive terms differ by a fixed number.

Example 7.2.9. The sequences $1,4,7,10,13,17,21,\ldots$ and $2,6,10,14,18,22,26,\ldots$ share the trait that the difference between two consecutive terms is a constant. In the first case, each term is three more than the prior term; in the second sequence, each term is four more than its predecessor.

□

Note that arithmetic sequences are defined recursively by $a_{n+1} = a_n + C$, where C is the constant. Also, since sequences are just functions with the special domain $\mathbb{N}$, we see that an arithmetic sequence is just a linear function on that domain.

Example 7.2.10. Find a formula for the nth term of an arithmetic sequence with first term 1 and difference 3.

Solution: Here we need to find a formula for the nth term without knowing what natural number n represents. We know that $a_1 = 1$ and $C = 3$, so $a_2 = 1+3 = 4$, $a_3 = 4+3 = 7$, and so on. Let's put this information into a table and try to organize our data.

n	a_n	
1	1	1
2	$1+3$	4
3	$(1+3)+3$	7
4	$(1+3+3)+3$	10
5	$(1+3+3+3)+3$	13

Notice that a_5 is a 1 with four 3's added, a_4 is a 1 with three 3's added, and so on. It looks reasonable that $a_n = 1+3(n-1)$, which simplifies to $a_n = 3n-2$. Let's check this: $3(5)-2 = 13$, which is what the fifth term should be.

□

Theorem 74. *The nth term for an arithmetic sequence with first term a and difference d is* $a_n = a+(n-1)d$.

Proof. We defer the proof, by induction on n, until another unit. □

Example 7.2.11. Find the first two terms of an arithmetic sequence if $a_3 = 11$ and $a_{30} = 146$.

Solution: We have that $a_3 = a+(3-1)d = 11$ which implies that $a+2d = 11$. We also have that $a_{30} = a+(30-1)d = 146$, hence, $a+29d = 146$. We now have a system of two equations in two unknowns. This gives $27d = 135 \implies d = 5$, so $a = 1$. Hence, the first two terms are $a_1 = 1$, and $a_2 = 6$.

□

Definition 7.2.6. A **geometric sequence** is a sequence in which successive terms have a constant ratio r.

These are much like arithmetic sequences, except that instead of *adding* the same number to each term, we *multiply* each term by the same number.

Example 7.2.12. The sequence $1, 2, 4, 8, 16, \ldots$ is a geometric sequence with initial term 1 and common ratio 2.

□

Example 7.2.13. The sequence $2, 6, 18, 54, 162, 468, 1458, \ldots$ is a geometric sequence with initial term 2 and common ratio 3.

□

As was the case with arithmetic sequences, geometric sequences are defined recursively. We will again make a table to try to find a general formula for the nth term of a geometric sequence with initial term a and common ratio r.

n	a_n
1	a
2	ar
3	ar^2
4	ar^3
5	ar^4

We apparently have the formula $a_n = ar^{n-1}$; we will prove that this is correct by induction in the next section.

Theorem 75. *If $a_1, a_2, a_3, \ldots$ is a geometric sequence with common ratio r and initial term $a = a_1$, then $a_n = ar^{n-1}$.*

Example 7.2.14. Above, we had $1 \cdot 2^{n-1}$ and $2 \cdot 3^{n-1}$.

□

Just as arithmetic sequences were linear functions, geometric sequences are exponential functions and, therefore, have application in growth and decay problems.

Example 7.2.15. Find the eighth term in the geometric sequence $2/3, 2/9, 2/27, \ldots$.

Solution: The initial term is $\frac{2}{3}$ and the common ratio is $\frac{1}{3}$, so the eighth term is $\frac{2}{3}\left(\frac{1}{3}\right)^{8-1} = \frac{2}{6561}$.

□

Section 7.2 Exercises

Write the first five terms of each sequence.

1. $\{3n\}$
2. $\{n^2 + 2n\}$

3. $\left\{\left(\frac{4}{3}\right)\right\}$

4. $\left\{\frac{3^n}{n}\right\}$

5. $\left\{\frac{3^n}{2n+3}\right\}$

6. $\left\{\frac{n^2}{2^n}\right\}$

Assuming the given pattern continues, write a formula for the nth term of each sequence suggested by the pattern.

7. $1, 3, 5, 7, 9, 11, \ldots$

8. $-3, 1, 5, 9, 13, 17, \ldots$

9. $5, 2, -1, -4, -7, -10, \ldots$

10. $\frac{1}{2}, 1, 2, 4, 8, \ldots$

11. $-3, 6, -12, 24, -48, 96, \ldots$

12. $\frac{1}{2}, \frac{1}{4}, \frac{1}{8}, \frac{1}{16}, \ldots$

13. $\frac{2}{3}, \frac{3}{4}, \frac{4}{5}, \frac{5}{6}, \ldots$

14. $2, 6, 12, 20, 30, 42, \ldots$

15. $1, -2, 3, -4, 5, -6, \ldots$

16. $1, \frac{1}{2}, 3, \frac{1}{4}, 5, \frac{1}{6}, \ldots$

17. $1, 3, 6, 10, 15, 21, 28, \ldots$

Find the indicated term in each of the given sequences.

18. Find the fifth term of the arithmetic sequence with first term 6 and difference 3.

19. Find the seventh term of the arithmetic sequence with first term -7 and difference 5.

20. Find the fifteenth term of the arithmetic sequence with first term 3 and difference 8.

21. Find the fifth term of the arithmetic sequence with first term 1 and difference π.

22. Find the fifth term of the geometric sequence with first term 2 and ratio 3.

23. Find the sixth term of the geometric sequence with first term $\frac{1}{2}$ and ratio 7.

24. Find the eighth term of the geometric sequence with first term 5 and ratio -3.

25. Find the first term of the arithmetic sequence with seventh term 12 and difference 3.

26. Find the first term of the arithmetic sequence with fifth term -9 and difference -2.

27. Find the first term of the arithmetic sequence with third term 8 and eighth term 23.

28. Find the first term of the arithmetic sequence with eighth term 8 and twentieth term 44.

29. Find the first term of the arithmetic sequence with fifth term -2 and thirteenth term 30.

30. Find x so that x, $x+2$, and $x+3$ are terms of a geometric sequence.

31. Find x so that $x-1$, x, and $x+2$ are terms of a geometric sequence.

32. Does there exist a sequence $\{a_n\}$ that is both arithmetic and geometric? Why or why not?

7.3 Operations on Functions

Functions give us a new kind of "numbers," so we once again need to define some notion of equality and introduce some operations.

Definition 7.3.1. Two functions f and g are **equal**, written $f = g$, if and only if

1. f and g have the same domain D, and
2. $f(x) = g(x)$ for all $x \in D$.

Example 7.3.1. Are the functions defined by $f(x) = x+1$ and $g(x) = \dfrac{x^2-1}{x-1}$ equal functions?

Solution: No, f and g are not equal because they do not have the same domain. The domain of f is $\mathbb{R}$, and the domain of g is $\{x \in \mathbb{R} | x \neq 1\}$. However, f and g **do** have the same action on the restricted domain $\{x \in \mathbb{R} | x \neq 1\}$.

□

When two functions have a common domain, it is sufficient to compare their action on each domain element to determine whether the functions are equal.

Example 7.3.2. The functions $f(x) = x$ and $g(x) = |x|$ have a common domain, $\mathbb{R}$, but they are not equal functions since they "disagree" on $(-\infty, 0)$. In particular,

$$f(-1) = -1 \neq 1 = g(-1).$$

□

Example 7.3.3. We have already seen that $f(x) = |x|$ and $g(x) = \sqrt{x^2}$ are equal functions; both have domain $\mathbb{R}$, and they agree at all elements of their domain.

□

We will now define some operations on the set of functions and investigate which field properties are satisfied by the operations.

Definition 7.3.2. Let $f : D_1 \to R_1$ and $g : D_2 \to R_2$ be functions with R_1 and R_2 subsets of the same field (think of $\mathbb{R}$ and/or $\mathbb{C}$). The **sum** of f and g is the function $f+g$, given by $(f+g)(x) = f(x)+g(x)$ for $x \in D_1 \cap D_2$.

Example 7.3.4. Let $f(x) = 4x-6$ and $g(x) = x^2-3x+2$. Then

$$(f+g)(x) = f(x)+g(x) = (4x-6)+(x^2-3x+2) = x^2+4x-4,$$

and

$$(f+g)\left(\frac{1}{2}\right) = f\left(\frac{1}{2}\right)+g\left(\frac{1}{2}\right) = \left(4\cdot\frac{1}{2}-6\right)+\left(\left(\frac{1}{2}\right)^2-3\cdot\frac{1}{2}+2\right) = -\frac{7}{4}.$$

□

Naturally, we want to know about field properties. We will have to be careful with the domain, but if we are, what do we get?

Theorem 76. *Addition of functions is associative and commutative.*

Proof. A function is determined by what it does to its domain. In particular, $f+g$ is determined by the values $(f+g)(x)$, where x is in the domain of $f+g$. Also, notice that $f+g$ and $g+f$ have the same domain since $D_1 \cap D_2 = D_2 \cap D_1$. The key in the proof below is the realization that $f(x)$ and $g(x)$ **are field elements** (members of $\mathbb{R}$). Consider:

$$\begin{aligned}(f+g)(x) &= f(x)+g(x) && \text{Definition of Function Addition}\\ &= g(x)+f(x) && f(x), g(x) \in \mathbb{R} \text{ and addition in a field commutes}\\ &= (g+f)(x) && \text{Definition of Function Addition.}\end{aligned}$$

Therefore, for any x in the domain of $f+g$, $(f+g)(x) = (g+f)(x)$, so $f+g = g+f$. The proof of associativity is similar and will be left to the reader. □

Theorem 77. *Let F be a field. The function $f(x) = 0$ with domain F is an identity for function addition.*

Proof. If h is any function, then $(f+h)(x) = f(x)+h(x) = 0+h(x) = h(x)$ for all x in the domain of h. Thus $f+h = h$. By commutativity of function addition, $h+f = h$. Therefore, f is an identity for function addition. □

Theorem 78. *The function $-f$ given by $(-f)(x) = -(f(x))$ is an additive inverse for the function f.*

Proof. Note that $-f$ has the same domain as f. For any x in the domain of f, we have

$$\begin{aligned}(f+(-f))(x) &= f(x)+(-f)(x) && \text{Definition of Function Addition}\\ &= f(x)+(-(f(x))) && \text{Definition of } (-f)(x)\\ &= 0 && \text{Additive Inverse (remember, } f(x) \in \mathbb{R}!\text{).}\end{aligned}$$

Since we know that addition of functions is commutative, we have $f+(-f) = 0 = (-f)+f$, so $-f$ is the additive inverse of f. □

Since every function has an additive inverse, we can now define function subtraction.

Definition 7.3.3. Let $f : D_1 \to R_1$ and $g : D_2 \to R_2$ be functions with R_1 and R_2 subsets of the same field. The **difference** of f and g is the function $f-g = f+(-g)$, given by $(f-g)(x) = f(x)-g(x)$ for $x \in D_1 \cap D_2$.

Example 7.3.5. Again using $f(x) = 4x-6$ and $g(x) = x^2-3x+2$, we have

$$(f-g)(x) = (4x-6)-(x^2-3x+2) = 4x-6-x^2+3x-2 = -x^2+7x-8$$

and

$$(f-g)(2) = -(2^2)+7(2)-8 = 2.$$

□

We'll now turn our attention to defining and investigating multiplication of functions.

Definition 7.3.4. Let $f : D_1 \to R_1$ and $g : D_2 \to R_2$ be functions with R_1 and R_2 subsets of the same field (again, think of $\mathbb{R}$ and/or $\mathbb{C}$). The **product** of f and g is the function fg, given by $(fg)(x) = f(x)g(x)$ for all $x \in D_1 \cap D_2$.

Example 7.3.6. Let $f(x) = 4x - 6$ and $g(x) = x^2 - 3x + 2$. Then

$$(fg)(x) = (4x-6)(x^2-3x+2)$$

and

$$(fg)(-1) = (4(-1)-6)((-1)^2-3(-1)+2) = (-10)(6) = -60.$$

□

Theorem 79. *Multiplication of functions is associative and commutative.*

We omit the proof, as it is very similar to that of the theorem above.

Theorem 80. *Let F be a field. The function $g(x) = 1$ with domain F is an identity for function multiplication.*

Proof. If h is any function, then $(gh)(x) = g(x)h(x) = 1 \cdot h(x) = h(x)$ for all x in the domain of h. Thus $gh = h$. Since function multiplication commutes, $hg = h$ as well, so $g(x) = 1$ is an identity for function multiplication. □

Theorem 81. *Let f be a function with domain D, and let $D' - \{x \in D | f(x) \neq 0\}$. Then the function g given by $g(x) = \dfrac{1}{f(x)}$ is a multiplicative inverse for f on D'.*

Proof. We need to restrict the domain to D' to prevent a division by zero. Now, let $x \in D'$. Then

$$\begin{aligned}
(f \cdot g)(x) &= f(x) \cdot g(x) && \text{Definition of function multiplication} \\
&= f(x)\frac{1}{f(x)} && \text{Definition of } g(x). \\
&= 1 && \text{Multiplicative inverses (remember that } f(x) \text{ is a real number!),}
\end{aligned}$$

which is the multiplicative identity for functions.

Since we know that multiplication of functions commutes, we also have $f \cdot g = 1 = g \cdot f$, so $g = \dfrac{1}{f}$ is the multiplicative inverse of f. □

In practice, the multiplicative inverse of a function is only marginally useful, except for in defining function division. Note: we **do not** use the -1 exponent for denoting the multiplicative inverse of a function.

Definition 7.3.5. Let $f : D_1 \to R_1$ and $g : D_2 \to R_2$ be functions with R_1 and R_2 subsets of the same field. The **quotient** of f and g is the function $\dfrac{f}{g}$, given by $\left(\dfrac{f}{g}\right)(x) = f(x)/g(x)$ for all $x \in D_1 \cap D_2 \cap \{x | g(x) \neq 0\}$ (since division by 0 is not defined).

Example 7.3.7. Let $f(x) = 4x - 6$ and $g(x) = x^2 - 3x + 2$. Then

$$\left(\frac{f}{g}\right)(x) = \frac{4x-6}{x^2-3x+2}$$

and

$$\left(\frac{f}{g}\right)(0) = \frac{4\cdot 0-6}{0^2-3\cdot 0+2} = -3.$$

The domain of $\frac{f}{g}$ is $\{x \in \mathbb{R} | x^2 - 3x + 2 \neq 0\}$. If $x^2 - 3x + 2 = 0$, then $(x-2)(x-1) = 0$, so $x = 2$ or $x = 1$ by the Zero Product Theorem. Therefore, the domain of $\frac{f}{g}$ is

$$\mathbb{R} \cap \mathbb{R} \cap \{x \in \mathbb{R} | x \neq 1,2\} = \{x \in \mathbb{R} | x \neq 1,2\}.$$

□

Theorem 82. *Function multiplication distributes over function addition.*

Proof. Let f, g, and h be functions with domains A, B, and C, respectively. Let $D = A \cap B \cap C$, so that D is the domain of both $f \cdot (g+h)$ and $fg + fh$. (Note that $f \cdot (g+h)$ does NOT refer to function evaluation in this context.) Let $x \in D$. Then

$$\begin{array}{rcll}
(f\cdot(g+h))(x) & = & f(x)((g+h)(x)) & \text{Definition of Function Multiplication} \\
& = & f(x)(g(x)+h(x)) & \text{Definition of Function Addition} \\
& = & f(x)g(x)+f(x)h(x) & \text{Distribution of multiplication over addition in } \mathbb{R} \\
& = & (fg)(x)+(fh)(x) & \text{Definition of Function Multiplication} \\
& = & (fg+fh)(x) & \text{Definition of Function Addition.}
\end{array}$$

Therefore $f(g+h)$ and $fg+fh$ have the same values for all $x \in D$, so they must be the same function. That is, $f \cdot (g+h) = fg + fh$. Similarly, $(g+h) \cdot f = gf + hf$ (or use commutativity of function multiplication). □

Therefore, as long as we are careful with domains, we have that functions satisfy the field axioms, so we may use the theorems we proved for fields in general. However, we really must pay attention to domains to avoid inappropriately using field theorems.

We have one more operation to define.

Definition 7.3.6. Let $f : D_1 \to R_1$ and $g : D_2 \to R_2$ be functions with R_1 and R_2 subsets of the same field. The **composition** of f by g is $f \circ g$, given by $(f \circ g)(x) = f(g(x))$. The domain of the composition is $\{x \in D_2 | g(x) \in D_1\}$.

Think of composition of functions as a two-step process. First, evaluate g at x. Then take the *output* from that, and substitute it into f. This is why the domain looks like it does: in evaluating $f \circ g$ at x, the first step is to evaluate g at x, so x had better be in D_2, the domain of g. The second step is to evaluate f at the *output* from g, so that output (which is $g(x)$) had better be in D_1, the domain of f.

Example 7.3.8. Let $f(x) = \sqrt{x}$ and $g(x) = x-4$. The domain of g is $\mathbb{R}$, and the domain of f is $\{x \in \mathbb{R} | x \geq 0\}$. The domain of $f \circ g$ is therefore $\{x \in \mathbb{R} | x-4 \geq 0\} = \{x \in \mathbb{R} | x \geq 4\}$.
$(f \circ g)(x) = \sqrt{x-4}$.

□

Example 7.3.9. It is interesting that composition of functions does NOT commute. Consider $f(x) = x^2$ and $g(x) = x+1$. $(f \circ g)(x) = f(g(x)) = f(x+1) = (x+1)^2 = x^2+2x+1$. On the other hand, $(g \circ f)(x) = g(f(x)) = g(x^2) = x^2+1$, which is not at all the same function!

□

Theorem 83. *Composition of functions is associative.*

We omit the proof of this theorem; it is relatively straight-forward as long as one is careful about the technicalities involved with the necessary domain restriction.

Theorem 84. *Let F be a field. The function $\iota(x) = x$ with domain F is an identity for composition.*

Note: ι is the Greek letter lower-case iota.

Proof. Let f be a function with domain $D \subseteq F$, and let $x \in D$. Then $(f \circ \iota)(x) = f(\iota(x)) = f(x)$, and $(\iota \circ f)(x) = \iota(f(x)) = f(x)$. Thus $f \circ \iota = f = \iota \circ f$, so ι is an identity for composition. □

At this point, one naturally wonders about the existence of inverses with respect to composition. Like multiplicative inverses for matrices, inverse functions (with respect to composition) do not always exist. We explore inverse functions in detail in a later section.

We end this section with a definition that will help us analyze graphical symmetry in the next section.

Definition 7.3.7. An **even function** is a function f that satisfies $f(-x) = f(x)$ for all x in the domain of f. An **odd function** is a function that satisfies $f(-x) = -f(x)$ for all x in the domain of f.

Example 7.3.10. The function defined by $f(x) = x^2+1$ is an even function because

$$f(-x) = (-x)^2+1 = x^2+1 = f(x).$$

The function defined by $g(x) = -x^3+x$ is an odd function because

$$g(-x) = -(-x)^3+(-x) = x^3-x = -(-x^3+x) = -g(x).$$

The function defined by $h(x) = x^2+x$ is neither even nor odd since

$$h(-x) = (-x)^2+(-x) = x^2-x,$$

which equals neither $h(x)$ nor $-h(x)$.

□

Section 7.3 Exercises

Perform the indicated operations and simplify. Find the domain for each function, as well. (Remember that the domain of sum, difference, etc. must be found *before* simplifying.

1. For $f(x) = \dfrac{1}{x^2+1}$ and $g(x) = \dfrac{1}{x-1}$, find $f+g$, $\dfrac{f}{g}$, and $(f-g)(4)$.

2. Let $f(x) = x^2 - 3x + 5, g(x) = 3x + 4$, and $h(x) = -2x^2 - 9$. Find $f+g+h$ and $\dfrac{f-h}{g}$.

3. For $f(x) = 2x - 3$ and $g(x) = x^2 + 2$, find fg and $f \circ g$.

4. **Profit** is defined as revenue (income) minus cost (expense). If a certain company's revenue function is described by $R(x) = -x^2 + 18x + 6$ and its cost function is $C(x) = 4x - 22$, find its profit function $P(x)$. (Here x is the number of units produced.)

We can define a function from the set $\mathbb{R}^2$ of 2×1 column matrices to itself by matrix multiplication. Let $A = \begin{bmatrix} 1 & 2 \\ -2 & 3 \end{bmatrix}$. Define $f : \mathbb{R}^2 \to \mathbb{R}^2$ by $f(X) = AX$.

5. Verify that if X is 2×1, then $f(X)$ is 2×1.

6. Compute $f\left(\begin{bmatrix} 3 \\ -1 \end{bmatrix}\right)$.

7. Compute $f\left(\begin{bmatrix} 12 \\ -8 \end{bmatrix}\right)$.

8. Show that f is a **linear operator**; that is, show that $f(X+Y) = f(X) + f(Y)$. (This makes this particular function extremely special; most functions do not behave this way – think about x^2, for example, or $x - 3$.)

9. Let $f(x) = \dfrac{1}{x-1}$ and $g(x) = \sqrt{x}$.

   (a) Find $(f \circ g)(x)$ (including its domain).

   (b) Find $(g \circ f)(x)$ (including its domain).

   (c) What is $(f \circ g)(4)$?

   (d) What is $(g \circ f)(3)$?

10. Let $f(x) = x^2$ and $g(x) = 2x - 1$.

    (a) Find $(f \circ g)(x)$ (including its domain).

    (b) Find $(g \circ f)(x)$ (including its domain).

    (c) What is $(f \circ g)(-2)$?

    (d) What is $(g \circ f)(-2)$?

Determine whether each function is even, odd, or neither.

11. $f(x) = 2x + 4$
12. $g(x) = |x|$
13. $g(x) = \frac{1}{x}$
14. $r(x) = x^2 + x$
15. $f(x) = x^3 + x + 1$
16. $h(x) = -16x^2 + 80$
17. $f(x) = \frac{x}{x^2+1}$

A **piece-wise defined function** is a function whose description is given separately for different parts of its domain. For example, the absolute value function

$$f(x) = \begin{cases} x & \text{if } x \geq 0 \\ -x & \text{if } x < 0 \end{cases}$$

is a piece-wise defined function; the "pieces" are $[0, \infty)$ and $(-\infty, 0)$. Many of the functions we encounter in day-to-day life are piece-wise defined.

18. A mail-order book company advertises shipping rates of \$5 for up to 6 books, and then \$0.75 for each additional book.
    (a) Determine a reasonable domain for this function.
    (b) Write a description of this function. (Note that it is piece-wise defined.)
    (c) Find the shipping cost for (i) 3 books and (ii) 12 books.
19. A car rental company advertises a flat fee of \$25 per day with the first 200 miles free, and then \$0.35 cents per mile thereafter. Assuming a one day rental, the cost to rent a car from them and drive it x miles is given by

$$C(x) = \begin{cases} 25 & \text{if } 0 \leq x \leq 200 \\ 25 + 0.35(x - 200) & \text{if } x > 200 \end{cases}$$

    (a) Determine $f(150)$
    (b) Determine $f(245)$
    (c) Explain why in the $x > 200$ piece of the function, the formula involves $x - 200$ instead of just x.
20. Describe the symmetry of the graph of an odd function.
21. Describe the symmetry of the graph of an even function.
22. Prove that the sum of two odd functions is odd.
23. Prove that the sum of two even functions is even.
24. Prove that the product of two odd functions is even.
25. Prove that the product of two even functions is even.

26. Prove that if f is an odd function, then so are $\frac{1}{f}$ and $-f$.

27. Prove that the product of an even function and an odd function is odd.

28. Prove that addition of functions is associative.

29. Explain the difference between f and $f(x)$, where f is a function.

30. Prove that multiplication of functions is associative.

31. Prove that multiplication of functions is commutative.

7.4 Graphing Functions

We return to graphing. The power of graphing is in the visual representation it can give of an otherwise mysterious relationship. Given the points $(-1,2),(4,12),(0,4),(1,5)$, can you see the relationship among them? Probably not! What if we graph them?

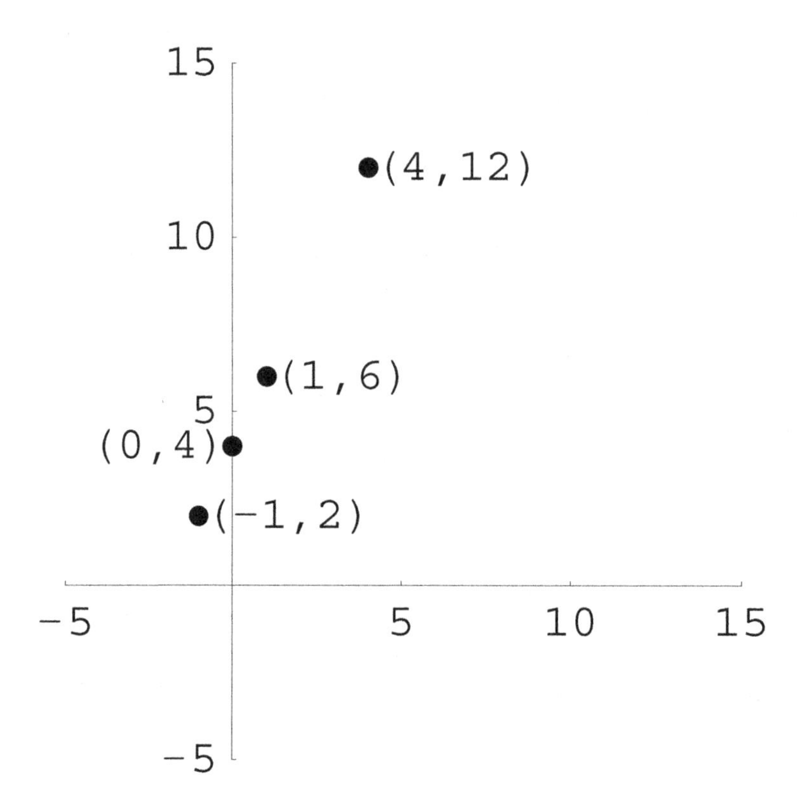

Viewing the graph, it is not hard to believe that these points all lie on a line, although that was not readily apparent from just looking at the ordered pairs. Scientists, business people, journalists, and, in fact, most professions regularly use graphical representations of data to help determine or illustrate relationships.

In this section we will build up a library of basic graphs we should be able to recognize, and then see how they may be transformed. You will want to memorize the graphs of the basic functions so that you can recognize them at a glance.

First, let's consider the conditions under which a given graph is the graph of a function $y = f(x)$. Recall that for a relation to be a function we need each x-value in the domain to be paired with exactly one y-value. Since equal x-values correspond graphically to points on a vertical line (consider the graph of $x = 2$), we have the following theorem.

Theorem 85 (Vertical Line Test). *A graph in the xy-coordinate plane is the graph of a function $y = f(x)$ if and only if every vertical line intersects the graph at most once.*

Example 7.4.1. The relationship defined by the equation $(x-1)^2 + y^2 = 9$ does not represent a function because the graph of this equation is a circle with center $(1,0)$ and radius 3, which fails the vertical line test.

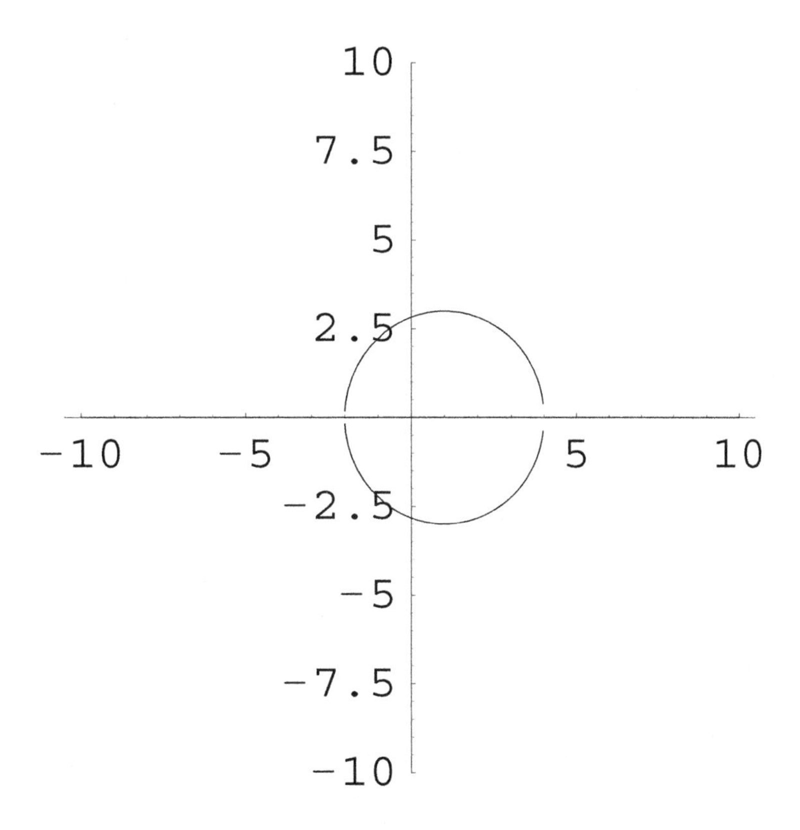
10
7.5
5
2.5
-10
-5
5
10
-2.5
-5
-7.5
-10

□

Example 7.4.2. Does the following graphical relationship represent a function? If so, find the domain and range of the function.

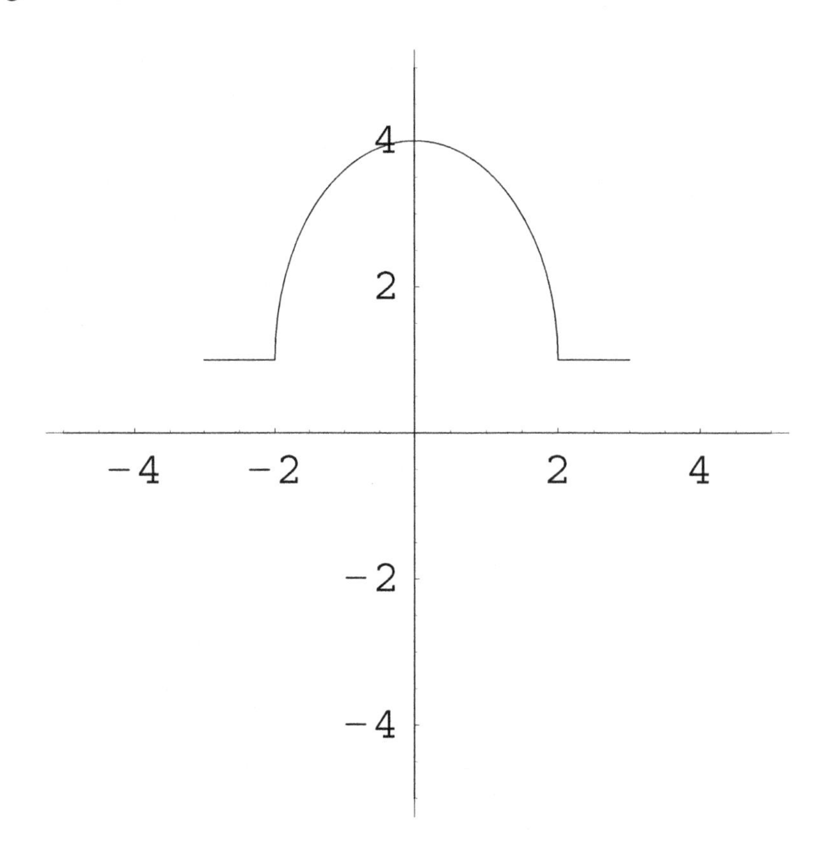

Solution: Since the graph passes the vertical line test, this relationship is a function, f. The domain is the set of x-values used as first coordinates and the range is the set of second coordinates of the ordered pairs comprising f. Hence, the domain appears to be $[-3,3]$, and the range is $[1,4]$.

□

For the vertical line test to be very useful to us, we need to have knowledge of the graph of a given relation. For a completely unknown function, we must plot enough ordered pairs belonging to the function to give us a general idea of what its graph looks like. Consider the following "basic" functions.

$f(x)$	Name	Returns...	Example	Domain
c	Constant	c regardless of x.	$f(-2)=c$	$\mathbb{R}$
x	Identity	the input.	$f(-2)=-2$	$\mathbb{R}$
$\lvert x\rvert$	Absolute Value	x if $x \geq 0$, and $-x$ if $x < 0$.	$f(-2)=2$	$\mathbb{R}$
$\sqrt{x}$	Square root	$a \geq 0$ such that $a^2 = x$	$f(4)=2$	$[0,\infty)$
x^2	Square	the square of x	$f(-2)=4$	$\mathbb{R}$
x^3	Cube	the cube of x	$f(-2)=-8$	$\mathbb{R}$
$\frac{1}{x}$	Reciprocal	the multiplicative inverse of x	$f(2/5)=5/2$	$\{x \in \mathbb{R} \lvert x \neq 0\}$

The graphs of these "basic" functions are given below.

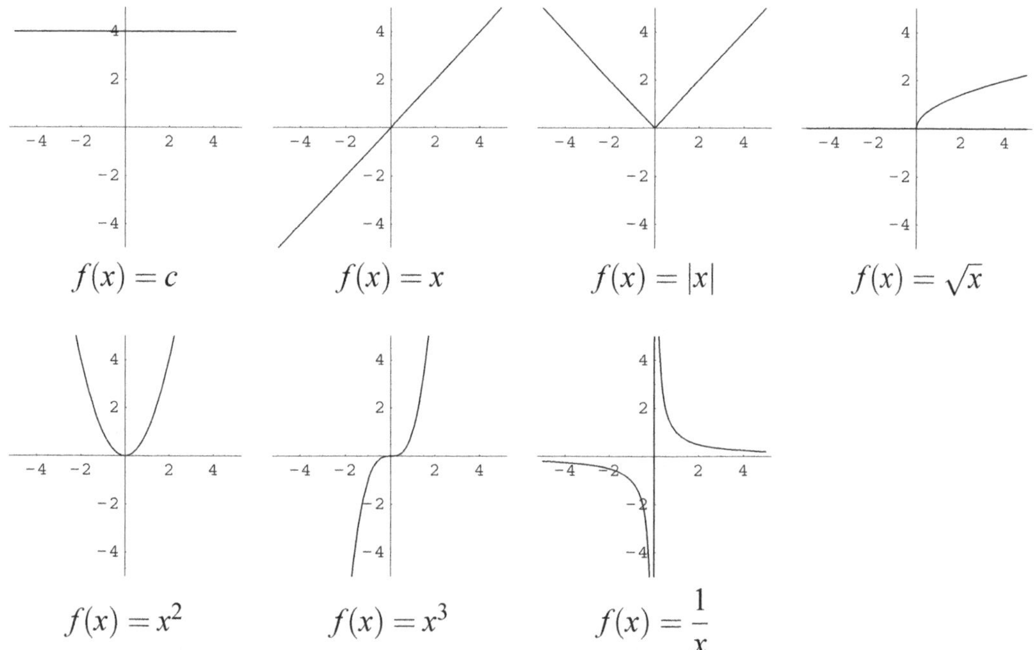

We have several trans formations we can do with these (and other) functions and still get functions of the same basic shape. Let $c > 0$, and let f be a given function (not necessarily one of those above). We will compare the graph of $f(x)$ to the graph of $g(x) = f(x+c)$.

Remember that in graphing a function, the x-coordinate represents a domain element and the y-coordinate (or "height") represents the function value at x. Notice that, for example, $f(0) = g(-c)$, so g has the same height at $-c$ as f does at 0. Also, $f(-2) = g(-2-c)$, so g has the same height at $-2-c$ as f does at -2. In general, g will have the same height at $x-c$ as f does at x. Geometrically, this just means that the graph of g follows *exactly the same* "ups and downs" as f does; that is, the graph of g has the same shape as the graph of f. However, while the graph of g attains the same heights as does the graph of f, it does so at $x-c$ instead of at x; that is, c units to the left of where f does.

In summary, the graph of $g(x) = f(x+c)$ is exactly the same shape as the graph of f, but translated to the left by c units. Similar arguments lead to the table below.

New function:	Effect on graph of $y = f(x)$
$h(x) = f(x) + c$	Vertical shift up c units
$h(x) = f(x) - c$	Vertical shift down c units
$h(x) = f(x+c)$	Horizontal shift left c units
$h(x) = f(x-c)$	Horizontal shift right c units
$h(x) = -f(x)$	Reflection across x-axis
$h(x) = f(-x)$	Reflection across y-axis
$h(x) = cf(x)$	Vertical stretch by a factor of c for $c > 1$; shrink for $0 < c < 1$
$h(x) = f(cx)$	Horizontal shrink by a factor of c for $c > 1$; stretch for $0 < c < 1$

One can see that there are two fundamentally different ways to modify a graph: one can modify the input into f, as $f(x+c)$ or $f(cx)$, or one can modify the output from f, as $f(x)+c$ or $cf(x)$. **The parentheses matter**; read very carefully and consider every symbol.

Example 7.4.3. For each function, identify the basic function and the transformation involved.

1. $h(x) = x^2 - 3$. The basic function is $f(x) = x^2$; h is a downward shift by 3 units.

2. $h(x) = |x+1|$. The basic function is $f(x) = |x|$; h is a shift to the left by 1 unit.

3. $h(x) = (x-4)^2 + 3$. The basic function is $f(x) = x^2$; we have two transformations here. If we put $g(x) = (x-4)^2$, then $h(x) = g(x) + 3$. To obtain the graph of g, shift the graph of f right 4 units. To obtain the graph of h, shift the graph of g up 3 units.

4. $h(x) = -x^2$. The basic function is $f(x) = x^2$; h is a reflection across the x-axis.

5. $h(x) = 7x^2$. The basic function is $f(x) = x^2$; h is a vertical stretch by a factor of 7.

6. $h(x) = \dfrac{x^2}{4}$. The basic function is $f(x) = x^2$; h is a vertical shrink by a factor of $1/4$.

7. $h(x) = \sqrt{-x}$. The basic function is $f(x) = \sqrt{x}$; h is a reflection across the y-axis.

8. $h(x) = \sqrt{2x}$. The basic function is $f(x) = \sqrt{x}$. $h(x) = f(2x)$, so its graph is a horizontal compression by a factor of 2 of that of f.

9. $h(x) = -|x| + 3$. The basic function is $f(x) = |x|$. Let $g(x) = -f(x)$, so $h(x) = g(x) + 3$. Therefore, we first reflect the graph of f across the x-axis, and then shift up 3 units.

10. $h(x) = \sqrt{-x+7} = \sqrt{-(x-7)}$. The basic function is $f(x) = \sqrt{x}$. Let $g(x) = \sqrt{-x} = f(-x)$. Then the graph of g is the reflection of the graph of f across the y-axis. Also, $h(x) = g(x-7) = f(-(x-7)) = \sqrt{-(x-7)} = \sqrt{-x+7}$. Therefore, the graph of h is the graph of g shifted right 7 units. We again had to perform two transformations.

11. $h(x) = \dfrac{-1}{x+2}$. The basic function is $f(x) = \dfrac{1}{x}$. Let $g(x) = -f(x)$ (a reflection across the x-axis), and then $h(x) = g(x+2)$ (a shift right by 2 units).

12. $h(x) = (2x+6)^3$. The basic function is $f(x) = x^3$. Let $g(x) = f(2x)$. Then $h(x) = g(x+3)$. Thus the graph of h is the graph of x^3 after a horizontal compression by a factor of 2 and a shift to the left 3 units. We could also have used $g(x) = f(x+6)$, a shift to the left by 6 units, followed by $h(x) = g(2x) = f(2x+6)$, a horizontal compression by a factor of 2. We can choose either way, but we must be careful not to mix them together.

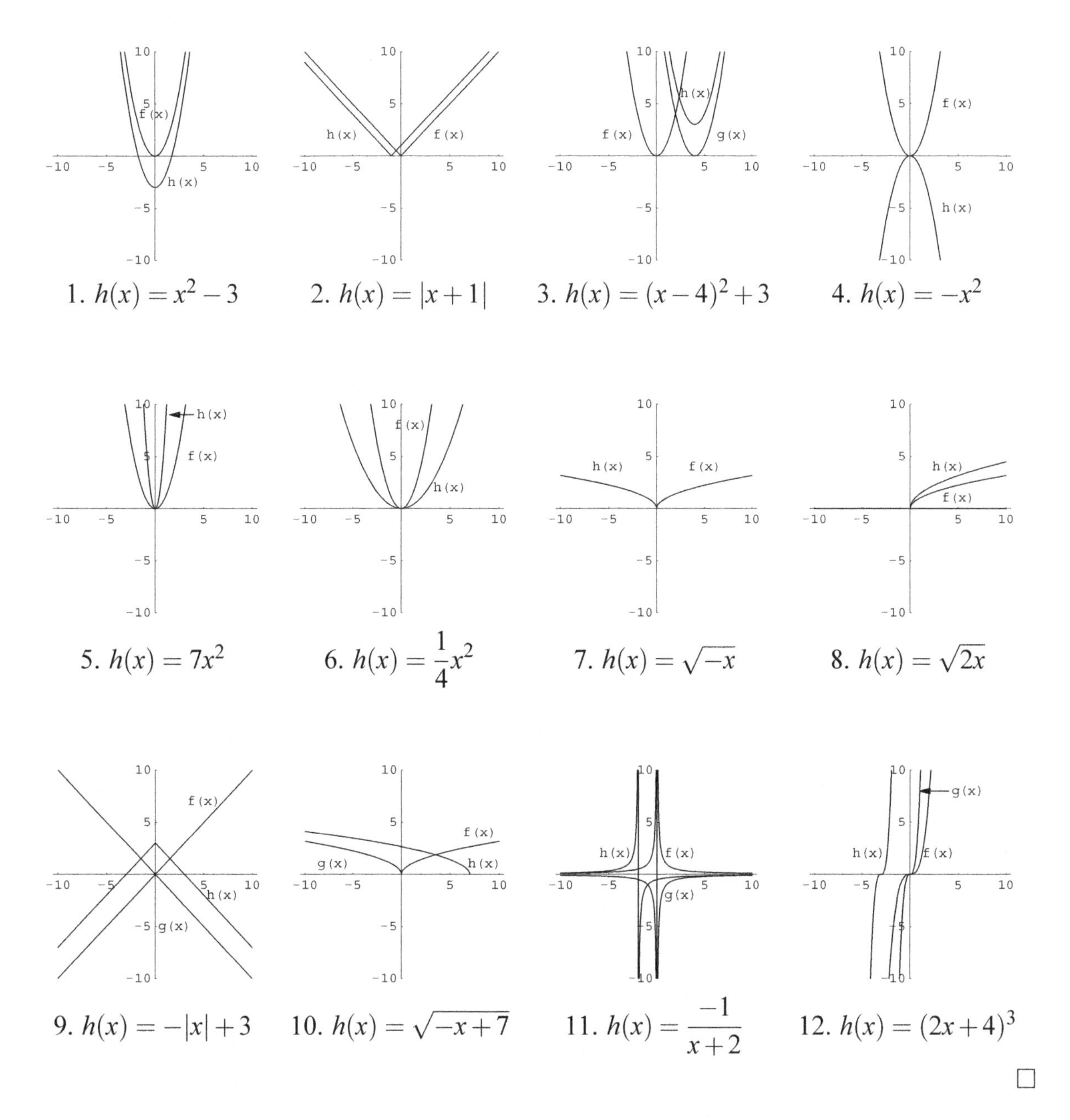

1. $h(x) = x^2 - 3$ 2. $h(x) = |x+1|$ 3. $h(x) = (x-4)^2 + 3$ 4. $h(x) = -x^2$

5. $h(x) = 7x^2$ 6. $h(x) = \frac{1}{4}x^2$ 7. $h(x) = \sqrt{-x}$ 8. $h(x) = \sqrt{2x}$

9. $h(x) = -|x| + 3$ 10. $h(x) = \sqrt{-x+7}$ 11. $h(x) = \frac{-1}{x+2}$ 12. $h(x) = (2x+4)^3$

□

Note that the graph touches the x-axis if and only if the y-coordinate is 0. The corresponding x-value is called a **zero** of f.

Definition 7.4.1. If $f(x) = 0$, then x is a **zero** of f.

Zeros are especially important in equation solving, one of the major purposes of algebra.

Section 7.4 Exercises

Determine whether the given graph is the graph of a function.

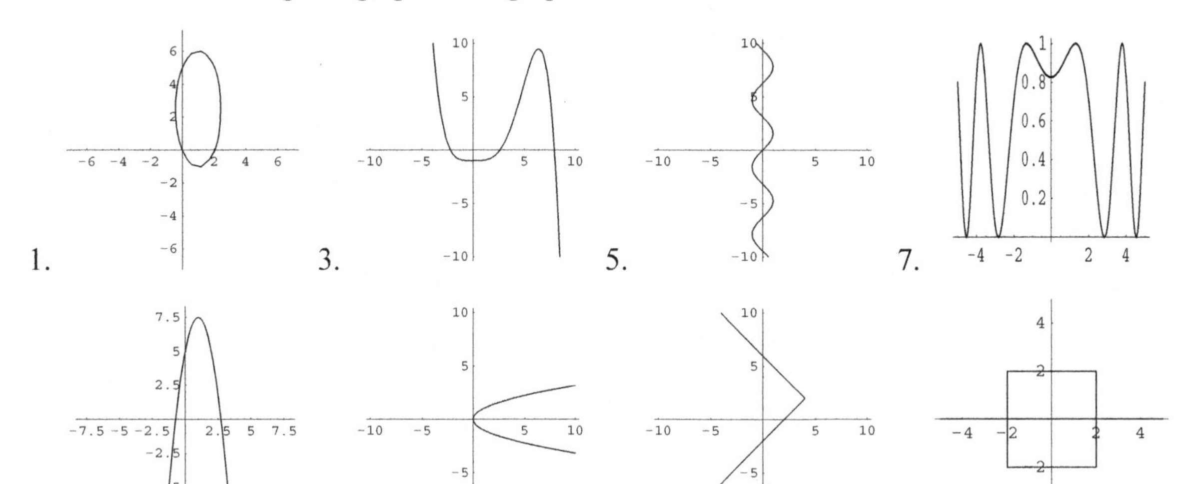

9. For each graph in problems 1 through 8 that represents a function, determine the domain and range from the graph as well as possible. (You may need to extrapolate beyond what you can see on the graph. Realize, though, that doing so is somewhat hazardous!)

10. For each graph in problem 1 that represents a function f, determine $f(2)$ from the graph, if possible. Realize that this will only be an approximation. If it is not possible, explain why not.

Sketch the graph of each function. If possible, do so by transforming the graph of a known function rather than plotting points.

11. $f(x) = x^2 + 1$

12. $f(x) = \dfrac{1}{x-2}$

13. $f(x) = \dfrac{x+1}{x}$

14. $f(x) = \sqrt{2x-5}$

15. $f(x) = 2|x+3| - 1$

16. $f(x) = -|x-2|$

17. $f(x) = \sqrt{-(x+1)}$

18. $f(x) = (3x)^3$

19. $f(x) = -\dfrac{1}{x-3} + 2$

20. Convince yourself that each of the transformations in the table "Transformations of Graphs" does what the table claims it will do.

21. The **floor** function is another basic function; floor(x), also written $\lfloor x \rfloor$ (pronounced "floor x") is the greatest integer that is less than or equal to x. For example, $\lfloor 2.75 \rfloor = 2$, $\lfloor \pi \rfloor = 3$, $\lfloor -4.1 \rfloor = -5$, and $\lfloor 461 \rfloor = 461$. Sketch a graph of the floor function.

22. The **ceiling** function is another basic function; ceil(x), also written $\lceil x \rceil$ (pronounced "ceiling x") is the least integer that is greater than or equal to x. For example, $\lceil 2.75 \rceil = 3$, $\lceil \pi \rceil = 4$, $\lceil -4.1 \rceil = -4$, and $\lceil 461 \rceil = 461$. Sketch a graph of the ceiling function.

Each graph below is a transformation of some basic function. Identify the basic function.

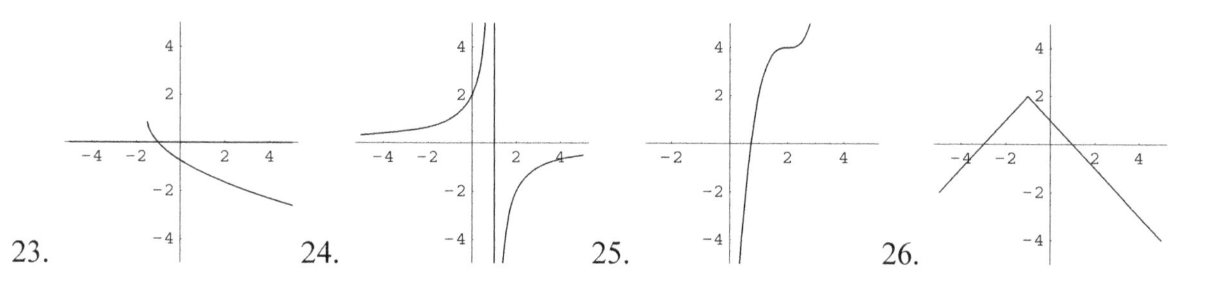

27. For each $x \in \mathbb{R} - \mathbb{Z}$, show that $\lceil x \rceil = \lfloor x+1 \rfloor$.

28. The following data was collected in an experiment in which the height h of a dropped ball was measured (in centimeters) at several times t. Sketch a graph of the height h versus time t, and decide whether the graph is a transformation of one of our basic graphs.

$h(t)$	202	201	200	196	191	185	178	169	159	148	135	122	107	90	73	54	34	12
t	$\frac{3}{60}$	$\frac{5}{60}$	$\frac{7}{60}$	$\frac{9}{60}$	$\frac{11}{60}$	$\frac{13}{60}$	$\frac{15}{60}$	$\frac{17}{60}$	$\frac{19}{60}$	$\frac{21}{60}$	$\frac{23}{60}$	$\frac{25}{60}$	$\frac{27}{60}$	$\frac{29}{60}$	$\frac{31}{60}$	$\frac{33}{60}$	$\frac{35}{60}$	$\frac{37}{60}$

7.5 Inverse Functions

We now return to the notion of inverse functions (with respect to function composition). Recalling the definition of an inverse element, the definition of composition, and the fact that the identity function for composition is given by $\iota(x) = x$, we have the following definition.

Definition 7.5.1. If f and g are two functions satisfying $f \circ g = \iota$ and $g \circ f = \iota$, then f and g are called **inverses** of each other. In this case, f and g are said to each have an inverse, f is the inverse of g, and g is the inverse of f. These relationships are denoted by $f = g^{-1}$ and $g = f^{-1}$, respectively.

NOTE: The f^{-1} notation refers to the inverse of f under **composition**, *not* the multiplicative inverse of f. This similarity in notation is unfortunate, but standard. It should be clear from the context whether we are talking about an inverse function or a multiplicative inverse of a number.

Since two functions are equal provided they have identical domains and the same action on each element of their domain, we have the following equivalent statement.

Theorem 86. *If f and g are two functions satisfying $(f \circ g)(x) = x$ for all x in the domain of g and $(g \circ f)(x) = x$ for all x in the domain of f, then f and g are called* **inverses** *of each other.*

Example 7.5.1. Verify that the functions $f(x) = 2x - 1$ and $g(x) = \dfrac{x+1}{2}$ are inverse functions.

Solution: First note that f and g have the same domain, namely $\mathbb{R}$. So, we compute the compositions $f \circ g(x)$ and $g \circ f(x)$ for all $x \in \mathbb{R}$ and see that we get x.

$$\begin{aligned}(f \circ g)(x) &= f(g(x)) \\ &= f\left(\frac{x+1}{2}\right) \\ &= 2\left(\frac{x+1}{2}\right) - 1 \\ &= (x+1) - 1 \\ &= x\end{aligned}$$

$$\begin{aligned}(g \circ f)(x) &= g(f(x)) \\ &= g(2x-1) \\ &= \frac{(2x-1)+1}{2} \\ &= \frac{2x}{2} \\ &= x\end{aligned}$$

□

Example 7.5.2. Verify that the functions $f(x) = \dfrac{1}{1+x}$ and $g(x) = \dfrac{1}{x} - 1$ are inverse functions.

Solution: We need to compute $f \circ g(x)$ for all $x \neq 0$ and $g \circ f(x)$ for all $x \neq -1$ and see that these both equal x.

$$\begin{aligned} f \circ g(x) &= f(g(x)) \\ &= f\left(\frac{1}{x} - 1\right) \\ &= \frac{1}{1 + \left(\frac{1}{x} - 1\right)} \\ &= \frac{1}{(1/x)} \\ &= x \end{aligned} \qquad \begin{aligned} g \circ f(x) &= g(f(x)) \\ &= g\left(\frac{1}{1+x}\right) \\ &= \frac{1}{\left(\frac{1}{1+x}\right)} - 1 \\ &= (1+x) - 1 \\ &= x \end{aligned}$$

□

Suppose the functions f and g are functions defined as sets of ordered pairs. Say that $f = \{(x,y)|y = f(x)\}$ and $g = \{(x',y')|y' = g(x')\}$. Then f and g are inverse functions if and only if $f(g(x')) = x'$ for all x' (in the domain of g) and $g(f(x)) = x$ for all x (in the domain of f). If $y = f(x)$, then $g(f(x)) = g(y) = x$. Similarly, if $y' = g(x')$, then $f(g(x')) = f(y') = x'$. Thus we have that inverse functions, as sets of ordered pairs, have "reversed" coordinates.

Theorem 87. *Let f and g be inverse functions. Then $y = f(x)$ if and only if $x = g(y)$. That is, $(x,y) \in f$ if and only if $(y,x) \in g$.*

It may be seen by direct substitution that if the above relationship holds, then f and g are indeed inverse functions. (This is what we did above.) Thus, for the function f to have an inverse *function*, it must be that the relation obtained by reversing all the ordered pairs of f is also a function.

That is, the function $f = \{(x,y)|y = f(x)\}$ has an inverse g if and only if the relation $g = \{(y,x)|y = f(x)\}$ is also a function. In effect, we need each y to be "used" exactly once. A function satisfying this new condition is called a **one-to-one** function.

Definition 7.5.2. A function f is called **one-to-one** provided whenever $f(a) = f(b)$, $a = b$. In other words, no element of the range is paired with more than one element of the domain.

Suppose that $c = f(a) = f(b)$. In terms of ordered pairs, the definition of one-to-one means that if (a,c) and (b,c) are both elements of f, then (c,a) and (c,b) can only be elements of another function if $a = b$; otherwise, c would be paired with *two* elements of the range. To verify that a given function is one-to-one, we prove the conditional: if $f(a) = f(b)$, then $a = b$.

Example 7.5.3. Verify that the function given by $f(x) = \sqrt{x}$ is one-to-one.

Solution: We prove the conditional: if $f(a) = f(b)$, then $a = b$.

$f(a)$	$= f(b)$	Hypothesis
$\sqrt{a}$	$= \sqrt{b}$	Definition of f
$(\sqrt{a})^2$	$= (\sqrt{b})^2$	______________
a	$= b$	______________

Therefore, $f(x) = \sqrt{x}$ is one-to-one.

□

Example 7.5.4. Verify that the function given by $g(x) = x^2$ is NOT one-to-one.

Solution: We show the conditional, "If $g(a) = g(b)$, then $a = b$" is false. To do this, we show that even when the hypothesis is true, the conclusion might be false (either algebraically or by a counter-example).

$$\begin{array}{rcll} g(a) & = & g(b) & \text{Hypothesis} \\ a^2 & = & b^2 & \text{Definition of } f \\ \sqrt{a^2} & = & \sqrt{b^2} & \underline{\hspace{5cm}} \\ |a| & = & |b| & \underline{\hspace{5cm}} \\ a & = & \pm b & \underline{\hspace{5cm}} \end{array}$$

Therefore, $g(x) = x^2$ is NOT one-to-one.

Alternatively, a counterexample where $g(a) = g(b)$ but $a \neq b$ will suffice. Take, for instance, $a = 2$ and $b = -2$.

□

The graph of a function can also be used to determine whether a function is one-to-one. A function is one-to-one provided each y-value "gets used" at most once. Since y-values correspond graphically to horizontal lines, we have the following theorem.

Theorem 88 (Horizontal Line Test). *A function is one-to-one if and only if every horizontal line intersects its graph in the xy-coordinate plane at most once.*

Example 7.5.5. Consider the functions $f(x) = \sqrt{x}$ and $g(x) = x^2$ from the examples above. They are graphed below. Notice that the horizontal line $y = 4$ intersects g in both points $(-2,4)$ and $(2,4)$, but no horizontal line will intersect the graph of f more than once.

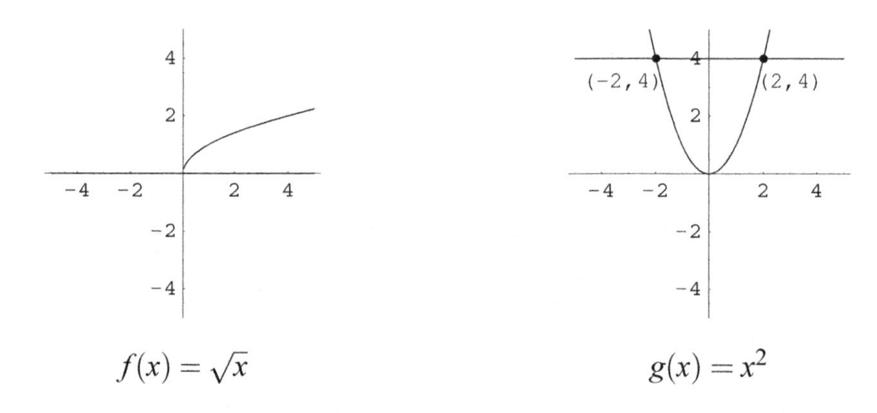

$f(x) = \sqrt{x}$ $\qquad$ $g(x) = x^2$

□

To formally reiterate the relationship between inverse functions and one-to-one functions, we have the following theorem.

Theorem 89. *A function f has an inverse function if and only if f is one-to-one.*

Note that it is possible to define an inverse for a function that is NOT one-to-one by restricting the domain so that the function is one-to-one on the new domain.

Example 7.5.6. The function $f(x) = x^2$ is one-to-one on $[0, \infty)$ and has inverse $f^{-1}(x) = \sqrt{x}$.

□

Recall that the graph of f looks like $\{(x,y)|y = f(x)\}$, so the graph of f^{-1} (if it exists) looks like $\{(y,x)|x = f^{-1}(y)\}$. The effect of this on the graph is to reflect the graph of f across the line $y = x$. Notice that reflecting this way will turn a horizontal line (which we used to perform the horizontal line test) into a vertical line (which we use to perform the vertical line test for functions). This confirms the theorem above; a function will have an inverse if and only if it is one-to-one.

The inverse of a given function f may be found algebraically via the following procedure.

1. Verify that f is one-to-one.
2. Find a rule for $f^{-1}(x)$.
   (a) Write $y = f(x)$. (This is the rule for f.)
   (b) Interchange x and y in the equation. (This creates an ordered pair (y,x) that also belongs to f since we have $x = f(y)$.)
   (c) Solve for y. (This gives $y = f^{-1}(x)$, so the ordered pair (x,y) belongs to f^{-1}.)
3. Verify that f and f^{-1} are indeed inverse functions.

Example 7.5.7. Find the inverse of $f(x) = 3x - 2$, provided the inverse exists.

1. Verify that f is one-to-one:

$$\begin{aligned} f(a) &= f(b) && \text{Hypothesis} \\ 3a-2 &= 3b-2 && \text{Definition of } f \\ 3a &= 3b && \text{Additive Cancellation} \\ a &= b && \text{Multiplicative Cancellation} \end{aligned}$$

2. Find a rule for f^{-1}:

$$\begin{aligned} y &= f(x) && \text{Write } y = f(x) \\ y &= 3x-2 && \text{Function evaluation} \\ x &= 3y-2 && \text{Interchange } x \text{ and } y \\ y &= \frac{x+2}{3} && \text{Solve for } y \\ f^{-1}(x) &= \frac{x+2}{3} && y = f^{-1}(x) \end{aligned}$$

3. Verify that f and f^{-1} are inverse functions:

$$\begin{aligned} f\circ f^{-1}(x) &= f(f^{-1}(x)) \\ &= f\left(\frac{x+2}{3}\right) \\ &= 3\left(\frac{x+2}{3}\right)-2 \\ &= (x+2)-2 \\ &= x \end{aligned}$$

$$\begin{aligned} f^{-1}\circ f(x) &= f^{-1}(f(x)) \\ &= f^{-1}(3x-2) \\ &= \frac{(3x-2)+2}{3} \\ &= \frac{3x}{3} \\ &= x \end{aligned}$$

□

We provide a slightly more complicated example for illustration purposes. The process is identical to that in the example above, only the specific calculations are different.

Example 7.5.8. Find the inverse of $f(x)=\dfrac{5x}{4x+1}$, provided the inverse exists.

1. Verify that f is one-to-one:

$$\begin{aligned} f(a) &= f(b) && \text{Hypothesis} \\ \frac{5a}{4a+1} &= \frac{5b}{4b+1} && \text{Definition of } f \\ 5a(4b+1) &= 5b(4a+1) && \text{Fraction equivalence} \\ a(4b+1) &= b(4a+1) && \text{Multiplicative Cancellation} \\ 4ab+a &= 4ab+b && \text{Distributive Law} \\ a &= b && \text{Additive Cancellation} \end{aligned}$$

2. Find a rule for f^{-1}:

$$\begin{aligned} y &= f(x) && \text{Write } y=f(x) \\ y &= \frac{5x}{4x+1} && \text{Function evaluation} \\ x &= \frac{5y}{4y+1} && \text{Interchange } x \text{ and } y \\ y &= \frac{-x}{4x-5} && \text{Solve for } y \\ f^{-1}(x) &= \frac{-x}{4x-5} && y=f^{-1}(x) \end{aligned}$$

3. Verify that f and f^{-1} are inverse functions:

$$
\begin{aligned}
f \circ f^{-1}(x) &= f(f^{-1}(x)) \\
&= f\left(\frac{-x}{4x-5}\right) \\
&= \frac{5\left(\frac{-x}{4x-5}\right)}{4\left(\frac{-x}{4x-5}\right)+1} \\
&= \frac{\left(\frac{-5x}{4x-5}\right)}{\left(\frac{-4x+(4x-5)}{4x-5}\right)} \\
&= \frac{\left(\frac{-5x}{4x-5}\right)}{\left(\frac{-5}{4x-5}\right)} \\
&= x
\end{aligned}
$$

$$
\begin{aligned}
f^{-1} \circ f(x) &= f^{-1}(f(x)) \\
&= f^{-1}\left(\frac{5x}{4x+1}\right) \\
&= \frac{-\left(\frac{5x}{4x+1}\right)}{4\left(\frac{5x}{4x+1}\right)-5} \\
&= \frac{\left(\frac{-5x}{4x+1}\right)}{\left(\frac{20x-5(4x+1)}{4x+1}\right)} \\
&= \frac{\left(\frac{-5x}{4x+1}\right)}{\left(\frac{-5}{4x+1}\right)} \\
&= x
\end{aligned}
$$

□

How do the graphs of a one-to-one function and its inverse compare? When we interchange x and y, the effect on the graph is to reflect it across the line $y = x$. We do not prove this fact here; it is a topic more for a course in analytic geometry. Instead, we will illustrate it with an example.

Example 7.5.9. Consider the function $f(x) = \frac{5x}{4x+1}$ from the previous example, with inverse $f^{-1}(x) = \frac{-x}{4x-5}$. The graphs are below, graphed with the line $y = x$.

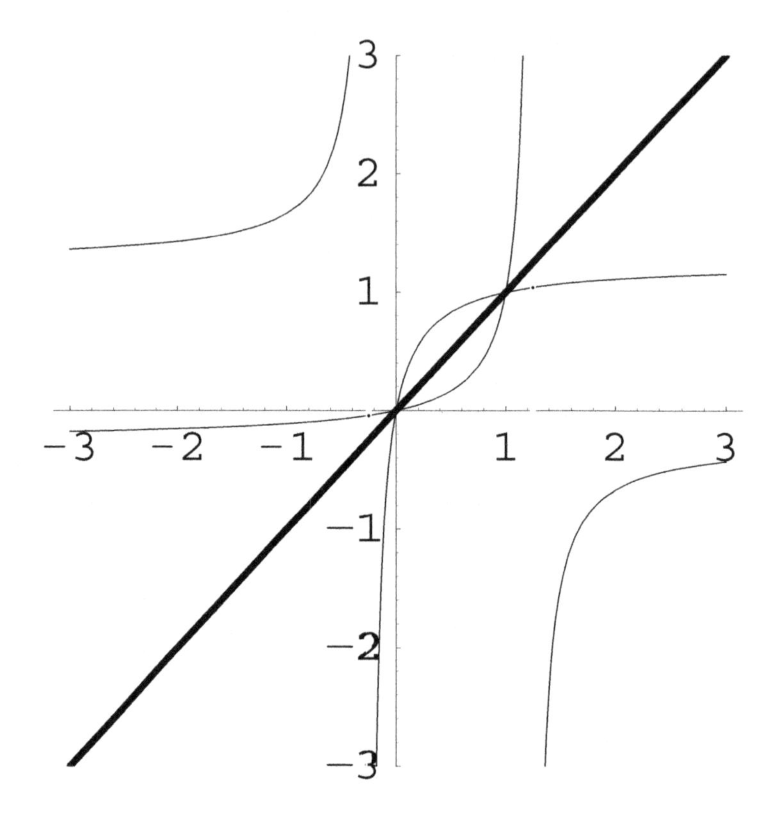

□

Notice that if f and g are inverses, then the domain of f is the range of g, and the range of f is the domain of g.

We have used the word "inverse" in several contexts. Originally, we spoke only of additive and multiplicative inverses, and we were very careful about saying which we were using. In some contexts, however, we just say "inverse" without specifying what kind. When we say "the" inverse of a matrix M, we mean the *multiplicative* inverse of M every time. When we say "the" inverse of a function f, we always mean the inverse of f with respect to *composition*. You will need to be alert for context so you can tell what is meant by "the inverse" of some object.

Section 7.5 Exercises

Verify that the given functions are inverses, or show that they are not.

1. $f(x) = \sqrt{2x+1} - 3$ for $x \in [-1/2, \infty)$ and $g(x) = \dfrac{(x+3)^2 - 1}{2}$ for $x \in [-3, \infty)$.
2. $f(x) = \dfrac{3x+1}{x-1}$ and $g(x) = \dfrac{x+1}{x-3}$
3. $f(x) = x - 5$ and $g(x) = x + 5$.
4. Determine which "basic" functions (from the section on graphing) are one-to-one.

Determine whether the given function is one-to-one. If a function is one-to-one, find its inverse.

5. $f(x) = (x+2)^3$

6. $g(x) = (x+3)^2$

7. $f = \{(1,3),(2,5),(0,4),(-2,2),(5,5)\}$

8. $h(t) = -16t^2 + 80t + 100$

9. $g = \{(1,2),(2,3),(0,4),(-2,-3),(5,5)\}$

10. $h(x) = \dfrac{x^2-1}{x+1}$

11. $f(x) = \dfrac{|x|}{x}$

12. $r(x) = \dfrac{2}{x+1}$

13. $g(x) = \dfrac{-5}{2x-3}$

14. $h(x)\dfrac{2x}{3x+4}$

15. $f(x) = \dfrac{3x+5}{6x-1}$

16. $r(x) = \dfrac{2x+7}{3x-2}$

Determine which functions are one-to-one. If a function is one-to-one, find its inverse graphically.

17. 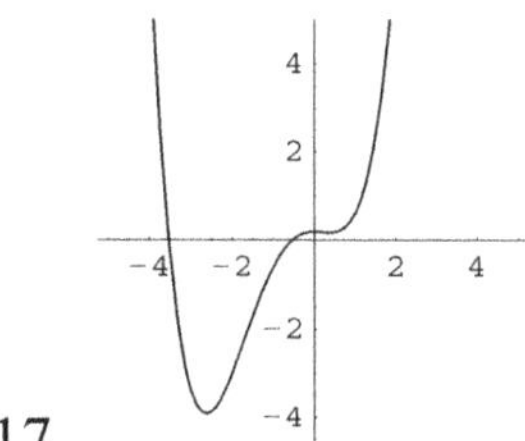

18. 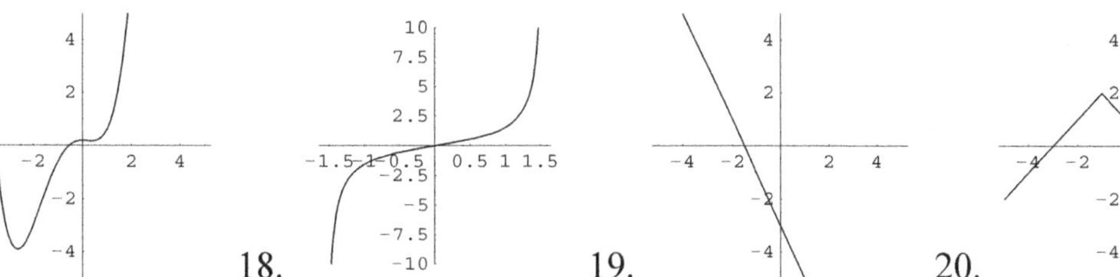

19. 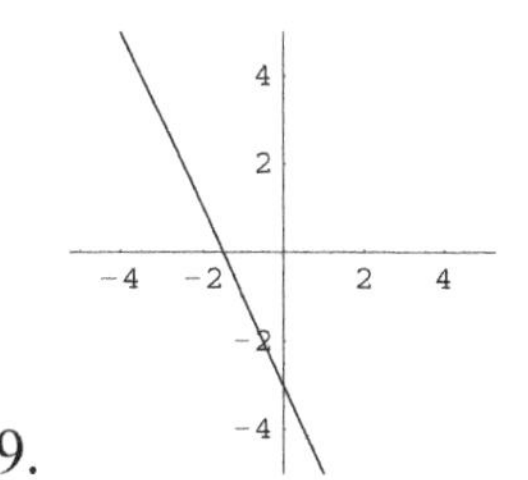

20. 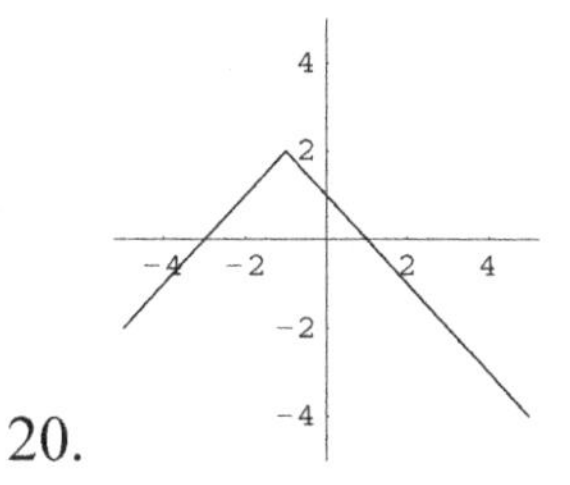

21. Explain the difference between the idea of a function and the idea of a one-to-one function.

22. In the previous section, we gave data associated with dropping a ball. The function

$$h(t) = -490.5t^2 + 203$$

gives an approximation for the height of this ball after t seconds.

(a) Verify that this function is one-to-one on the interval $[0,\infty)$.

(b) Find an inverse for h on the interval $[0,\infty)$.

(c) What does the inverse function tell you?

Chapter 8

Exponential and Logarithmic Functions

This unit defines and investigates exponential and logarithmic functions. We motivate exponential functions by their "similarity" to monomials as well as their wide variety of application. We introduce logarithmic functions as the inverse functions of exponential functions and exploit our previous knowledge of inverse functions to investigate these functions. In particular, we use this inverse relationship for the purpose of solving exponential and logarithmic equations

Objectives

- To define exponential and logarithmic functions
- To investigate the properties of exponential and logarithmic functions
- To introduce some applications of exponential and logarithmic functions
- To solve exponential and logarithmic equations

Terms

- exponential function
- logarithmic function

8.1 Exponential Functions

Up to this point, we have been concerned solely with polynomial functions like $f(x) = x^3$ which have a constant exponent and variable base. We now want to consider functions like 3^x which have a variable exponent and constant base. What makes such functions important is the wide variety of applications they have: compound interest, radioactive decay, learning curves, charge in a capacitor, etc.

Definition 8.1.1. The **exponential function** f with **base** a is $f(x) = a^x$, where $a > 0$, $a \neq 1, x \in \mathbb{R}$. The domain of f is $\mathbb{R}$, and the range is $(0, \infty)$. (0 itself is not in the range of f.)

Example 8.1.1. Consider $f(x) = 2^x$. We will create a table of values for x and $f(x)$, and then sketch a graph of f.

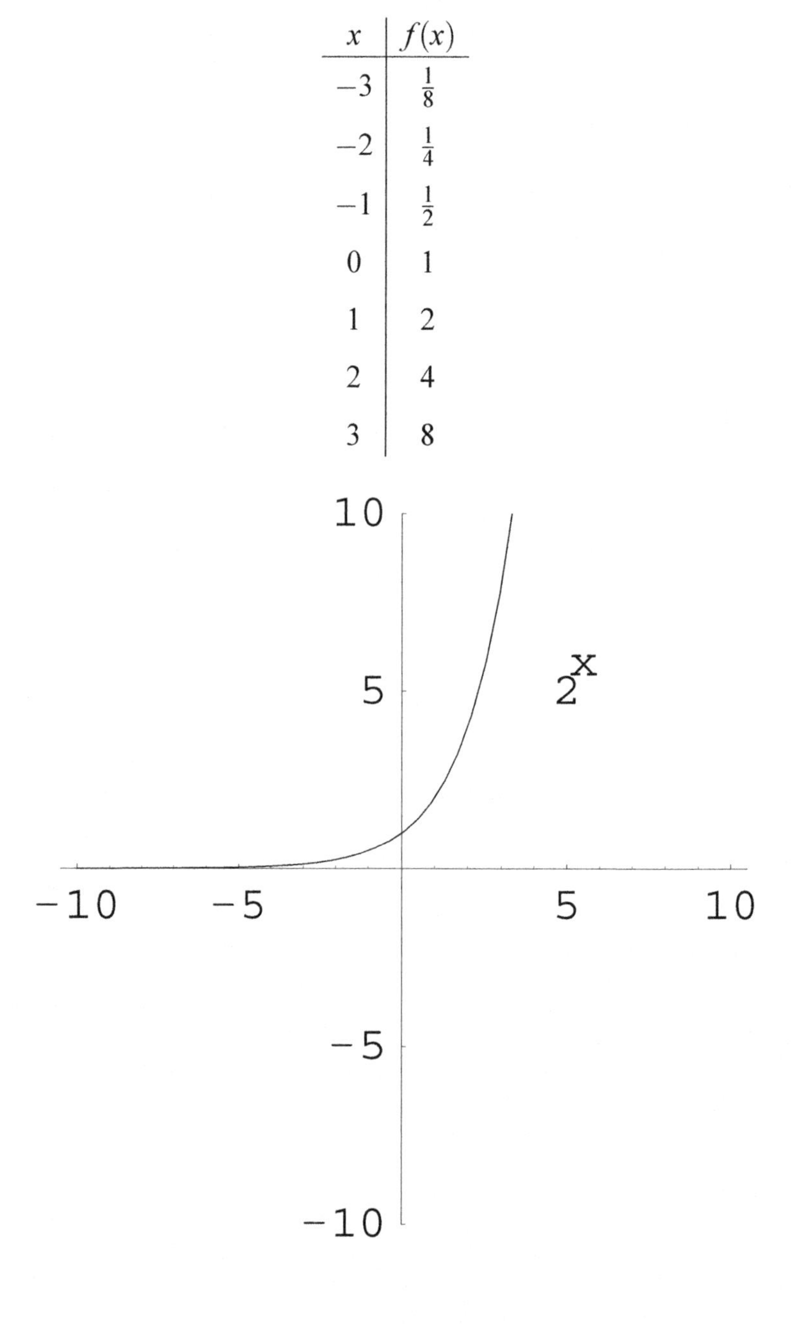

x	$f(x)$
-3	$\frac{1}{8}$
-2	$\frac{1}{4}$
-1	$\frac{1}{2}$
0	1
1	2
2	4
3	8

Observe that f is increasing and one-to-one. Also, f is strictly positive, as we noted earlier when we said that the range of a^x is $(0,\infty)$. Notice too that $f(0)=1$ and $f(1)=2$.

Now consider $g(x)=\left(\frac{1}{2}\right)^x$. This looks like a completely new function, but $g(x)=f(-x)$ since $\left(\frac{1}{2}\right)^x=\frac{1}{2^x}=2^{-x}$. Therefore, the graph of g is just the reflection of the graph of f across the y-axis!

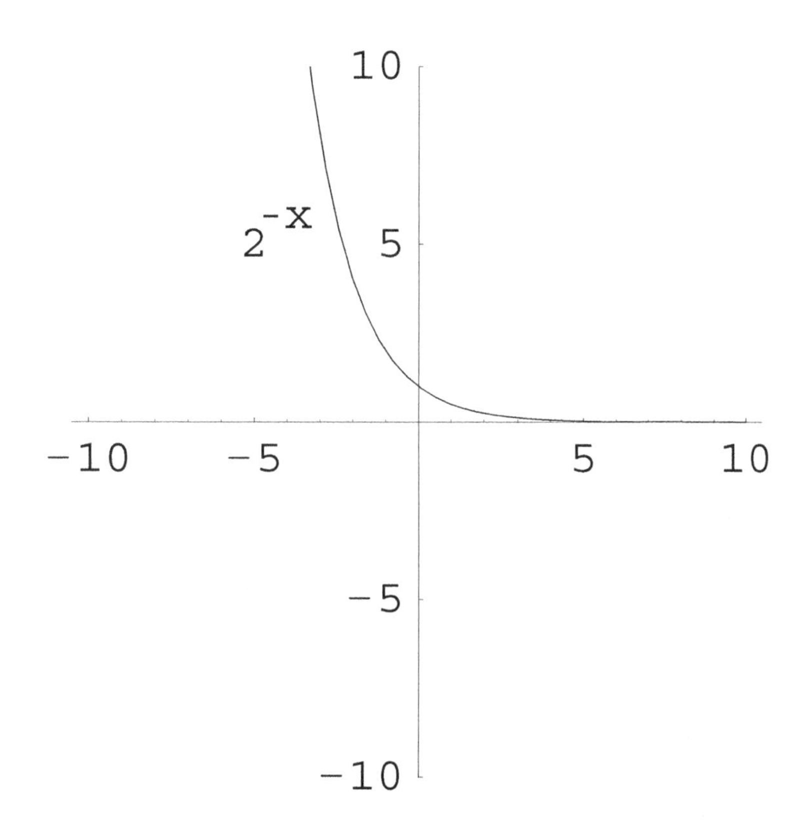

□

The observations above lead us to the following theorem.

Theorem 90. *Let $a>0$ with $a\neq 1$.*

1. *If $a>1$, the graph of $f(x)=a^x$ will rise to the right.*

2. *If $a<1$, then the graph of $f(x)=a^x$ will fall to the right.*

3. *The graph of $g(x)=\left(\frac{1}{a}\right)^x$ is the reflection of the graph of $y=a^x$ across the y-axis.*

4. *If $f(x)=a^x$, then $f(0)=1$ and $f(1)=a$.*

Naturally, a larger value of a will cause the graph to rise more rapidly. For purposes of comparison, we show below the graphs of $f(x)=2^x, 3^x$, and 10^x.

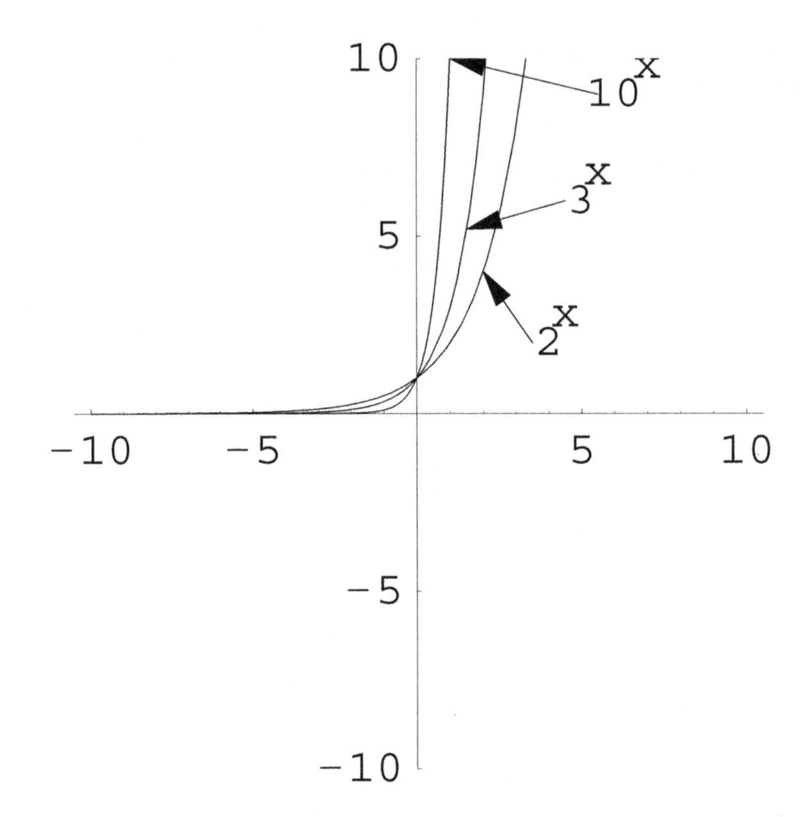

We may also transform these graphs according to the same principles we use to transform other graphs.

Example 8.1.2. $f(x) = 3^{x+1} - 2$ has the same graph as $y = 3^x$, but is shifted one unit left and two units down.

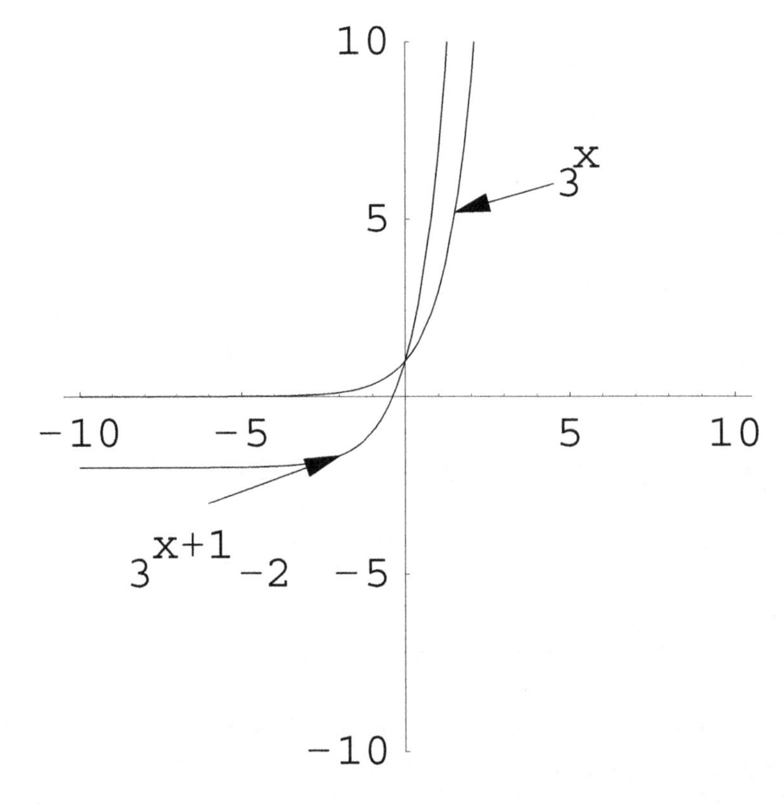

□

Exponential functions are important because of their wide application in both business and the sciences.

Example 8.1.3. **Interest**. Suppose you invest P dollars at an annual interest rate of r (expressed as a decimal), and interest is compounded n times per year. Let $A(t)$ denote your account balance after t years have elapsed. Find a closed formula for $A(t)$ in terms of P, r, n, and t.

Solution. First, since r is an annual rate, the rate for a single compounding period is $\frac{r}{n}$. (For example, if interest is compounded monthly, then your rate for a month is $\frac{r}{12}$.)

Initially, your account holds P dollars. After one compounding period, you add to this interest in the amount of $\frac{r}{n}(P)$, so that your balance is

$$A\left(\frac{1}{n}\right) = P + \frac{r}{n}(P) = P\left(1+\frac{r}{n}\right).$$

This is your beginning balance for the next period. (We use $t = \frac{1}{n}$ since one period out of n has elapsed.)

The amount of money at the beginning of the period is irrelevant to the above computation: if you begin the period with x dollars, you end the period with $\left(1+\frac{r}{n}\right)x$ dollars. Therefore, since you begin the second period with $P\left(1+\frac{r}{n}\right)$ dollars, you end it with

$$A\left(\frac{2}{n}\right) = P\left(1+\frac{r}{n}\right)\left(1+\frac{r}{n}\right) = P\left(1+\frac{r}{n}\right)^2$$

dollars.

This is what you begin the third compounding period with, so you end the third compounding period with

$$A\left(\frac{3}{n}\right) = P\left(1+\frac{r}{n}\right)^2\left(1+\frac{r}{n}\right) = P\left(1+\frac{r}{n}\right)^3$$

dollars.

Notice that in each case, the exponent on $\left(1+\frac{r}{n}\right)$ is n times the argument of A. For example, the exponent on $\left(1+\frac{r}{n}\right)$ in $A(\frac{3}{n})$ is 3. In general, then, we have

$$A(t) = P\left(1+\frac{r}{n}\right)^{nt}.$$

□

Example 8.1.4. Suppose \$1000 is invested at 8.2% for 5 years. Find the account balance if it is compounded annually, quarterly, monthly, and daily.

Solution: Recall that $A(t) = P\left(1+\dfrac{r}{n}\right)^{nt}$, where P is the amount invested, r is the interest rate (as a decimal), t is time elapsed in years, and n is the number of compounding periods per year. Thus, we have $P = 1000, r = 0.082$ and $t = 5$ with n varying in each case.

Annual compounding means once per year, so we have $n = 1$, and we get
$A(5) = 1000(1+0.082)^5 = 1000(1.082)^5 \approx 1482.98$ dollars.

Quarterly compounding means four times per year, and we get

$$A(5) = 1000\left(1+\frac{0.082}{4}\right)^{4(5)} \approx 1500.58 \text{ dollars.}$$

Monthly compounding means 12 times per year, and we get

$$A(5) = 1000\left(1+\frac{0.082}{12}\right)^{12(5)} \approx 1504.72 \text{ dollars.}$$

Finally, daily compounding means 365 times per year, so we get

$$A(5) = 1000\left(1+\frac{0.082}{365}\right)^{365(5)} \approx 1506.75 \text{ dollars.}$$

It is reasonable that the interest grows more rapidly when the compounding is more frequent. What is perhaps surprising is that compounding annually versus compounding daily made less than \$25 difference over the course of five years. Just to satisfy our curiosity, let's see how much difference it makes if we let interest accrue for 25 years.

Compounding annually: $A(25) = 1000(1.082)^{25} \approx 7172.68$ dollars.

Compounding daily: $A(25) = 1000\left(1+\dfrac{0.082}{365}\right)^{365(25)} \approx 7766.11$ dollars.

Now the difference is nearly \$600.

□

Example 8.1.5. **Radioactive decay**. Let $A(t)$ be the amount in grams of a radioactive substance. Let A_0 be the initial amount (the amount at $t = 0$), so that $A(0) = A_0$. Let k be the half-life (the time it takes for half of the substance to decay). After one half-life, we have $A(k) = \frac{1}{2}A_0$. After two half-lives, we have $A(2k) = \frac{1}{2}\left(\frac{1}{2}A_0\right) = \frac{1}{4}A_0$. Notice that it doesn't matter how much you start with, after a half-life elapses, you have exactly half that much left. After t years have gone by, $\frac{t}{k}$ half-lives have gone by. For example, if the half-life is 4 years, then after three years, $\frac{3}{4}$ of a half-life has elapsed. Thus we have

$$A(t) = A_0\left(\frac{1}{2}\right)^{t/k}.$$

□

Example 8.1.6. Suppose that a certain material has a half life of 25 years, and there are

$$A(t) = 10\left(\frac{1}{2}\right)^{t/25}$$

grams remaining after t years. Find the initial amount and the amount after 80 years.

Solution: The initial amount is 10g, as we can read directly off of the formula for $A(t)$. Alternatively, "initial amount" means the amount when no time has gone by, at $t = 0$, so we can simply compute $A(0) = 10(1/2)^0 = 10$ grams. Thus, after 80 years, we have $A(80) = 10\left(\dfrac{1}{2}\right)^{80/25} \approx$ 1.088 grams left.

□

Section 8.1 Exercises

Use your calculator to approximate each of the following to the nearest ten-thousandth, when possible.

1. $(2.3)^5$
2. $(1.4)^{-2}$
3. $(3.18)^{2.35}$
4. $(\sqrt{2})^{-3}$
5. $(\sqrt{\pi})^{-4}$
6. $(-3)^{\sqrt{2}}$
7. $-3^{\sqrt{2}}$
8. 100^{π}
9. $-100^{3\pi}$
10. $(5000)^{\frac{3}{25}}$
11. $\left(\frac{2}{3}\right)^{-\sqrt{2}}$
12. $\left(\frac{\sqrt{2}}{5}\right)^{7}$
13. $\left(\frac{1}{9}\right)^{-10}$

Sketch the graph of each exponential function, and explicitly evaluate each function at least four values of x.

14. $f(x) = \left(\frac{1}{3}\right)^x$
15. $f(x) = 4^{-x}$
16. $f(x) = 5^x$
17. $f(x) = \left(\frac{2}{3}\right)^{-x}$
18. $f(x) = \pi^x$
19. $f(x) = \left(\frac{5}{4}\right)^{-x}$
20. $f(x) = 3^{x-1} + 3$
21. $f(x) = 2^{x+4} - 2$
22. Use the graphs of $f(x) = 3^x$ and $g(x) = 4^x$ to solve the inequality $3^x < 4^x$.
23. Use the graphs $f(x) = \left(\frac{1}{3}\right)^x$ and $g(x) = \left(\frac{1}{4}\right)^x$ to solve the inequality $\left(\frac{1}{4}\right)^x \leq \left(\frac{1}{3}\right)^x$.
24. Compute the value of an account of \$12,000 after four years if the interest rate is 7% and is compounded

    (a) monthly.
    (b) quarterly.
    (c) semiannually.
    (d) weekly.
    (e) daily.
    (f) hourly.

25. Compute the *interest* earned on a CD of \$1500 after 18 months if the interest rate is 8% and is compounded monthly.
26. Compute the *interest* earned on an investment of \$5000 after 18 months if the interest rate is 8.25% and is compounded daily.
27. If you had \$5,000 to invest for 4 years with the goal of greatest return on your investment, would you rather invest in an account paying

(a) 7.2% compounded quarterly,

(b) 7.15% compounded daily, or

(c) 7.1% compounded hourly?

28. Many credit cards have interest rates of around 19%. Compute the interest on a balance of $1000 after one year if the interest is compounded daily. (Assume that you do not make any payments; if you paid enough in the previous month, some cards will allow you to skip payments.)

29. A certain radioactive substance has a half-life of 1200 years. If a sample of the substance is 1000 grams, how much of the substance will be left in 10,000 years?

30. The half-life of uranium-235 is approximately 710,000,000 years. Of a 10-gram sample, how much will remain after 2000 years?

31. Estimate each quantity to the nearest hundred-thousandths.

(a) $\left(1+\frac{1}{1}\right)^{1}$

(b) $\left(1+\frac{1}{10}\right)^{10}$

(c) $\left(1+\frac{1}{100}\right)^{100}$

(d) $\left(1+\frac{1}{1000000}\right)^{1000000}$

(e) Make a conjecture about the value of $\left(1+\frac{1}{x}\right)^{x}$ as x grows very large.

8.2 Logarithmic Functions

In the previous section we observed that exponential functions are one-to-one; this implies that they have inverse functions under composition.

Definition 8.2.1. The **logarithmic function** g with **base** a is the inverse of the function $f(x) = a^x$ for $a > 0, a \neq 1$. We write $g(x) = \log_a(x)$. That is,

$$y = \log_a(x) \text{ if and only if } a^y = x.$$

The domain of $\log_a$ is $(0, \infty)$, and the range is $(-\infty, \infty)$.

One way to think about logarithms is to note that the number $\log_a(x)$ answers the question, "To what power must one raise a to get x?"

Example 8.2.1. Using the definition above, we see that

1. $\log_2(8) = 3$, since $2^3 = 8$;
2. $\log_3(81) = 4$, since $3^4 = 81$;
3. $\log_{16}(4) = \frac{1}{2}$, since $16^{\frac{1}{2}} = 4$; and
4. $\log_{10}\left(\frac{1}{100}\right) = -2$, since $10^{-2} = \frac{1}{100}$.

□

Note that the domain of the function $\log_a(x)$ is the range of the function a^x, and the range of $\log_a(x)$ is the domain of a^x. We will sometimes write $\log_a x$ for $\log_a(x)$. Also, the symbol $\log_a$ represents a *function*, while $\log_a(x)$ represents a *number*; they are two entirely different kinds of objects. This is a somewhat subtle difference that may seem unimportant; however, if you take the time to understand it, it will help you avoid some serious computational errors. (See the unit on functions for more detail.)

Finally, since the functions a^x and $\log_a(x)$ are inverse functions, if we compose them, we get the identity function. Thus,

$$\log_a(a^x) = x \text{ and } a^{\log_a(x)} = x.$$

This is just the definition of inverse functions, but these identities are extremely important and will be essential when we begin solving equations.

Example 8.2.2. We have that

1. $\log_2(2^x) = x$,
2. $\log_a(a^3) = 3$,
3. $4^{\log_4(x)} = x$, while
4. $2^{\log_2(-5)} \neq -5$ since -5 is not in the domain of $g(x) = \log_2(x)$.

□

The graph of $\log_a(x)$ is not too hard to find since we already have graphs for exponential functions a^x; we need only reflect the graph of a^x across the line $y = x$. We may, as always, manipulate the graphs according to principles we have discussed. Thus, below, the graph of $\log_{10}(x-1)$ is just the graph of $\log_{10}(x)$ shifted to the right 1 unit.

Example 8.2.3.

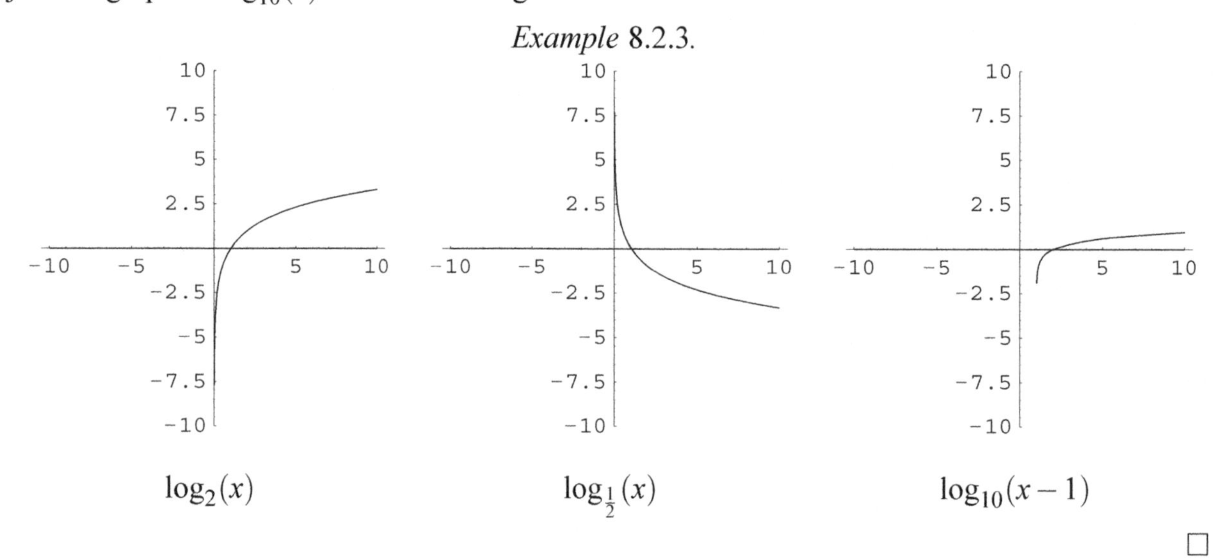

$\log_2(x)$ $\log_{\frac{1}{2}}(x)$ $\log_{10}(x-1)$

□

Example 8.2.4. Find the domain of $f(x) = \log_4(3x+5)$.

Solution: The domain of $\log_a(x)$ is $(0,\infty)$ regardless of a, so we must have $3x+5 > 0$. Thus $x > -\frac{5}{3}$, and the domain of f is $(-\frac{5}{3},\infty)$.

□

The logarithm has properties corresponding to the properties of exponential functions. This is pretty reasonable since we defined logarithms based on exponentials. We summarize the properties of logarithms in the following theorem.

Theorem 91. *Let* $a > 0$, $a \neq 1$, *let* $b,x,y \in \mathbb{R}$, *and let* $u,v > 0$.

	Laws of Exponents	*Laws of Logarithms*
1	$a^0 = 1$	$\log_a(1) = 0$
2	$a^1 = a$	$\log_a(a) = 1$
3	$a^{\log_a x} = x$	$\log_a(a^x) = x.$
4	$a^x a^y = a^{x+y}$	$\log_a(uv) = \log_a u + \log_a v$
5	$\frac{a^x}{a^y} = a^{x-y}$	$\log_a \frac{u}{v} = \log_a u - \log_a v$
6	$(a^x)^y = a^{xy}$	$\log_a(u^b) = b\log_a u.$

Proof. We are already familiar with the properties of exponential functions; our task is to prove the properties of logarithmic functions.

1. To what power must we raise a to get 1? We raise a to the zero power: $a^0 = 1$. Thus $\log_a(1) = 0$.

2. To what power must we raise a to get a? We raise a to the first power: $a^1 = a$. Thus $\log_a(a) = 1$.

3. To what power must we raise a to get a^x? We raise a to the x power! Thus $\log_a(a^x) = x$.

4. Let $x = \log_a u$ and $y = \log_a v$. Then $a^x = u$ and $a^y = v$, so

$$\begin{aligned}\log_a(uv) &= \log_a(a^x a^y)\\ &= \log_a(a^{x+y})\\ &= x+y\\ &= \log_a u + \log_a v.\end{aligned}$$

   This is a template for the way these proofs go: we must rewrite our logarithm statements in terms of exponentials.

5. We leave the proof of this as an exercise.

6. Let $x = \log_a u$, so that $u = a^x$. Then

$$\begin{aligned}\log_a(u^b) &= \log_a[(a^x)^b]\\ &= \log_a(a^{bx})\\ &= bx\\ &= b\log_a u\end{aligned}$$

□

Notice that $\log_a(u^b)$ and $(\log_a u)^b$ are two different things. The placement of the parentheses makes a big difference here; the theorem does not apply to $(\log_a u)^b$.

Example 8.2.5. Suppose you are told that $\log_a(2) \approx 0.3562$ and $\log_a(3) \approx 0.5646$. Find

1. $\log_a(6)$,
2. $\log_a(1.5)$, and
3. $\log_a(9)$.

Solution:

1. $\log_a(6) = \log_a(2\cdot 3) = \log_a(2) + \log_a(3) \approx 0.3562 + 0.5646 = 0.9208$.

2. $\log_a(1.5) = \log_a\left(\frac{3}{2}\right) = \log_a 3 - \log_a 2 \approx 0.5646 - 0.3562 = 0.2084$.

3. $\log_a(9) = \log_a(3^2) = 2\log_a(3) \approx 2(.5646) = 1.1292$.

□

Example 8.2.6. Expand $\log_{10}\frac{5}{x^3y}$.

Solution:

$$\begin{aligned}\log_{10}\frac{5}{x^3y} &= \log_{10}(5)-\log_{10}(x^3y)\\ &= \log_{10}(5)-(\log_{10}(x^3)+\log_{10}(y))\\ &= \log_{10}5-3\log_{10}x-\log_{10}y.\end{aligned}$$

□

Example 8.2.7. Express $2\log_3(x+3)+\log_3(x)-\log_3 7$ as a single logarithm.

Solution:

$$\begin{aligned}2\log_3(x+3)+\log_3(x)-\log_3 7 &= \log_3(x+3)^2+\log_3\left(\frac{x}{7}\right)\\ &= \log_3\left(\frac{x(x+3)^2}{7}\right).\end{aligned}$$

□

Example 8.2.8. Condense the logarithmic expression $\frac{1}{2}\log_{10}(x)+3\log_{10}(x+1)$.

Solution:

$$\begin{aligned}\frac{1}{2}\log_{10}(x)+3\log_{10}(x+1) &= \log_{10}(x^{\frac{1}{2}})+\log_{10}(x+1)^3\\ &= \log_{10}(\sqrt{x}\cdot(x+1)^3).\end{aligned}$$

□

It is perhaps worth noting that there are no formulas for rewriting a logarithm of a sum or difference.

Section 8.2 Exercises

Compute each logarithm.

1. $\log_8(512)$
2. $\log_{\frac{1}{3}}(27)$
3. $\log_5(625)$
4. $\log_2(16^5)$
5. $\log_6\left(\frac{1}{36}\right)$
6. $\log_{10}(0.00001)$

Sketch the graph of each logarithmic function.

7. $f(x)=\log_4(x)$
8. $f(x)=-\log_2(x)$
9. $f(x)=\log_5(x+2)$
10. $f(x)=-\log_2(x-3)$

11. $f(x) = \log_3(x-1)+3$

12. $f(x) = \log_{2/3}(x)+4$

Expand each logarithm.

13. $\log_4(3x^4)$

14. $\log_5(2x)$

15. $\log_6\left(\frac{x}{3}\right)$

16. $\log_3(5xy)$

17. $\log_5(x(x+1)^3)$

18. $\log_{12}((x+2)^3(x^2+1)^4)$

19. $\log_2\left(\frac{1}{2}\right)$

20. $\log_7\left(\frac{xy}{2z}\right)$

21. $\log_3\left(\frac{2x^2(x-4)^3}{(x+1)^4}\right)$

22. $\log_2\left(\frac{1}{x}\right)$

Condense each expression into a single logarithm.

23. $\log_3(x)+\log_3(2)$

24. $\log_5(z)-\log_5(y)$

25. $3\log_2(x+y)$

26. $-4\log_3(2x)$

27. $\frac{3}{2}\log_7(x-5)$

28. $-\frac{2}{3}\log_7(x+5)$

29. $\log_2(3x)-\log_2(x+3)+\log_2(x)$

30. $\log_5(4x)+3\log_5(x-1)$

31. $2\log_{10}(3x+2y)-2\log_{10}(6x+4y)$

32. $\log_6(x^2)-\log_6(2x)+\log_6\left(\frac{2}{x}\right)$

If $\log_a(10)\approx 2.3026$ and $\log_a(8)\approx 2.0794$, estimate the following:

33. $\log_a(80)$

34. $\log_a\left(\frac{4}{5}\right)$

35. $\log_a(640)$

36. $\log_a(1.25)$

37. $\log_a(512)$

38. $\log_a(6.4)$

39. Use the graphs $f(x)=\log_3 x$ and $g(x)=\log_4 x$ to solve the inequality $\log_3 x < \log_4 x$.

40. Use the graphs $f(x)=\log_{\frac{1}{3}}(x)$ and $g(x)=\log_{\frac{1}{4}}(x)$ to solve the inequality $\log_{\frac{1}{4}}(x)\leq \log_{\frac{1}{3}}(x)$.

41. Prove part 5 of Theorem 91.

8.3 The Natural Exponential and Logarithm Functions

So far, we have been using any positive number (except 1) as a base for an exponential or logarithmic function. Now we want to focus in on one particular base. This base is very special; we choose it for properties that one needs to take calculus to fully appreciate, but one can still see its utility without going so far.

Definition 8.3.1. The irrational number e is approximately $2.71828\ldots$. It is the base of the **natural exponential function** and the **natural logarithm function**.

When you encounter e in your reading or in exercises, just remember that it is a symbol we use to indicate a specific, very special number, in the same way we use π to represent a specific, very special number.

Definition 8.3.2. The function $f(x) = e^x$ is the **natural exponential function**.

You will need to use your calculator to graph it or evaluate it. One interesting fact is that as x grows infinitely large, $\left(1+\frac{1}{x}\right)^x$ comes arbitrarily close to e. This is explored more fully in an exercise.

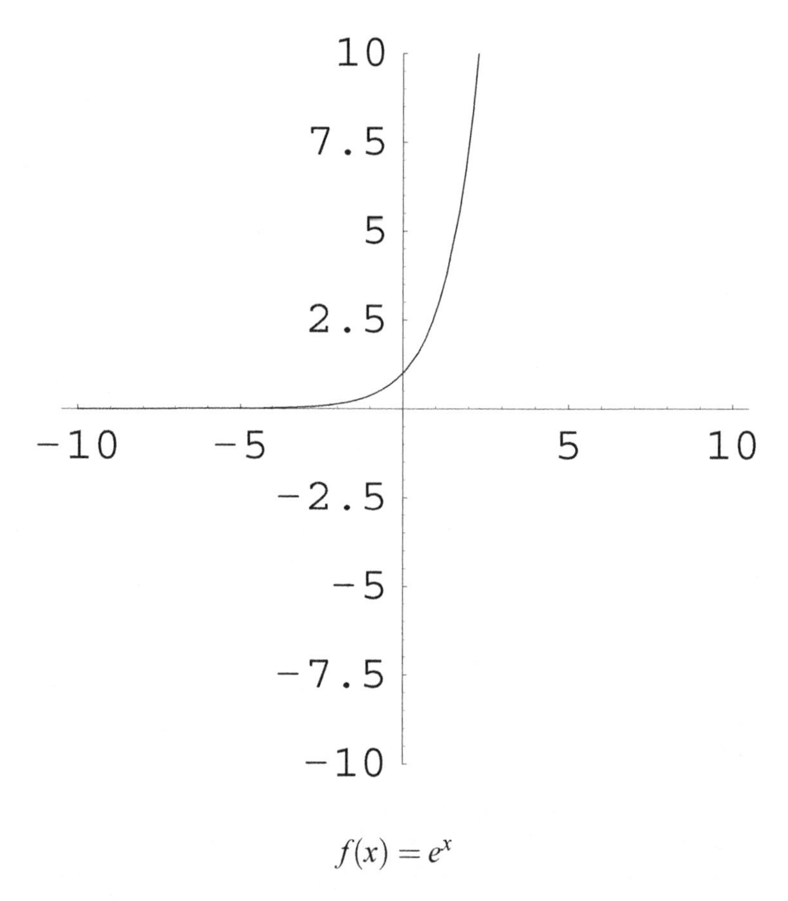

$f(x) = e^x$

Example 8.3.1. Approximate each of the following.

1. $e^3 \approx 20.08553692$

2. $e^{\frac{1}{2}} = \sqrt{e} \approx 1.648721271$

3. $\frac{3}{2}(e^{\pi}) \approx 34.71103895$

4. $e^{-2} \approx .1353352832$

□

Since the function $f(x) = e^x$ is one-to-one, we also need the inverse to e^x.

Definition 8.3.3. The **natural logarithm function** is defined by $f(x) = \log_e x$ for $x > 0$. It is usually written $\ln x$. Its graph is given below.

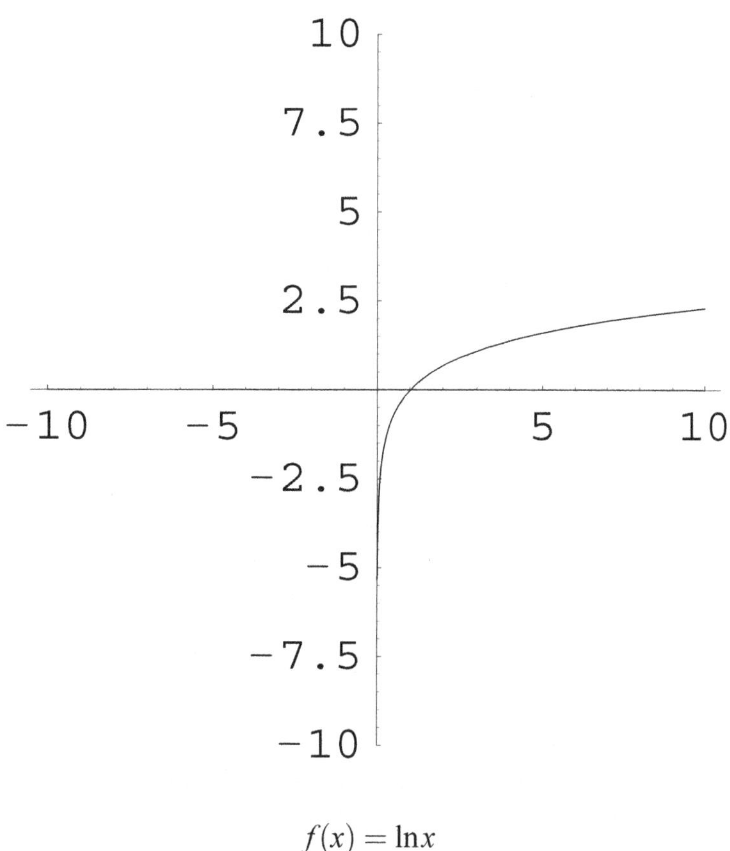

$f(x) = \ln x$

Of course, the natural logarithm obeys the same laws as the other logarithm functions. We have merely singled out a special base.

Example 8.3.2. Evaluate each of the following.

1. $\ln\left(\frac{1}{e}\right) = \log_e\left(\frac{1}{e}\right) = -1$

2. $\ln(e^2) = \log_e(e^2) = 2$

3. $\ln 6 \approx 1.7918$. (This required an approximation on a calculator.)

4. $\ln(x+2)^3 = 3\ln(x+2)$

5. $3\ln(x) + 5\ln(y) = \ln(x^3) + \ln(y^5) = \ln(x^3 \cdot y^5)$

□

Definition 8.3.4. The **common logarithm function** is given by $f(x) = \log_{10} x$ for $x > 0$. It is usually written simply as $\log x$.

Example 8.3.3. 1. $\log 10 = 1$.

2. $\log 100 = 2$.

3. $\log\left(\frac{1}{1000}\right) = -3$

4. $\log 12 \approx 1.0792$. (This required an approximation on a calculator.)

□

Your calculator is probably only set up to do logarithms to the bases 10 and e. What do we do if we need to compute a logarithm to another base? Somehow, we must express our logarithms in either base e or base 10. The theorem below tells us how to do this.

Theorem 92 (Change of Base Formula)**.** *Let a, b, x be positive real numbers with $a, b \neq 1$. Then* $\log_a x = \dfrac{\log_b x}{\log_b a}$.

Proof. Let $y = \log_a x$, so that $a^y = x$. Then

$$\begin{aligned} \log_b x &= \log_b a^y \\ &= y \log_b a \\ &= (\log_a x)(\log_b a). \end{aligned}$$

Therefore, since $\log_b x = (\log_a x)(\log_b a)$, we have $\log_a x = \dfrac{\log_b x}{\log_b a}$. □

Example 8.3.4. Approximate $\log_5 12$.

Solution: We have $a = 5$ and $x = 12$. We will use $b = 10$; that is, we will convert this to a base 10 logarithm: $\log_5 12 = \dfrac{\log 12}{\log 5} \approx 1.5440$.

We could have chosen to convert this to base e, in which case we would have found $\log_5 12 = \dfrac{\ln 12}{\ln 5} \approx 1.5440$, as before.

□

Note that the natural logarithm is not the same as the common logarithm, but either may be used in the change of base formula given above.

Section 8.3 Exercises

Evaluate each logarithm. (Use your calculator as necessary to approximate the values.)

1. $\log 1000$

2. $\ln e^{-3}$
3. $\ln 13$
4. $\log 103$
5. $\log_8 14$
6. $\log_{13} 169$
7. $\log_{1/3} 42$
8. $\ln(\ln 31)$

9. Sketch the graph of e^{3x}.
10. Sketch the graph of $f(x) = 2\ln(x+3) - 1$.

Expand each logarithm.

11. $\log(2x)$
12. $\ln\left(\frac{x}{3}\right)$
13. $\log(5x^2y)$
14. $\ln(12x^4)$
15. $\log(x(x+1)^3)$
16. $\ln((x+5)^2(x^2+1)^3)$
17. $\ln\left(\frac{1}{e}\right)$
18. $\ln\left(\frac{e^x}{2x}\right)$
19. $\log\left(\frac{2x^2(x-4)^3}{(x+1)^4}\right)$
20. $\ln\left(\frac{1}{x}\right)$

Condense each expression into a single logarithm.

21. $\log(3x) + \log(2)$
22. $\ln(z) - \ln(y)$
23. $3\log(x+y)$
24. $-4\ln(2x)$
25. $\frac{3}{2}\ln(x-5)$
26. $-\frac{2}{3}\log(x+5)$
27. $\log(3x) - \log(x+2) + \log(x)$
28. $\ln(4x) + 3\ln(x-1)$
29. $2\log(3x+2y) - 2\log(6x+4y)$
30. $\ln(x^3) - \ln(3x) + \log\left(\frac{3}{x}\right)$

31. Compute each quantity.

    (a) $\left(1+\frac{1}{1}\right)^1$

    (b) $\left(1+\frac{1}{10}\right)^{10}$

    (c) $\left(1+\frac{1}{100}\right)^{100}$

    (d) $\left(1+\frac{1}{1000000}\right)^{1000000}$

(e) $\left(1+\dfrac{1}{10^{14}}\right)^{10^{14}}$ Can you explain what went wrong?

(f) Graph $f(x) = \left(1+\dfrac{1}{x}\right)^{x}$ on your calculator. With your window set to $[0, 1000000000000]$, use the trace feature to determine the y-values on the graph. What do you notice?

8.4 Exponential and Logarithmic Equations

We want, as usual, to be able to solve equations involving our new functions. We will need to use the properties of exponents and logarithms to do this. It is important to note that if f is a function and $x = y$, then $f(x) = f(y)$. In particular, we need to know that if $x = y$, then $a^x = a^y$ and $\log_a x = \log_a y$. This allows us to solve equations, as you will see in the following example.

Example 8.4.1. Solve each equation.

1. $\log_3(4x-7) = 2$
2. $2\log_5 x = \log_5 9$
3. $3^{x+1} = 81$
4. $4^x - 2^x - 12 = 0$
5. $5^{x-2} = 3^{3x+2}$

Solution:

1.

$$\begin{array}{rcll} \log_3(4x-7) & = & 2 & \text{Given} \\ 3^{\log_3(4x-7)} & = & 3^2 & \text{Apply } f(x) = 3^x \text{ to both sides} \\ 4x-7 & = & 9 & \text{Inverse Functions} \\ 4x & = & 16 & \text{Add 7 to both sides} \\ x & = & 4 & \text{Divide by 4 on both sides} \end{array}$$

Check that $x = 4$ is a solution.

2.

$$\begin{array}{rcll} 2\log_5 x & = & \log_5 9 & \text{Given} \\ \log_5 x^2 & = & \log_5 9 & \text{Property of Logarithms} \\ 5^{\log_5 x^2} & = & 5^{\log_5 9} & \text{Apply } f(x) = 5^x \text{ to both sides} \\ x^2 & = & 9 & \text{Inverse Functions} \\ x & = & \pm 3 & \underline{\hspace{6cm}} \end{array}$$

However, a solution must solve the *original* equation, and $x = -3$ does not. Remember that the domain of any logarithm is $(0,\infty)$, so $\log_a(-3)$ does not make sense for any a. You should check that $x = 3$ is a solution.

3.

$$\begin{array}{rcll} \log_3(3^{x+1}) & = & \log_3(81) & \text{Given} \\ x+1 & = & 4 & a^{\log_a x} = x \\ x & = & 3 & \text{Subtract 1 from both sides} \end{array}$$

Check that $x = 3$ is a solution.

4. This one looks very difficult at first, but if we rewrite it cleverly as $(2^x)^2 - 2^x - 12 = 0$, then we can recognize it as a disguised quadratic.

$$\begin{array}{rcll} 4^x - 2^x - 12 & = & 0 & \text{Given} \\ (2^x)^2 - 2^x - 12 & = & 0 & 4^x = (2^2)^x = 2^{2x} = 2^{x2} = (2^x)^2 \\ z^2 - z - 12 & = & 0 & \text{Substitute } z = 2^x \\ (z-4)(z+3) & = & 0 & \text{Factor} \\ z - 4 = 0 & \text{or} & z + 3 = 0 & \text{Zero Product Theorem} \\ z = 4 & \text{or} & z = -3 & \text{Add equals to equals} \\ 2^x = 4 & \text{or} & 2^x = -3 & \text{Substitute } z = 2^x \\ x = 2 & & & \text{Inspection} \end{array}$$

Note that since the range of a^x is $(0, \infty)$ for any a, $2^x = -3$ is not possible. Check that $x = 2$ is a solution.

5. Up until now, we chose the base for the logarithm to use by seeing what the base on the exponentials was; this time, however, we have two different bases! Should we apply $\log_5$ to both sides, or $\log_3$? Actually, it doesn't matter what base we decide to use. (Why?) For convenience, we will take the natural logarithm of both sides.

$$\begin{array}{rcll} 5^{x-2} & = & 3^{3x+2} & \text{Given} \\ \ln(5^{x-2}) & = & \ln(3^{3x+2}) & \text{Apply } f(x) = \ln x \text{ to both sides} \\ (x-2)\ln 5 & = & (3x+2)\ln 3 & \text{Property of Logarithms} \\ x\ln 5 - 2\ln 5 & = & 3x\ln 3 + 2\ln 3 & \text{Distributive Law} \\ x\ln 5 - 3x\ln 3 & = & 2\ln 5 + 2\ln 3 & \text{Add equals to both sides} \\ x(\ln 5 - 3\ln 3) & = & \ln 5^2 + \ln 3^2 & \text{Distributive Law} \\ \\ x & = & \dfrac{\ln 5^2 + \ln 3^2}{\ln 5 - 3\ln 3} & \text{Divide on both sides} \\ x & = & \dfrac{\ln(25 \cdot 9)}{\ln 5 - \ln 27} & \text{Properties of Logarithms} \\ x & = & \dfrac{\ln 225}{\ln\left(\frac{5}{27}\right)} & \text{Properties of Logarithms} \\ x & \approx & -3.2116 & \text{Calculator approximation} \end{array}$$

□

We summarize the above methods below.
To solve an **exponential equation:**

1. Isolate the exponential expression.

2. Take the appropriate logarithm of both sides.

3. Solve for the variable.

We outline an analogous procedure to solve logarithmic equations.

To solve a **logarithmic equation:**

1. Isolate the logarithmic expression (write each side as a single logarithm).
2. Exponentiate both sides. (That is, raise both sides to the power which is the base of the logarithms you're dealing with, or a convenient base if the problem involves more than one base.)
3. Solve for the variable.

Example 8.4.2. 1. Solve $\ln x + \ln 7 = 2$.

Solution:

$$\begin{array}{rcll} \ln x + \ln 7 &=& 2 & \text{Given} \\ \ln(7x) &=& 2 & \text{Property of Logarithms} \\ e^{\ln(7x)} &=& e^2 & \text{Exponentiate both sides} \\ 7x &=& e^2 & \text{Inverse Functions} \\ x &=& \frac{e^2}{7} & \text{Divide by 7} \\ x &\approx& 1.05579 & \text{Calculator approximation} \end{array}$$

Check that this is indeed a solution to the original equation.

2. $\log_2(x+5) - \log_2(x-2) = 3$

Solution:

$$\begin{array}{rcll} \log_2(x+5) - \log_2(x-2) &=& 3 & \text{Given} \\ \log_2 \dfrac{x+5}{x-2} &=& 3 & \text{Property of Logarithms} \\ \dfrac{x+5}{x-2} &=& 2^3 & \text{Apply } f(x) = 2^x \text{ to both sides, inverse functions} \\ \dfrac{x+5}{x-2} &=& 8 & 2^3 = 8 \\ x+5 &=& 8(x-2) & \text{Fraction Equivalence} \\ x+5 &=& 8x-16 & \text{Gen. Distributive Law} \\ 21 &=& 7x & \text{Add/subtract equals} \\ x &=& 3 & \text{Divide by 7, symmetry of equality} \end{array}$$

Check this.

3. $\ln x + \ln(2-x) = 0$.

Solution:

$$\begin{array}{rcll} \ln x+\ln(2-x) &=& 0 & \text{Given} \\ \ln(x(2-x)) &=& 0 & \text{Property of Logarithms} \\ 2x-x^2 &=& e^0 & \text{Exponentiate both sides} \\ 2x-x^2 &=& 1 & e^0=1 \\ -x^2+2x-1 &=& 0 & \text{Subtract 1, Summand Permutation} \\ x^2-2x+1 &=& 0 & \text{Symmetry of equality} \\ (x-1)^2 &=& 0 & \text{Factor} \\ (x-1) &=& 0 & \text{Zero Product Theorem} \\ x &=& 1 & \text{Add 1 to both sides} \end{array}$$

Check that this is a solution.

□

Example 8.4.3. Carbon 14 is a radioactive isotope of carbon 12, with a half-life of about 5700 years. When an organism dies, its carbon 14 is at the same level as that of the environment, but then it begins to decay. We can determine an object's age by knowing the percentage of carbon-14 that remains. (This actually involves some big, not necessarily very good, assumptions that we will not go into here.) We know from section 8.1 that

$$A(t) = A_0\left(\frac{1}{2}\right)^{t/5700},$$

where $A(t)$ is the amount of carbon 14 remaining after t years, and $A(0)$ is the amount of carbon 14 present at time 0 (when the organism died).

□

Example 8.4.4. Charcoal from an ancient tree that was burned during a volcanic eruption has only 45% of the standard amount of carbon 14. When did the volcano erupt?

We know that $A(t) = A_0\left(\frac{1}{2}\right)^{t/5700}$. We also know that $A(t) = 0.45A_0$, although we don't know what t is. (Note that $0.45A_0$ is 45% of A_0.) Thus

$$\begin{array}{rcll} A(t) &=& A_0\left(\frac{1}{2}\right)^{t/5700} & \text{Given} \\ A(t) &=& 0.45A_0 & \text{Given} \\ A_0\left(\frac{1}{2}\right)^{t/5700} &=& 0.45A_0 & \text{Substitution} \\ \left(\frac{1}{2}\right)^{t/5700} &=& 0.45 & \text{Divide by } A_0 \neq 0 \\ \ln\left(\left(\frac{1}{2}\right)^{t/5700}\right) &=& \ln(0.45) & \text{Take ln of both sides} \\ \frac{t}{5700}\ln\left(\frac{1}{2}\right) &=& \ln 0.45 & \text{Property of Logarithms} \\ t &=& 5700\left(\frac{\ln 0.45}{\ln 0.50}\right) & \text{Multiply by equals} \\ t &\approx& 6566 & \text{Calculator approximation} \end{array}$$

Thus, the eruption occurred approximately 6566 years ago.
This technique is known as **carbon dating**.

□

Section 8.4 Exercises

Solve each exponential equation. After finding an exact solution, give an approximation to the nearest thousandths.

1. $3^{x+2} = 27$
2. $2^{3x+1} = 17$
3. $5^{-x/2} = 125$
4. $1 + e^{x} = 13$
5. $e^{3x+2} = 5$
6. $3e^{x+2} = 75$
7. $\left(1 + \frac{.08}{12}\right)^{12x} = 2$
8. $\frac{1000}{1+e^{x}} = \frac{1}{2}$
9. $36^{x} - 3 \cdot 6^{x} = -2$
10. $4^{2x+1} = 5^{x-4}$

Solve each logarithmic equation. After finding an exact solution, give an approximation to the nearest thousandth.

11. $\ln x = 7.2$
12. $\log(x+2) = 75$
13. $\ln(x+2) = -2.6$
14. $\ln(x) - \ln(5) = 13$
15. $3.5\ln(x) = 8$
16. $\log_2(x) + \log_2(x+2) = 4$
17. $2\log(x) + \log(4) = 2$
18. $\log_4(x+1) - \log_4(x) = 1$
19. $\ln(x+1) + \ln(x-2) = \ln(x^2)$
20. $\log_2(x+1) - \log_2(x-2) = 3$
21. An amateur archeologist claims to have discovered a dinosaur fossil that is only 3000 years old. It contains very little carbon-14; the best you can tell, it has less than 0.01% of the normal amount. (Your instruments are very crude.) What do you tell the archeologist, and why?
22. A skin coat has only 39% of the normal amount of carbon-14. How long ago was the animal killed to make the coat?
23. How long will it take to produce \$10,000 from a \$7,000 investment at 8% compounded monthly?
24. What interest rate, compounded monthly, will produce \$10,000 from a \$7,000 investment in 5 years?

Chapter 9

Polynomial Functions

In this unit, we define and investigate polynomial functions as a special class of functions. We begin with a review of linear and quadratic functions before discussing polynomials of arbitrary degree. We present the "traditional" theory of equations in a concise format and tie the procedures to the basic algebraic properties previously presented.

Objectives

- To define polynomial functions
- To explore, in detail, the properties of linear and quadratic functions
- To investigate general properties of polynomial functions
- To increase knowledge about the graphs of polynomial functions
- To introduce techniques for solving polynomial equations

Terms

- polynomial function
- degree
- leading coefficient
- linear function
  - slope, parallel, perpendicular
- quadratic function
  - standard form, general form, completing the square
  - parabola, vertex
  - discriminant
- Leading Coefficient Test
- Division Algorithm (for Polynomial Functions)
- Remainder Theorem
- Factor Theorem
- Fundamental Theorem of Algebra
- Rational Root Test
- Synthetic Division

9.1 Linear Functions

Definition 9.1.1. A function f defined by $f(x) = mx + b$ for some constants m and b is called a **linear function**. The number m is called the **slope** of the line, and b is the **y-intercept**.

The number b is called the y-intercept because $f(0) = b$. This means that when $x = 0$ (that is, when the point is on the y-axis), $y = b$. Thus, the y-intercept is where the graph crosses the y-axis. (Some prefer to reserve the term y-intercept for the *point* $(0, b)$; we will not make this distinction.)

We have dealt with linear equations already. Just remember that you MUST use the algebraic principles we have developed. There is no magic to it!

Consider the following example:

Example 9.1.1. Peggy leaves for work in her car, and drives at 45 mph. Her husband Bob notices that Peggy has left her lunch behind. He hops in his car ten minutes later and chases her at 55 mph. How far will Bob go before he catches Peggy?

Solution: Let t be how long Peggy has driven. Then her distance is $45t$, and Bob's distance is $55(t-10)$. We need to know when these are equal. We solve: $45t = 55(t-10) \implies 45t = 55t - 550 \implies -10t = -550 \implies t = 55$. Therefore, Peggy has driven for 55 minutes and Bob has driven for 45 minutes. This means that Bob drives 55 mph for 3/4 of an hour, so he goes 41.25 miles.

□

There is nothing new here; the goal is for you to be able to think your way through setting up such problems. Solving them should then be a simple exercise in the algebraic principles we have studied.

We examined graphs in a previous section. Suppose that $f(x) = mx + b$. If $m = 0$, then $f(x) = b$, so we get a horizontal line. If $m \neq 0$, then $f(x) = m\left(x + \frac{b}{m}\right)$. We can now recognize this as a transformation of the line $y = x$. However, it is usually easier to obtain the graph of $f(x) = mx + b$ by plotting two points and connecting them with a line. A good choice for one point is the y-intercept.

Example 9.1.2. Sketch the graph of $f(x) = -3x + 5$.

Solution: The y-intercept is 5, so the point $(0, 5)$ is on the graph of f. Since $f(1) = 2$, we have that $(1, 2)$ is also on the graph of f. Since two points determine a line, this is sufficient for us to draw the rest of the graph.

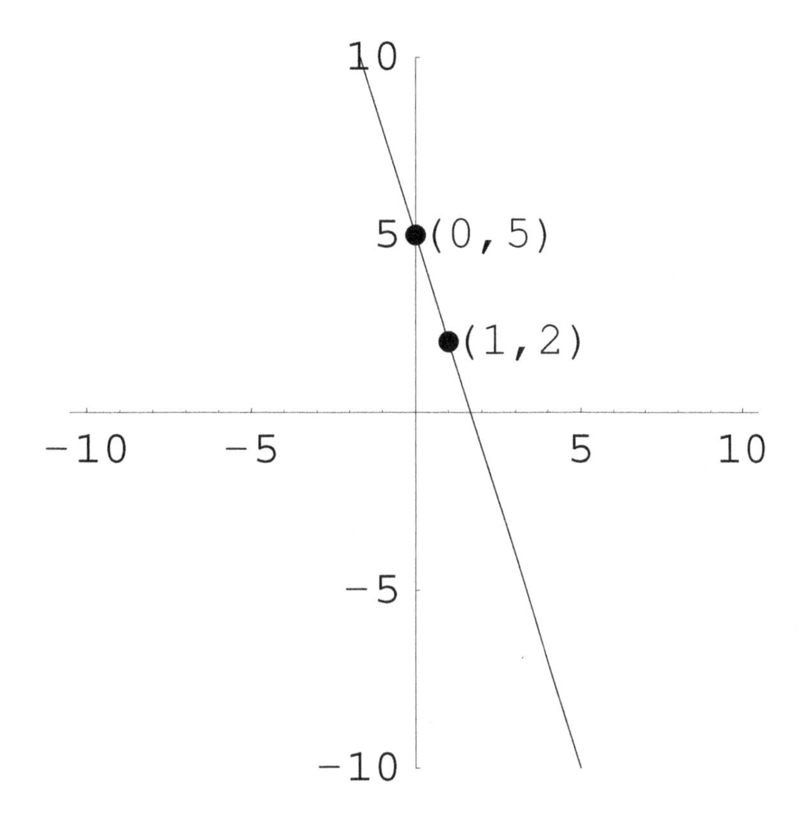

□

Definition 9.1.2. Let $P_1 = (x_1, y_1)$ and $P_2 = (x_2, y_2)$ be two distinct points with $x_1 \neq x_2$. The **slope** of the nonvertical line L containing P_1 and P_2 is given by

$$m = \frac{y_2 - y_1}{x_2 - x_1}, \quad x_1 \neq x_2.$$

If $x_1 = x_2$, then L is a vertical line and the slope m of L is undefined. (It is sometimes said that a vertical line has *no slope*. This is not to be confused with a horizontal line having *zero slope.*)

It is not too hard to show that this definition of slope is equivalent to the definition we gave previously.

Example 9.1.3. The slope of the line through the points $\left(1, -\frac{1}{3}\right)$ and $\left(\frac{4}{9}, -2\right)$ is

$$m = \frac{\left(-2 + \frac{1}{3}\right)}{\left(\frac{4}{9} - 1\right)} = \frac{\left(-\frac{5}{3}\right)}{\left(-\frac{5}{9}\right)} = 3.$$

If (x, y) is another point on this line, then $\frac{y - (-2)}{x - \left(\frac{4}{9}\right)} = 3$ as well. Thus, an equation of the line is $y + 2 = 3\left(x - \frac{4}{9}\right)$, or $y = 3x + \frac{2}{3}$.

□

Definition 9.1.3. Two distinct nonvertical lines L_1 and L_2 are **parallel** if and only if their slopes m_1 and m_2 are equal. (Any two distinct vertical lines are parallel.)

We are accustomed to thinking of parallel lines as being lines that do not intersect; we will show now that that definition corresponds to our definition above. If $y = mx + b_1$ and $y = mx + b_2$ are two distinct lines (so that $b_1 \neq b_2$), then they cannot intersect: if (x,y) is a point on both lines, then $y = mx + b_1$ (since (x,y) is on the first line), and $y = mx + b_2$ (since (x,y) is on the second line). Therefore, $mx + b_1 = mx + b_2$, so $b_1 = b_2$ by additive cancellation. But we already said that $b_1 \neq b_2$, so this is a contradiction. Since the only possible error was the assumption that the two lines intersect, they must not intersect, so they are parallel in the geometrical sense.

On the other hand, if two lines have different slopes, say, m_1 and m_2, then the point

$$\left(\frac{b_2 - b_1}{m_1 - m_2}, \frac{m_1 b_2}{m_1 - m_2}\right)$$

lies on both lines $y = m_1 x + b_1$ and $y = m_2 x + b_2$. (Check that.)

This means that lines intersect if and only if they have different slopes.

Example 9.1.4. Prove that the lines $2x - 3y + 7 = 0$ and $4x - 6y - 3 = 0$ are parallel.

Solution: Solving both equations for y we get $y = \frac{2}{3}x + \frac{7}{3}$ and $y = \frac{4}{6}x - \frac{3}{6} = \frac{2}{3}x - \frac{1}{2}$. Since these lines both have slope $\frac{2}{3}$, they are parallel lines.

□

Definition 9.1.4. Two nonvertical lines L_1 and L_2 are **perpendicular** if and only if the product of their slopes m_1 and m_2 is -1. Alternatively, two nonvertical lines are perpendicular if and only if their slopes are negative reciprocals of each other. (A vertical line is perpendicular to a horizontal line.)

It can also be shown that this definition of perpendicular corresponds to the usual definition ("intersect at right angles") by means of the Pythagorean theorem, but that is more appropriate for a course in analytic geometry.

Example 9.1.5. Find an equation of the line perpendicular to $2x - 3y + 7 = 0$ and passing through the point $(1,3)$.

Solution: Since $y = \frac{2}{3}x - \frac{7}{3}$, we know the slope of the given line is $m_1 = \frac{2}{3}$. The desired perpendicular line has slope $m_2 = -\frac{3}{2}$ and passes through the point $(1,3)$. Thus, the desired line has equation $y - 3 = -\frac{3}{2}(x - 1)$. (This can be found from the fact that if (x,y) lies on the line, then $\frac{y-3}{x-1} = -\frac{3}{2}$.)

□

The following review of equations of lines may prove helpful.

Description	Equation
Vertical Line	$x = a$
Horizontal Line	$y = b$
Slope-Intercept Form	$y = mx + b$
Point-Slope Form	$y - y_1 = m(x - x_1)$
Two-Point Form	$y - y_1 = \frac{y_2 - y_1}{x_2 - x_1}(x - x_1)$
General Form	$Ax + By + C = 0$

Section 9.1 Exercises

Determine whether the given pair of lines is parallel, perpendicular, or neither.

1. $y = 3x - 5$; $y = 3x + 2$
2. $6x - 4y + 8 = 0$; $9x - 6y + 2 = 0$
3. $2x + y = 4$; $4x - 8y = 0$
4. $4x + y - 2 = 6$; $y = 4x + 9$

Graph the line through each pair of points.

5. $(-1,4)$ and $(2,2)$
6. $(-2/3,5)$ and $(-2/3,-2)$
7. $(4,-1)$ and $(-2,-1)$
8. $(2.4,1.1)$ and $(0.5,4.7)$

Find an equation of each line by first finding two points lying on the line.

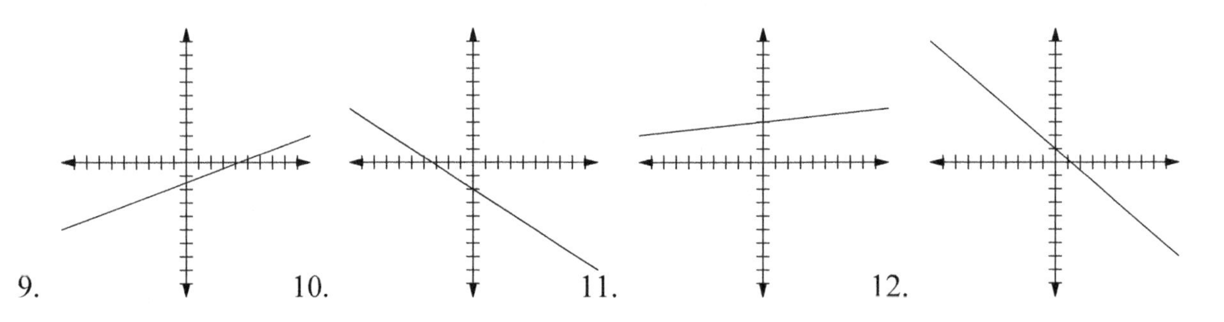

13. Find an equation of the line that is perpendicular to the line $2x + 5y - 3 = 0$ and that passes through the point $(12,-7)$.

14. Find an equation of the line that is perpendicular to the line $y = 6x - 8$ and that passes through the point $(2,-4)$.

15. Find an equation of the line that is parallel to the line $2x + 5y - 3 = 0$ and that passes through the point $(12,-7)$.

16. Find an equation of the line that is parallel to the line $y = 6x - 8$ and that passes through the point $(2,-4)$.

17. JarCo, Inc. has observed that they will sell 2400 jars if the price of a jar is \$0.55, and that for every \$0.05 they raise the price, they sell 125 fewer jars. Find a linear function expressing the number of jars they sell as a function of the price they charge.

18. The distance from Houston to Dallas is approximately 250 miles. Marcie lives in Houston and her friend Bryson lives in Dallas; they leave their respective homes at the same time and drive toward the others' city.

    (a) If Bryson drives 65 miles per hour, what is his distance from Houston t hours after he has left Dallas?

(b) If Marcie drives 50 miles per hour, what is her distance from Houston t hours after she has left Houston?

(c) If Marcie and Bryson leave their homes at 8:00 in the morning, at what time will they be at the same place on the highway?

9.2 Quadratic Functions

Definition 9.2.1. A **quadratic function** is a function of the form $f(x) = ax^2 + bx + c$, where a, b, c are constants and $a \neq 0$. A quadratic function f is in **standard form** if it is written as $f(x) = a(x-h)^2 + k$.

To put a function in standard form, we must **complete the square**. Recall from our earlier work (proving theorems in fields) that $(a+b)^2 = a^2 + 2ab + b^2$. Therefore, if we have $a^2 + 2ab + c$, we can "complete the square" by rewriting this number as

$$(a^2 + 2ab + b^2) - b^2 + c = (a+b)^2 - b^2 + c.$$

Since we both added and subtracted b^2, we really only added "a clever 0." Therefore, the final number is equal to the original number (because 0 is the additive identity). Notice that what we added (and subtracted) was the *square of half of the coefficient of a*. That is, the coefficient of a was $2b$, and we added (and subtracted) b^2.

Example 9.2.1. Put $f(x) = x^2 - 6x + 2$ in standard form. We have $f(x) = x^2 + 2(-3)x + 2$. Thus our a is x, and our b is -3. We get

$$f(x) = (x^2 + 2(-3)x + (-3)^2) - (-3)^2 + 2 = (x + (-3))^2 - 9 + 2 = (x-3)^2 - 7.$$

□

Example 9.2.2. Put $f(x) = 2x^2 - 3x + 2$ in standard form. It is simpler to complete the square if the leading coefficient is 1. We rewrite f as

$$\begin{aligned} f(x) &= 2\left(x^2 - \frac{3}{2}x\right) + 2 \\ &= 2\left(x^2 - \frac{3}{2}x + \left(-\frac{3}{4}\right)^2 - \left(-\frac{3}{4}\right)^2\right) + 2 \\ &= 2\left(x^2 - \frac{3}{2}x + \left(-\frac{3}{4}\right)^2\right) - 2 \cdot \left(-\frac{3}{4}\right)^2 + 2 \\ &= 2\left(x - \frac{3}{4}\right)^2 + \frac{7}{8}. \end{aligned}$$

□

These results also indicate certain features of the graph of a quadratic function, which is called a **parabola**, as noted in an earlier section. We can see in the example above that the graph of f is just a transformation of the graph of $y = x^2$. We have a translation to the right by $\frac{3}{4}$, followed by a vertical stretch by a factor of 2, and finally an upward shift by $\frac{7}{8}$. Thus, we can find the graph of f from prior knowledge.

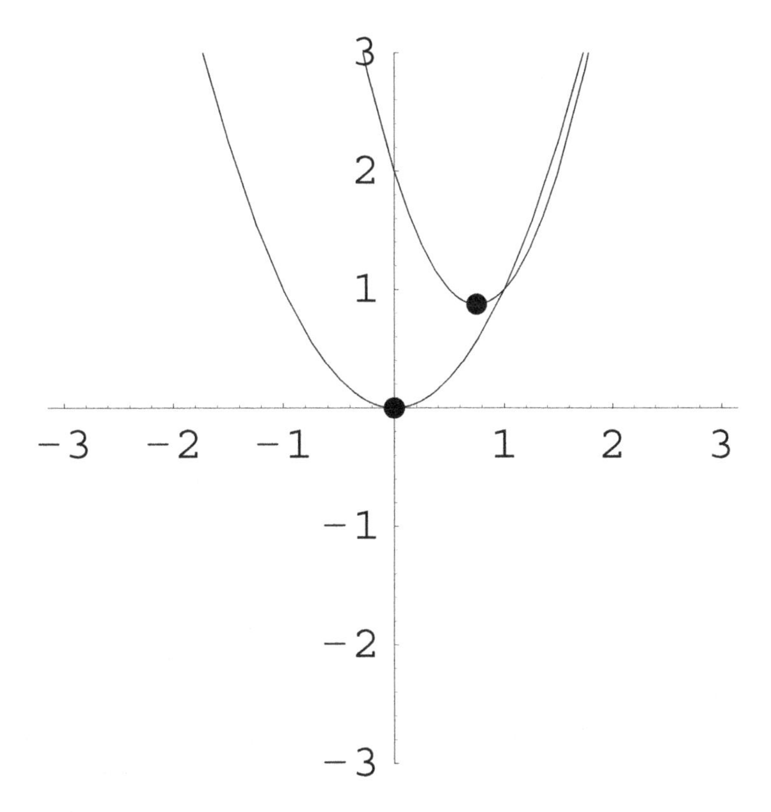

Furthermore, we may ask what happens to the **vertex** of the parabola when we perform this transformation. (The vertex is the highest or lowest point of the parabola.)

The answer is that the vertex is shifted right by $\frac{3}{4}$ (as is every other point), the stretch does not affect the vertex (since the vertex is located at $y = 0$, and $2 \cdot 0 = 0$), and then the vertex is shifted up $\frac{7}{8}$ (as is every other point). The net result is that the vertex of f is located at $(\frac{3}{4}, \frac{7}{8})$.

We will generalize this idea in the following example.

Example 9.2.3. Consider a generic quadratic function $f(x) = ax^2 + bx + c$, with $a \neq 0$. We will complete the square in exactly the same way we did in Example 2.

$$\begin{aligned}
f(x) &= ax^2 + bx + c \\
&= a\left(x^2 + 2\left(\frac{b}{2a}\right)x\right) + c \\
&= a\left(x^2 + \left(\frac{b}{a}\right)x + \left(\frac{b}{2a}\right)^2 - \left(\frac{b}{2a}\right)^2\right) + c \\
&= a\left(x^2 + \left(\frac{b}{a}\right)x + \left(\frac{b}{2a}\right)^2\right) - a\left(\frac{b}{2a}\right)^2 + c \\
&= a\left(x + \frac{b}{2a}\right)^2 - \frac{b^2}{4a} + c \\
&= a\left(x - \frac{-b}{2a}\right)^2 - \frac{b^2}{4a} + c.
\end{aligned}$$

We can see here again a transformation of the graph of $y = x^2$. The graph is first shifted to the right by $\frac{-b}{2a}$. (Note that if $\frac{-b}{2a} < 0$, then this is really a shift to the left.) In particular, the vertex of the graph is translated to the point $\left(\frac{-b}{2a}, 0\right)$.

Next, the graph is stretched or compressed vertically by a factor of a. If it happens that $a < 0$, the graph is also reflected across the x-axis. Neither operation will affect the location of the vertex since it is on the x-axis.

Finally, the graph is shifted up by $-\frac{b^2}{4a} + c$. (This will be a downward shift if this quantity is negative.) This will move the vertex from $\left(\frac{-b}{2a}, 0\right)$ to $\left(\frac{-b}{2a}, -\dfrac{b^2}{4a} + c\right)$.

Thus, the vertex of the parabola $y = ax^2 + bx + c$ has x-coordinate $\frac{-b}{2a}$. There is no point in memorizing the expression for the y-coordinate, as it can always be obtained by substituting the x coordinate into the function. (See the example below.) Furthermore, we saw that if $a < 0$, the parabola will be reflected across the x-axis, and therefore open downward. It is useful to note that henceforth we will not need to complete the square to determine this information: because we used general coefficients a, b, and c, we can apply the formula to any quadratic function as long as $a \neq 0$.

□

Parabolas are especially significant since projectiles travel in parabolic paths. Next time you see one of those huge sprinklers spraying crops, look carefully at the path of the water — it is a parabola (or nearly; our parabolic model does not account for wind resistance). Parabolas also possess many properties that make them useful in engineering; a course in analytic geometry will include such topics.

Example 9.2.4. A juggler in a parade juggles chainsaws on one of the floats. The height of a chainsaw above the ground x seconds after the juggler releases it is given by

$$h(x) = -16x^2 + 30x + 5$$

feet. Should the juggler rest while his float goes under a bridge with 15 feet of clearance?

Solution: We need to find the highest point reached by his chainsaw; that is, we need to find the vertex of the parabola the chainsaw describes as it is juggled. Notice that we have $a = -16$, so the parabola opens downward. This tells us that the parabola will have a maximum — a good thing if this problem is going to make sense!

The vertex of the parabola occurs at $x = \frac{-30}{2(-16)} = 0.9375$ seconds. This is not the maximum height, however, it is the **time** at which the maximum height is reached. To determine the height, evaluate $h(0.9375) = -16(0.9375)^2 + 30(0.9375) + 5 = 19.0625$ feet. The juggler should probably pause while the float goes under the bridge.

□

Sometimes we are not given an equation describing the function; we may only be given a few points that lie on its graph, and told to find the equation ourselves. This happens very frequently in the sciences. Scientists collect data in the course of an experiment, and they want to find some general rule which can be used to predict the outcomes of future experiments. Consider the following example.

Example 9.2.5. Find an equation of the quadratic function f whose graph passes through the points $(1,3),(-1,4)$, and $(2,1)$.

Solution: If $(1,3)$ is on the graph of f, that means (by definition) that $f(1)=3$. Similarly, $f(-1)=4$ and $f(2)=1$. What does f itself look like? We are told that f is a quadratic function, so f must have the form $f(x)=ax^2+bx+c$ for some a,b, and c which we need to determine.

Since $f(1)=3$, we have

$$3=f(1)=a(1)^2+b(1)+c=a+b+c.$$

Since $f(-1)=4$, we have

$$4=f(-1)=a(-1)^2+b(-1)+c=a-b+c$$

Finally, since $f(2)=1$, we have

$$1=f(2)=a(2)^2+b(2)+c$$

Together, these give us the system of **linear** equations

$$\begin{aligned} a+b+c&=3\\ a-b+c&=4\\ 4a+2b+c&=1 \end{aligned}$$

We may write this as the matrix equation

$$\begin{bmatrix} 1 & 1 & 1\\ 1 & -1 & 1\\ 4 & 2 & 1 \end{bmatrix}\begin{bmatrix} a\\ b\\ c \end{bmatrix}=\begin{bmatrix} 3\\ 4\\ 1 \end{bmatrix}.$$

Thus

$$\begin{aligned} \begin{bmatrix} a\\ b\\ c \end{bmatrix}&=\begin{bmatrix} 1 & 1 & 1\\ 1 & -1 & 1\\ 4 & 2 & 1 \end{bmatrix}^{-1}\begin{bmatrix} 3\\ 4\\ 1 \end{bmatrix}\\ &=\begin{bmatrix} -1/2\\ -1/2\\ 4 \end{bmatrix}. \end{aligned}$$

That is, $a=-\frac{1}{2}, b=-\frac{1}{2}$, and $c=4$. We finally arrive at $h(x)=-\dfrac{1}{2}x^2-\dfrac{1}{2}x+4$. You should check that all three of the given points lie on this parabola.

□

Section 9.2 Exercises

Complete the square to write each quadratic function in standard form.

1. $f(x) = x^2 + 6x - 12$
2. $f(x) = x^2 + 5x + 3$
3. $f(x) = -2x^2 - 8x + 15$
4. $f(x) = 5x^2 - 3x + 2$

Find the vertex of each parabola.

5. $f(x) = x^2 + 8x + 15$
6. $f(x) = 3x^2 - 7x - 1$
7. $f(x) = -16x^2 + 48x + 80$
8. $f(x) = 2x^2 + 13$

Find an equation of a parabola through the given points.

9. $(-1,2),(3,5),(4,2)$
10. $(-2,3),(0,0),(1,4)$
11. $(-100,20),(0,50),(100,20)$
12. $(0,203),(0.25,178),(0.45,107)$. (These are three points from the dropped ball exercises in the function sections. Compare your parabola to the one given there.)
13. A punted football follows a parabolic path. Its height at $t = 0$ s is 3 feet, its height at $t = 2$ s is 95 feet, and its height at $t = 4$ s is 59 feet.
    (a) Find an equation of the path the football follows.
    (b) How high does the football go?
    (c) What is the hang-time of the football?

9.3 Solving Quadratic Equations

We found in the previous section that a **general** quadratic function $f(x) = ax^2 + bx + c$ may be put into the **standard** form

$$f(x) = a\left(x + \frac{b}{2a}\right)^2 - \frac{b^2}{4a} + c.$$

We wish now to find the zeros of our general quadratic; that is, we want to know for which values of x we have $f(x) = 0$. We have four basic algebraic techniques:

1. Factoring
2. Extracting Square Roots
3. Completing the Square
4. The Quadratic Formula

The method of factoring uses the Zero-Product Theorem: If $ab = 0$, then $a = 0$ or $b = 0$. That is, if we have a product (in a field) equal to zero, then we must conclude that at least one of the factors is zero.

Example 9.3.1. Solve the (quadratic) equation $x^2 - 3x - 10 = 0$ by factoring.

Solution:

$$x^2 - 3x - 10 = 0$$
$$(x-5)(x+2) = 0$$
$$\begin{array}{ccc} (x-5) = 0 & \text{or} & (x+2) = 0 \\ x = 5 & \text{or} & x = -2 \end{array}$$

□

Example 9.3.2. Solve the equation $9x^2 - 6x + 3 = 2$.

Solution:

$$\begin{aligned} 9x^2 - 6x + 3 &= 2 \\ 9x^2 - 6x + 1 &= 0 \\ (3x-1)(3x-1) &= 0 \\ (3x-1) &= 0 \\ x &= \frac{1}{3} \end{aligned}$$

□

The second method, extracting square roots, refers to "taking the square-root of both sides." Solving the equation $x^2 = a$ means finding all values of x that square to a. In this case, $x = \pm\sqrt{a}$, since $(\sqrt{a})^2 = a$ and $(-\sqrt{a})^2 = a$. Note that finding all x such that $x^2 = a$ is a different problem from finding $\sqrt{a}$ since our convention is that the symbol $\sqrt{\ }$ means the *principal* square root; therefore, when taking the square root of both sides of an equation, we must remember the $\pm$. (Why is putting $\pm$ on one side of the equation sufficient?) Note also that the method of extracting square roots is a special case of factoring, as shown in the examples below.

Example 9.3.3. Solve the equation $4x^2 = 12$.

Solution 1:

$$\begin{aligned} 4x^2 &= 12 \\ x^2 &= 3 \\ x &= \pm\sqrt{3} \end{aligned}$$

Solution 2:

$$4x^2 = 12$$
$$4x^2 - 12 = 0$$
$$4(x^2 - 3) = 0$$
$$x^2 - 3 = 0$$
$$(x - \sqrt{3})(x + \sqrt{3}) = 0$$

$$\begin{array}{ccc} (x - \sqrt{3}) = 0 & \text{or} & (x + \sqrt{3}) = 0 \\ x = \sqrt{3} & \text{or} & x = -\sqrt{3} \end{array}$$

□

Example 9.3.4. Solve the equation $(x - 5)^2 = 15$.

Solution:

$$\begin{aligned} (x - 5)^2 &= 15 \\ x - 5 &= \pm\sqrt{15} \\ x &= 5 \pm \sqrt{15} \end{aligned}$$

□

We see from the above example that if we have a perfect square involving the "variable side" of the equation, then we might extract square roots to solve the equation. In fact, we may force a perfect square involving the variable on one side of the equation by **completing the square**.

Example 9.3.5. Solve the equation $x^2 - 6x + 2 = 0$.

Solution: It becomes apparent rather quickly that factoring this trinomial is not likely to work. Thus we consider forcing a perfect-square trinomial on one side of the equation. Recall that we may complete the square on a quadratic expression of the form $x^2 + bx$ by "adding" $\left(\frac{b}{2}\right)^2$.

$$\begin{aligned} x^2 - 6x + 2 &= 0 \\ x^2 - 6x &= -2 \\ x^2 - 6x + (-3)^2 &= -2 + (-3)^2 \\ x^2 - 6x + 9 &= -2 + 9 \\ (x - 3)^2 &= 7 \\ x - 3 &= \pm\sqrt{7} \\ x &= 3 \pm \sqrt{7} \end{aligned}$$

□

Example 9.3.6. Solve the equation $3x^2 - 4x - 5 = 0$.

Solution: It again becomes apparent rather quickly that factoring is not a viable option. Thus we consider completing the square, recalling that it is convenient to have a leading coefficient of 1. Hence, we'll divide both sides of the equation by 3 before attempting to complete the square.

$$\begin{aligned}
3x^2 - 4x - 5 &= 0 \\
3x^2 - 4x &= 5 \\
x^2 - \frac{4}{3}x &= \frac{5}{3} \\
x^2 - \frac{4}{3}x + \left(\frac{2}{3}\right)^2 &= \frac{5}{3} + \left(\frac{4}{9}\right) \\
\left(x - \frac{2}{3}\right)^2 &= \frac{19}{9} \\
x - \frac{2}{3} &= \pm\frac{\sqrt{19}}{3} \\
x &= \frac{2}{3} \pm \frac{\sqrt{19}}{3} \\
x &= \frac{2 \pm \sqrt{19}}{3}
\end{aligned}$$

□

In general, we may solve the equation $ax^2 + bx + c = 0$ by completing the square and extracting square roots as follows:

$$a\left(x+\frac{b}{2a}\right)^2-\frac{b^2}{4a}+c=0 \tag{9.1}$$

$$a\left(x+\frac{b}{2a}\right)^2=\frac{b^2}{4a}-c \tag{9.2}$$

$$a\left(x+\frac{b}{2a}\right)^2=\frac{b^2-4ac}{4a} \tag{9.3}$$

$$\left(x+\frac{b}{2a}\right)^2=\frac{b^2-4ac}{4a^2} \tag{9.4}$$

$$x+\frac{b}{2a}=\pm\sqrt{\frac{b^2-4ac}{4a^2}} \tag{9.5}$$

$$x+\frac{b}{2a}=\frac{\pm\sqrt{b^2-4ac}}{|2a|} \tag{9.6}$$

$$x+\frac{b}{2a}=\frac{\pm\sqrt{b^2-4ac}}{2a} \tag{9.7}$$

$$x=-\frac{b}{2a}\pm\frac{\sqrt{b^2-4ac}}{2a} \tag{9.8}$$

$$x=\frac{-b\pm\sqrt{b^2-4ac}}{2a} \tag{9.9}$$

Note that in moving from equation (6) to equation (7), we replaced $|2a|$ with $2a$. We can justify this as follows: if $a>0$, then $|2a|=2a$. If $a<0$, then $|2a|=-2a$, so we would have $\frac{\pm\sqrt{b^2-4ac}}{-2a}=\frac{\mp\sqrt{b^2-4ac}}{2a}$. But these are the same roots we had to begin with! Thus, we cover both cases ($a>0$ and $a<0$) by just using the $\pm$.

This computation leads to the following theorem and our fourth algebraic solution technique for quadratic equations.

Theorem 93. *If x is a solution of the equation $ax^2+bx+c=0$ $(a\neq 0)$, then*

$$x=\frac{-b\pm\sqrt{b^2-4ac}}{2a}.$$

Logically speaking, this does not mean that $x=\frac{-b+\sqrt{b^2-4ac}}{2a}$ and $x=\frac{-b-\sqrt{b^2-4ac}}{2a}$ will be solutions, but the reader may verify as an exercise the following theorem.

Theorem 94. *The values $x=\frac{-b\pm\sqrt{b^2-4ac}}{2a}$ are solutions to the quadratic equation $ax^2+bx+c=0$ where $a\neq 0$.*

Combining these two theorems leads to the theorem known as the Quadratic Formula.

Theorem 95 (The Quadratic Formula). *The solutions of the quadratic equation in general form* $ax^2 + bx + c = 0$, $a \neq 0$, *are given by*

$$x = \frac{-b \pm \sqrt{b^2 - 4ac}}{2a}.$$

The quantity $b^2 - 4ac$ is called the **discriminant** of the quadratic expression $ax^2 + bx + c$; it carries a substantial amount of information about the behavior of the function. We will investigate some of its properties after some examples.

Example 9.3.7. Solve each of the following quadratic equations.

1. $x^2 + 4x + 5 = 0$
2. $x^2 + 6x + 9 = 0$
3. $2x^2 - 3x = 5$

Solution.

1.

$$\begin{array}{rcll} x^2 + 4x + 5 & = & 0 & \text{Given} \\ a = 1, b = 4, & \text{and} & c = 5 & \text{Inspection} \\ x & = & \dfrac{-4 \pm \sqrt{4^2 - 4(1)(5)}}{2(1)} & \text{Quadratic Formula} \\ x & = & \dfrac{-4 \pm \sqrt{-4}}{2} & \text{Simplification} \\ x & = & \dfrac{-4 \pm 2i}{2} & \text{Simplification} \\ x & = & -2 \pm i & \text{Simplification} \end{array}$$

The solutions are $x = -2 + i$ and $x = -2 - i$. Notice that these are complex conjugates of each other.

We may factor the original polynomial as $x^2 + 4x + 5 = (x - (-2 + i))(x - (-2 - i))$.

2.

$$\begin{array}{rcll} x^2 + 6x + 9 & = & 0 & \text{Given} \\ a = 1, b = 6, & \text{and} & c = 9 & \text{Inspection} \\ x & = & \dfrac{-6 \pm \sqrt{6^2 - 4(1)(9)}}{2(1)} & \text{Quadratic Formula} \\ x & = & \dfrac{-6 \pm 0}{2} & \text{Simplification} \\ x & = & -3 & \text{Simplification} \end{array}$$

There is only one zero! Notice that $x^2 + 6x + 9 = (x + 3)^2$; had we factored at the beginning instead of blindly using the quadratic formula, we would have very quickly discovered that $x = -3$ is the only solution.

3.

$$\begin{array}{rcll} 2x^2-3x &=& 5 & \text{Given} \\ 2x^2-3x-5 &=& 0 & \text{Subtract 5} \\ (2x-5)(x+1) &=& 0 & \text{Factoring} \\ 2x-5=0 & \text{or} & x+1=0 & \text{Zero Product Theorem} \\ x=\frac{5}{2} & \text{or} & x=-1 & \text{Add/multiply by equals} \end{array}$$

The quadratic formula is not always the best tool; always be on the lookout for easier methods.

For practice, we will also present a solution using the quadratic formula.

$$\begin{array}{rcll} 2x^2-3x &=& 5 & \text{Given} \\ 2x^2-3x-5 &=& 0 & \text{Subtract 5} \\ a=2, b=-3, & \text{and} & c=-5 & \text{Inspection} \\ x &=& \dfrac{-(-3)\pm\sqrt{(-3)^2-4(2)(-5)}}{2(2)} & \text{Quadratic Formula} \\ x &=& \dfrac{3\pm\sqrt{49}}{2(2)} & \text{Simplification} \\ x &=& \dfrac{3\pm 7}{4} & \text{Simplification} \\ x=\frac{10}{4} & \text{or} & x=-\frac{4}{4} & \text{Simplification} \\ x=\frac{5}{2} & \text{or} & x=-1 & \text{Simplification} \end{array}$$

We may factor our polynomial as $2x^2-3x-5=(2x-5)(x+1)$.

We have the following graphs:

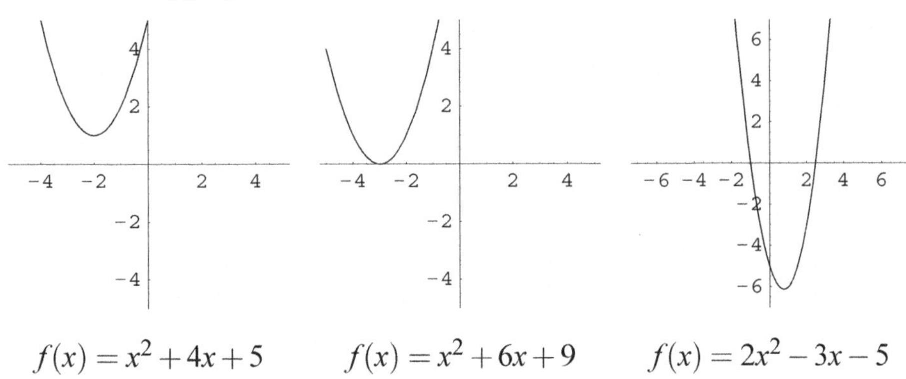

$f(x)=x^2+4x+5$ $\quad$ $f(x)=x^2+6x+9$ $\quad$ $f(x)=2x^2-3x-5$

□

Consider the behavior of the discriminant in each example above. In the first example, $b^2-4ac<0$, and we had two different nonreal complex solutions. This makes sense, since we must take the square root of the discriminant.

In the second example, the discriminant was zero, and we found only one solution, which was real. Again, this makes sense, since the reason we might get two solutions is that we take plus *or* minus the square root of the discriminant. However, if the discriminant is zero, adding it or subtracting it will give the same result, so there is only one solution.

In the third example, the discriminant was positive, and there were two distinct real roots. This, too, is reasonable since we both add and subtract the square root of the discriminant, which will be a real number as long as the discriminant is positive.

The third example also illustrates a general principle: the solutions will be rational if and only if the discriminant is a perfect square (of a rational number). In this case, as in the example, the original quadratic would have factored with rational coefficients. Thus, you can use the discriminant to tell you whether or not to look for a factorization.

In summary, we have the following characterizations of the two solutions to the quadratic equation $ax^2+bx+c=0$:

- If $b^2-4ac<0$, then the quadratic equation has no real solutions (and the two complex solutions are complex conjugates), and the graph of the corresponding parabola does not intersect the x-axis.

- If $b^2-4ac=0$, then the quadratic equation has one REAL (repeated) solution, and the graph of the corresponding parabola is tangent to the x-axis at the vertex.

- If $b^2-4ac>0$, then the quadratic equation has two distinct REAL solutions, and the graph of the corresponding parabola intersects the x-axis in two distinct points.

Example 9.3.8. Simplify $\dfrac{5x^2+17x+12}{x^2+x+1}$.

Solution: Before we go to the trouble of factoring, let's check the discriminant of the denominator: $1^2-4(1)(1)=-3<0$, so x^2+x+1 will not factor over the rationals. Therefore, this quotient is already simplified.

□

Some quadratic functions may also be "disguised."

Example 9.3.9. Consider $x^4-4x^2-5=0$. This is not a quadratic equation in x, but it *is* quadratic in x^2: $(x^2)^2-4(x^2)-5=0$.

$$\begin{array}{rclll}
x^4-4x^2-5 & = & 0 & & \text{Given}\\
(x^2)^2-4(x^2)-5 & = & 0 & & \text{Laws of Exponents}\\
u^2-4u-5 & = & 0 & & \text{Substitute } u=x^2\\
(u-5)(u+1) & = & 0 & & \text{Factor}\\
u-5=0 & \text{or} & u+1=0 & & \text{Zero Product Theorem}\\
u=5 & \text{or} & u=-1 & & \text{Add equals}\\
x^2=5 & \text{or} & x^2=-1 & & \text{Substitute } u=x^2\\
x=\pm\sqrt{5} & \text{or} & x=\pm i & & \text{Extract square root}
\end{array}$$

□

Example 9.3.10. Solve $x+\sqrt{x}=1$.

Solution:

$$\begin{aligned} x+\sqrt{x} &= 1 && \text{Given} \\ x+\sqrt{x}-1 &= 0 && \text{Subtract 1} \\ u^2+u-1 &= 0 && \text{Substitute } u=\sqrt{x} \\ u &= \frac{-1\pm\sqrt{1-4(1)(-1)}}{2} && \text{Quadratic Formula} \\ u &= \frac{-1\pm\sqrt{5}}{2} && \text{Simplification} \\ \sqrt{x} &= \frac{-1\pm\sqrt{5}}{2} && \text{Substitute } u=\sqrt{x} \\ x &= \left(\frac{-1+\sqrt{5}}{2}\right)^2 && \text{Square both sides} \\ x &= \frac{3\pm\sqrt{5}}{2} && \text{Simplification} \end{aligned}$$

Note that $\frac{3+\sqrt{5}}{2}$ is clearly not a solution since it (and its square root) are both greater than one. One may verify that $x=\frac{3-\sqrt{5}}{2}$ is in fact a solution. Be on the lookout for quadratics in disguise!

□

We close this section with an application of quadratic equations.

Example 9.3.11. If Roy and Joe work together, they can mow a field in 4 hours. Roy working alone will take 3 hours longer than Joe would working alone. How long would each take to mow the field alone?

Solution: Let r_1 and r_2 be the rates of Roy and Joe, respectively. Then their combined rate is r_1+r_2. If we multiply this rate by the time it takes to complete the mowing, we will have one complete job: $(r_1+r_2)4=1$. Also, if t is how long it takes Joe alone, then Roy will need $t+3$ hours to complete the same amount of work, so $r_2t=r_1(t+3)=1$. Therefore, $r_2=\frac{1}{t}$ and $r_1=\frac{1}{t+3}$. This gives us

$$4\left(\frac{1}{t}+\frac{1}{t+3}\right)=1.$$

Multiply both sides by $t(t+3)$ to clear denominators. The resulting equation (after distributing and simplifying) is

$$4(t+3)+4t=t(t+3)$$

or $8t+12=t^2+3t$. Thus we have $t^2-5t-12=0$. This has solutions $t=\frac{5\pm\sqrt{73}}{2}$. Only the positive root makes sense in this context, so Joe will take approximately 6.77 hours on his own, and Roy will take about 9.77 hours alone.

□

Section 9.3 Exercises

Solve each equation by completing the square.

1. $x^2+6x+2=-1$
2. $x^2-5x-5=0$
3. $3x^2+8x+20=14$
4. $-4x^2-5x=22$

Solve each equation by using the quadratic formula.

5. $x^2+5x-8=0$
6. $x^2+x+1=0$
7. $-3x^2-2x+4=0$
8. $\frac{3}{5}x^2+2x-1=\frac{4}{5}$

Solve each equation by an appropriate method.

9. $x^2-12=4$
10. $(x-3)^2=5$
11. $x^2+8x+12=0$
12. $3x^2-5x-1=14$
13. $x^2+2x=4$
14. $3x^2-12x+14=2$
15. $(x-3)(x-5)=(x-5)$
16. $x^8-6x^4-9=0$
17. $x^5-3x^3+\frac{1}{4}x=0$
18. $2x-\sqrt{x}=1$
19. $\left(\frac{x+2}{x-1}\right)^2-\frac{x+2}{x-1}-2=0$
20. The sum of a number and its multiplicative inverse is -1. What are the possibilities for the number?
21. If it takes Azuka and Barney 7 hours to paint a house together, and it takes Barney 14 hours to paint the house by himself, how long will it take Azuka working alone?
22. An arrow is shot into the air; its height is given by $h(t)=-4.9t^2+30t+2$ meters above the ground.
    (a) Determine the maximum height of the arrow and when it occurs.
    (b) When does the arrow hit the ground?
    (c) At what times is the arrow 40 meters in the air? Why is there more than one answer?
23. Verify that $(1+\sqrt{2})$ is a square root of $3+2\sqrt{2}$.

9.4 Polynomial Functions and Equations

A general polynomial function is a function of the form

$$f(x) = a_n x^n + a_{n-1} x^{n-1} + \ldots + a_2 x^2 + a_1 x + a_0,$$

where $a_n \neq 0$. In this case, f has degree n, written $\deg f = n$. The numbers $a_0, a_1, \ldots, a_n$ are called **coefficients**, $a_n x^n$ is the **leading term**, and a_n is the **leading coefficient**. We have already considered some special cases: constant functions (degree 0), linear functions (degree 1), and quadratic functions (degree 2).

The graph of a polynomial is a smooth curve with no sharp turns or breaks. The monomial x^n "looks like" x^2 if n is even and x^3 if n is odd. The difference is that when $n > 3$, the graph will be flatter on $[-1,1]$ and steeper on $(-\infty,-1) \cup (1,\infty)$.

Example 9.4.1.

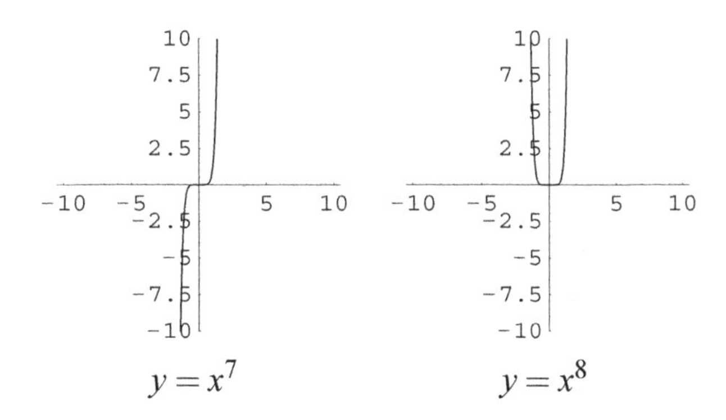

□

One can determine the shape of a more general polynomial's graph by considering the leading term. For example, consider the polynomial $f(x) = -3x^8 - 4x^5 + 6x^3 - 2$. If x is an extremely large positive number, then $-3x^8$ will be an enormous negative number. By comparison, $-4x^5, 6x^3$, and -2 will all be "small" numbers. Take $x = 1000$. This is not a particularly "large" x, but it will serve to illustrate the idea. We get

$$-3(1000)^8 = -3000000000000000000000000,$$

and

$$f(1000) = -2999999996000006000000002.$$

The relative sizes of these are nearly the same! This idea is summarized in the theorem below.

Theorem 96 (Leading Coefficient Test). *The polynomial $f(x) = a_n x^n + \ldots + a_0$ has the following end behavior.*

1. *If n is even and $a_n > 0$, then the end behavior is $(\nwarrow, \nearrow)$. (Like x^2)*

   If n is even and $a_n < 0$, then then the end behavior is $(\swarrow, \searrow)$. (Like $-x^2$)

2. *If n is odd and $a_n > 0$, then then the end behavior is (↙, ↗). (Like x^3)*
   If n is odd and $a_n < 0$, then then the end behavior is (↖, ↘). (Like $-x^3$)

Example 9.4.2. Consider the graph of $f(x) = -3x^8 - 4x^3 + 12$. The theorem tells us that this should have the shape of an inverted "U" if we zoom out far enough. The graph below bears this out.

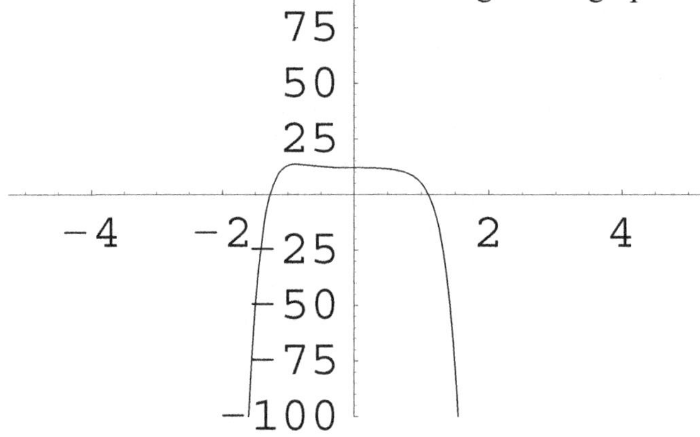

□

Such an analysis can tell you, for example, whether your graphing calculator has its window set appropriately.

We now come to an extremely important theorem about dividing polynomials. Its statement may not look familiar, but it is very likely that you are familiar with its application.

Theorem 97 (The Division Algorithm). *If $f(x)$ and $p(x)$ are polynomials, then there are unique polynomials $q(x)$ and $r(x)$ such that $f(x) = p(x)q(x) + r(x)$, where $r(x) = 0$ or $0 \leq \deg r(x) < \deg p(x)$.*

Example 9.4.3. Divide $f(x) = x^4 - 3x^2 + 2x + 1$ by $x^3 - 4$.

Solution: We employ polynomial long division. Notice that we must insert $0x^3$ into $f(x)$ as a place holder, in the same way that we use a 0 in 108 as a place holder. Also, recall the strategy of polynomial long division: all it really is is a bookkeeping system.

$$\begin{array}{r|l} & x \\ \hline x^3 - 4 & x^4 + 0x^3 - 3x^2 + 2x + 1 \\ & x^4 \qquad\qquad\qquad - 4x \\ \hline & \qquad\qquad - 3x^2 + 6x + 1 \end{array}$$

Thus, in terms of the division algorithm, $f(x) = (x^3 - 4)(x) + (-3x^2 + 6x + 1)$, with $p(x) = (x^3 - 4), q(x) = x$, and $r(x) = -3x^2 + 6x + 1$. The letter q stands for "quotient," and the letter r stands for "remainder."

How do we know when to stop our long division process? The division algorithm tells us we can quit as soon as our remainder has degree less than that of p. In this example, p has degree 3, so we stop when we get to $-3x^2 + 6x + 5$, which has degree 2.

□

The same reasoning says that if we divide a polynomial by a linear polynomial, the remainder will be a constant.

Example 9.4.4. Divide $x^2 - 3x + 5$ by $x + 2$.

$$\begin{array}{r|rrrrr} & x & - & 5 & & \\ \hline x+2 & x^2 & - & 3x & + & 5 \\ & x^2 & + & 2x & & \\ \hline & & - & 5x & + & 5 \\ & & - & 5x & - & 10 \\ \hline & & & & & 15 \end{array}$$

Thus the quotient is $x - 5$ and the remainder is 15. In terms of the division algorithm, we have $f(x) = (x+2)(x-5) + 15$.

□

It is reasonable to ask whether the constant has any particular significance; in fact, it does.

Theorem 98 (Remainder Theorem). *If $f(x)$ is divided by $x - a$, the remainder is $f(a)$.*

Proof. By the division algorithm, we may write $f(x) = (x-a)q(x) + R$, where $R = 0$ or R has degree 0; in either case, R is a constant (a specific number that doesn't depend on x). Now evaluate both sides of this equation at $x = a$: $f(a) = (a-a)q(a) + R = R$. □

Example 9.4.5. We may verify this by considering again Example 9.4.4. We have $x + 2$ as a divisor. The Remainder Theorem has as hypothesis that f is divided by $x - a$, so in order to use the theorem, we must first write $x + 2$ in this form. We have $x + 2 = x - (-2)$. Thus $a = -2$ (not 2). Now evaluate: $f(-2) = (-2)^2 - 3(-2) + 5 = 15$, the same as our remainder!

□

The following theorem is closely related to the Remainder Theorem.

Theorem 99 (Factor Theorem). *Let f be a polynomial. Then $x - a$ is a factor of f if and only if $f(a) = 0$.*

Proof. First, suppose that $x - a$ is a factor of $f(x)$. Then $f(x) = (x-a)q(x)$, with no remainder. Therefore, $f(a) = (a-a)q(x) = 0$.

Next, suppose that $f(a) = 0$. By the Remainder Theorem, if $f(x)$ is divided by $x - a$, the remainder is $f(a)$; in this case, 0. Therefore we may write $f(x) = (x-a)q(x) + 0 = (x-a)q(x)$, so $x - a$ is a factor of $f(x)$. □

Example 9.4.6. In Example 9.4.4, $x + 2$ is not a factor of $f(x) = x^2 - 3x + 5$ since $f(-2) = 15 \neq 0$. We already knew this since we performed polynomial long division and found a nonzero remainder; however, if we had had the Factor Theorem, we would not have needed to do the long division to determine this.

□

Example 9.4.7. Suppose that we wish to factor the polynomial $f(x) = x^4 + 4x^3 - x^2 - 16x - 12$. It does not take long to decide that $x = -1$ is a zero of f, so the Factor Theorem tells us that $x+1$ will be a factor of $f(x)$. Therefore, it is worth our while to divide $f(x)$ by $x+1$.

$$
\begin{array}{rrrrrrrrrr}
 & x^3 & + & 3x^2 & - & 4x & - & 12 & & \\
\cline{2-10}
x+1 \;) & x^4 & + & 4x^3 & - & x^2 & - & 16x & - & 12 \\
 & x^4 & + & x^3 & & & & & & \\
\cline{2-10}
 & & & 3x^3 & - & x^2 & - & 16x & - & 12 \\
 & & & 3x^3 & + & 3x^2 & & & & \\
\cline{4-10}
 & & & & - & 4x^2 & - & 16x & - & 12 \\
 & & & & - & 4x^2 & - & 4x & & \\
\cline{5-10}
 & & & & & & - & 12x & - & 12 \\
 & & & & & & - & 12x & - & 12 \\
\cline{7-10}
 & & & & & & & & & 0
\end{array}
$$

We see that the remainder is 0, as promised. We now have

$$f(x) = (x+1)(x^3 + 3x^2 - 4x - 12).$$

Notice that $q(x) = x^3 + 3x^2 - 4x - 12$ has degree 3, one less than the degree of f.

We may now factor q by grouping:

$$q(x) = x^2(x+3) - 4(x+3) = (x^2 - 4)(x+3) = (x-2)(x+2)(x+3).$$

Therefore, $f(x) = (x-2)(x+1)(x+2)(x+3)$. As a bonus, we can see that f has exactly four zeros: $2, -1, -2$, and -3.

□

Notice that we did not have to use polynomial long division repeatedly; once we found one factor, we were able to factor the rest by other methods. Be on the lookout for such things!

Also, it is not a coincidence that f had four zeros and that its degree is four. We have the following theorem.

Theorem 100. *A polynomial function of degree $n > 0$ has at most n real zeroes.*

Proof. We prove the theorem by mathematical induction. We take as S_n the statement "A polynomial function of degree $n > 0$ has at most n real zeros." Our goal is to prove that S_n is a true statement for all $n \in \mathbb{N}$.

Base Case: If $n = 1$, we have $f(x) = ax + b$ for some $a \neq 0$. Therefore, if $f(x) = 0$, then $ax + b = 0$, so $x = -\frac{b}{a}$. We see that we can have at most one zero if f has degree 1. Thus, S_1 is a true statement.

Induction Step: We may assume that S_k is a true statement for some $k \geq 1$; that is, any polynomial of degree k has at most k real zeros. We must show that S_{k+1} is also a true statement.

Thus, let $f(x)$ be a polynomial of degree $k+1$. If f does not have any real zeros, then f certainly does not have more than $k+1$ zeros, so S_{k+1} is a true statement. If f does have a real zero, say a, then we may write $f(x) = (x-a)q(x)$, where $q(x)$ has degree k. (We need $(x-a)q(x)$ to have degree $k+1$.) Now, if b is a zero of f, then by the Zero Product Theorem, we must have either $b-a=0$ or $q(b)=0$. Since q has at most k real zeros by the induction hypothesis and $x-a$ has at most one real zero by the base case, f can have at most $k+1$ real zeros, completing the proof.

□

We have developed several number systems in the course of this text, but most have what may be considered a serious "flaw." Consider:

1. The polynomial $f(x) = 2x+4$ has natural numbers for coefficients, but its only zero is **not** a natural number. We need more numbers to solve the equation $2x+4=0$.

2. If we allow integer coefficients (and solutions) instead of only natural numbers, we can solve $2x+4=0$, but we still cannot solve $2x-1=0$ in the integers. We need still a larger number system to solve this equation.

3. If we move into the rational numbers, we can then solve $2x-1=0$, but we are still unable to solve the equation $x^2-2=0$ with rational numbers even though the coefficients are rational. (Recall that $\sqrt{2}$ is not rational.)

4. If we allow all real numbers as coefficients and as solutions, we can solve $x^2-2=0$, but we cannot solve $x^2+1=0$ in the real numbers, although the coefficients are real numbers. We need yet again a larger number system in order to solve this equation.

5. We will now allow any complex number to be a coefficient or solution. Is this number system large enough that every polynomial equation having complex coefficients has a complex zero? There is no obvious reason for this to be the case; nevertheless, it is the case.

Theorem 101 (Fundamental Theorem of Algebra)**.** *Every nonconstant polynomial with complex coefficients has at least one (complex) zero.*

NOTE: Remember that $\mathbb{R} \subseteq \mathbb{C}$, so polynomials with real coefficients will also have (possible nonreal complex) zeros, as well.

Proof. The first proof of this important and remarkable theorem is due to Karl Friedrich Gauss, as are many other important theorems. Its proof is beyond the scope of this course; if you are interested in seeing a proof, you will find one in any course in Complex Analysis. □

In fact, we can go a little farther with this result.

Theorem 102. *A polynomial of degree n has exactly n zeros, some of which may be repeated.*

Proof. This may also be proved by induction; we leave the proof as an exercise. □

What does it mean for a zero to be repeated?

Definition 9.4.1. If a linear factor $x-a$ occurs m times, then a is a zero of **multiplicity** m.

Example 9.4.8. $(x-3)^4(3x+2)(5x-12)^5$ has solution set $\{3, -\frac{2}{3}, \frac{12}{5}\}$. However, 3 has multiplicity 4, $-\frac{2}{3}$ has multiplicity 1, and $\frac{12}{5}$ has multiplicity 5. We could also say that 2 has multiplicity 0.

□

Section 9.4 Exercises

Determine the end behavior of each polynomial. Also determine the maximum possible number of real zeros.

1. $p(x) = x^2 + 5$
2. $f(x) = 2x - 7$
3. $x(t) = -12t^5 - 7t^3 + 2t - 1$
4. $g(x) = 5x^4 + 3x^3 - x - 100000000001$

Express each quotient in terms of the division algorithm. (You may need to perform polynomial long division in order to do this.)

5. $(x^5 - 3x^4 - 2x^3 + x^2 + 4x + 1) \div (x^3 - 2)$
6. $(2x^4 - x - 1) \div (x - 1)$
7. $(3x^3 - 4x^2 + 2x + 2) \div (3x + 3)$
8. $(3x^3 - 7x^2 + x - 2) \div (x^2 + 3)$

Determine the remainder in each case *without* performing long division, and whether the divisor is a factor of the dividend.

9. $(2.4x^3 + x^2 - 3.2x) \div (x - 1.1)$
10. $(2x^6 + 6x^5 - 3x^4 - 9x^3 - 4x - 12) \div (x + 3)$
11. $(4x^4 - 3x^3 - x - 4) \div (x + 4)$
12. $(-2x^5 - 8x^3 - x + 12) \div (x + 2)$
13. $(x^4 - 2x^3 - 8x^2 - 3x + 2) \div (x + 1)$
14. $(-4x^2 - 2x - 4) \div (x - 2)$

Determine the multiplicity of each zero for each polynomial.

15. $(x-2)(x+3)^4(3x-5)^3$
16. $16(x+2)^4(3x-1)^{13}$
17. $(x+1)(x+2)^2(x+3)^3(x+4)^4$
18. Use Mathematical Induction to prove Theorem 102.

9.5 Rational Roots

In this section, we explore some techniques for finding zeros of polynomials. From the results of the previous section, we know that we can find zeros by factoring completely. We summarize those results in the following theorem.

Theorem 103. *Let $a \in \mathbb{R}$ and let f be a polynomial function. The following are equivalent.*

1. *$x = a$ is a zero of f.*
2. *$x = a$ is a solution of $f(x) = 0$.*
3. *$(x - a)$ is a factor of the polynomial $f(x)$.*
4. *$(a, 0)$ is an x-intercept of the graph of f.*

Example 9.5.1. Find all real zeroes of $f(x) = x^4 + x^3 - x^2 + x - 2$.

Solution: We can see that $x = 1$ is a zero, so according to the theorem above, $x - 1$ must be a factor. We get $f(x) = (x-1)(x^3 + 2x^2 + x + 2)$ from polynomial long division. Now we can factor by grouping, and we see that $f(x) = (x-1)(x+2)(x^2+1)$. Since $x^2 + 1$ does not factor over $\mathbb{R}$, the real roots are $x = 1, -2$. However, $x^2 + 1$ does factor over $\mathbb{C}$ as $x^2 + 1 = (x+i)(x-i)$, so we also have the nonreal complex zeros i and $-i$. We have $f(x) = (x-1)(x+2)(x-i)(x+i)$.

□

How did we know that 1 is a zero? We have a technique that we will discuss below. First, we introduce a simplified division algorithm for when the divisor is of the form $x - a$. This process is called **synthetic division**.

Example 9.5.2. Divide $4x^4 - 7x^3 + x - 5$ by $x + 2$. NOTE: we must have the divisor in the form $x - a$, so $a = -2$. Just as polynomial long division is really a bookkeeping system, so is synthetic division. We will compare the two methods in this example. The major difference between the two is that in polynomial long division, we use the powers of x (x^4, x^3, etc.) to keep track of like terms. In synthetic division, we use only the coefficients, and keep track of like terms according to their **positions** in an array, just as we use in our decimal number system.

We set up both processes in a similar way:

$$x+2 \overline{\big) \; 4x^4 \; - \; 7x^3 \; + \; 0x^2 \; + \; x \; - \; 5} \qquad\qquad -2 \overline{\big) \; 4 \quad -7 \quad 0 \quad 1 \quad -5}$$

Notice that in both cases we must include the 0 that holds the place of the x^2 term. It is especially important that we do this in synthetic division, since it is the *position* that tells us the corresponding power of x.

It is also important to notice that in polynomial long division, we have $x + 2$, whereas in synthetic division, we have a -2. We will need to account for this change in sign in the synthetic division process.

In ordinary long division, we keep track of our quotient above the line over the dividend; in synthetic division, we will keep track of the quotient at the bottom, instead. We start the long division process with a $4x^3$; correspondingly, in synthetic division, we record only the coefficient 4. **Note:** This is only valid if the divisor has the form $x-a$. If the divisor looks like $cx-a$, this method cannot be used!

$$\begin{array}{r|rrrrrrrrr} & 4x^3 & & & & & & & & \\ \hline x+2 & 4x^4 & - & 7x^3 & + & 0x^2 & + & x & - & 5 \end{array} \qquad \begin{array}{r|rrrrr} -2 & 4 & -7 & 0 & 1 & -5 \\ & & & & & \\ \hline & 4 & & & & \end{array}$$

In long division, we next record $4x^3(x+2) = 4x^4 + 8x^3$ in the second row. Where did the $8x^3$ come from? It came from multiplying the $4x^3$ by the 2; the coefficient is obtained by multiplying the 2 in $x+2$ by the 4 in $4x^3$. In synthetic division, we have recorded a -2 from $x+2 = x-(-2)$ and a 4 from $4x^4$. We will multiply these and record the product (-8) in the second row.

$$\begin{array}{r|rrrrrrrrr} & 4x^3 & & & & & & & & \\ \hline x+2 & 4x^4 & - & 7x^3 & + & 0x^2 & + & x & - & 5 \\ & 4x^4 & + & 8x^3 & & & & & & \end{array} \qquad \begin{array}{r|rrrrr} -2 & 4 & -7 & 0 & 1 & -5 \\ & & -8 & & & \\ \hline & 4 & & & & \end{array}$$

Notice the difference in sign for the two 8's. In polynomial long division, we *subtract* the $4x^4 + 8x^3$ from the previous line. In order to account for that sign difference in synthetic division, we *add* the 8 to the previous line.

$$\begin{array}{r|rrrrrrrrr} & 4x^3 & & & & & & & & \\ \hline x+2 & 4x^4 & - & 7x^3 & + & 0x^2 & + & x & - & 5 \\ & 4x^4 & + & 8x^3 & & & & & & \\ \hline & & & -15x^3 & + & 0x^2 & + & x & - & 5 \end{array} \qquad \begin{array}{r|rrrrr} -2 & 4 & -7 & 0 & 1 & -5 \\ & & -8 & & & \\ \hline & 4 & -15 & & & \end{array}$$

Notice that in both cases we end up with a -15 recorded. In the long division version, we next record a $-15x^2$ above the top bar, but in synthetic division, we already have the -15 recorded at the bottom. In long division, we multiply that $-15x^2$ by $x+2$ and get $-15x^3 - 30x^2$, which we write underneath the last line. Where did the $-30x^2$ come from? It came from multiplying the $-15x^2$ by the 2; accordingly, in synthetic division, we multiply the -15 by the -2, and record that in the second line.

$$\begin{array}{r|rrrrrrrrr} & 4x^3 & - & 15x^2 & & & & & & \\ \hline x+2 & 4x^4 & - & 7x^3 & + & 0x^2 & + & x & - & 5 \\ & 4x^4 & + & 8x^3 & & & & & & \\ \hline & & - & 15x^3 & + & 0x^2 & + & x & - & 5 \\ & & - & 15x^3 & - & 30x^2 & & & & \end{array} \qquad \begin{array}{r|rrrrr} -2 & 4 & -7 & 0 & 1 & -5 \\ & & -8 & 30 & & \\ \hline & 4 & -15 & & & \end{array}$$

Again, we subtract in polynomial long division, and add in synthetic division:

$$
\begin{array}{rrrrrrrrrr}
 & 4x^3 & - & 15x^2 & & & & & & \\ \cline{2-10}
x+2 \,\big) & 4x^4 & - & 7x^3 & + & 0x^2 & + & x & - & 5 \\
 & 4x^4 & + & 8x^3 & & & & & & \\ \cline{2-10}
 & & - & 15x^3 & + & 0x^2 & + & x & - & 5 \\
 & & - & 15x^3 & - & 30x^2 & & & & \\ \cline{3-10}
 & & & & & 30x^2 & + & x & - & 5
\end{array}
\qquad
\begin{array}{r|rrrrr}
-2 & 4 & -7 & 0 & 1 & -5 \\
 & & -8 & 30 & & \\ \hline
 & 4 & -15 & 30 & &
\end{array}
$$

In long division, we now record a $30x$ on the top line; this 30 is already in the bottom line of our synthetic division. We multiply this $30x$ by $x+2$ to get $30x^2+60x$. Similarly, in synthetic division, we multiply the -2 by the 30 (getting -60) and record that in the second line. We subtract the bottom line from the previous in long division, and add the top two lines in synthetic division, as we have been doing.

$$
\begin{array}{rrrrrrrrrr}
 & 4x^3 & - & 15x^2 & + & 30x & & & & \\ \cline{2-10}
x+2 \,\big) & 4x^4 & - & 7x^3 & + & 0x^2 & + & x & - & 5 \\
 & 4x^4 & + & 8x^3 & & & & & & \\ \cline{2-10}
 & & - & 15x^3 & + & 0x^2 & + & x & - & 5 \\
 & & - & 15x^3 & - & 30x^2 & & & & \\ \cline{3-10}
 & & & & & 30x^2 & + & x & - & 5 \\
 & & & & & 30x^2 & + & 60x & & \\ \cline{6-10}
 & & & & & & - & 59x & - & 5
\end{array}
\qquad
\begin{array}{r|rrrrr}
-2 & 4 & -7 & 0 & 1 & -5 \\
 & & -8 & 30 & -60 & \\ \hline
 & 4 & -15 & 30 & -59 &
\end{array}
$$

Again, notice that the -59's match. Finally, we record a $-59x$ on the top line for our long division (again observing that the -59 is already recorded in our synthetic division), multiply by $x+2$ (multiplying -59 by -2 in our synthetic division), and subtract (add in our synthetic division).

$$
\begin{array}{rrrrrrrrrr}
 & 4x^3 & - & 15x^2 & + & 30x & - & 59 & & \\ \cline{2-10}
x+2 \,\big) & 4x^4 & - & 7x^3 & + & 0x^2 & + & x & - & 5 \\
 & 4x^4 & + & 8x^3 & & & & & & \\ \cline{2-10}
 & & - & 15x^3 & + & 0x^2 & + & x & - & 5 \\
 & & - & 15x^3 & - & 30x^2 & & & & \\ \cline{3-10}
 & & & & & 30x^2 & + & x & - & 5 \\
 & & & & & 30x^2 & + & 60x & & \\ \cline{6-10}
 & & & & & & - & 59x & - & 5 \\
 & & & & & & - & 59x & - & 108 \\ \cline{7-10}
 & & & & & & & & & 103
\end{array}
\qquad
\begin{array}{r|rrrrr}
-2 & 4 & -7 & 0 & 1 & -5 \\
 & & -8 & 30 & -60 & 108 \\ \hline
 & 4 & -15 & 30 & -59 & 103
\end{array}
$$

Polynomial long division gives us the result

$$(4x^4-7x^3+x-5)\div(x+2)=4x^3-15x^2+30x-59+\frac{103}{x+2};$$

that is, in terms of the division algorithm,

$$4x^4 - 7x^3 + x - 5 = (x+2)(4x^3 - 15x^2 + 30x - 59) + 103.$$

What does synthetic division tell us? We read the coefficients of the quotient from the bottom row:

$$(4x^4 - 7x^3 + x - 5) \div (x+2) = 4x^3 - 15x^2 + 30x - 59 + \frac{103}{x+2},$$

the same thing we had before! How do we know what power of x to begin with? We must begin with a power on x one less than the degree of the dividend, since we are dividing by a polynomial of degree one. After that, the powers simply descend until we reach the last entry, which is the remainder.

□

Example 9.5.3. Divide $5x^3 - 7x^2 + 2x$ by $x - 3$ using synthetic division.
Solution: First, we write this in the form

$$5 \;\big)\overline{\;3 \quad -7 \quad 2 \quad 0\;}$$

to get things started. Remember, when the divisor has the form $x - a$, we record a itself as the first entry in the first row. In this case, we have $x - 5$, so we record the 5. Next, we bring down the 3.

$$\begin{array}{r|rrrr} 5 & 3 & -7 & 2 & 0 \\ & & & & \\ \hline & 3 & & & \end{array}$$

Now multiply that 3 by the 5, and record that in the second row of the next column.

$$\begin{array}{r|rrrr} 5 & 3 & -7 & 2 & 0 \\ & & 15 & & \\ \hline & 3 & & & \end{array}$$

Add the -7 and the 15:

$$\begin{array}{r|rrrr} 5 & 3 & -7 & 2 & 0 \\ & & 15 & & \\ \hline & 3 & 8 & & \end{array}$$

Multiply the 8 by the 5, and record the result in the second row of the next column.

$$\begin{array}{r|rrrr} 5 & 3 & -7 & 2 & 0 \\ & & 15 & 40 & \\ \hline & 3 & 8 & & \end{array}$$

Add the 2 and the 40:

$$\begin{array}{r|rrrr} 5 & 3 & -7 & 2 & 0 \\ & & 15 & 40 & \\ \hline & 3 & 8 & 42 & \end{array}$$

Multiply the 42 by the 5, and record the result in the second row of the next column.

$$\begin{array}{r|rrrr} 5 & 3 & -7 & 2 & 0 \\ & & 15 & 40 & 210 \\ \hline & 3 & 8 & 42 & \end{array}$$

Finally, add the 210 and the 0.

$$\begin{array}{r|rrrr} 5 & 3 & -7 & 2 & 0 \\ & & 15 & 40 & 210 \\ \hline & 3 & 8 & 42 & 210 \end{array}$$

Therefore, we have a quotient of $3x^2 + 8x + 42$ with a remainder of 210, which we may also write in any of the equivalent forms

$$\frac{3x^3 - 7x^2 + 2x}{x-5} = 3x^2 + 8x + 42 + \frac{210}{x-5},$$

$$(3x^3 - 7x^2 + 2x) \div (x-5) = 3x^2 + 8x + 42 + \frac{210}{x-5},$$

or

$$3x^3 - 7x^2 + 2x = (x-5)(3x^2 + 8x + 42) + 210.$$

The last, of course, is written in terms of the division algorithm.

□

Notice that a is a zero if and only if that last entry in row 3 is a zero; this is also an easy way to determine whether a given number is a zero! Of course, we have to have a number to check to begin with, so we need some system for finding possible zeros. We have one in the following theorem.

Theorem 104 (Rational Root Test)**.** *Let f be a polynomial of degree $n \geq 1$ with integer coefficients, say $f(x) = a_n x^n + \ldots + a_0$, $a_n \neq 0$. Then the rational zeros of f have the form $\frac{p}{q}$, where p is a factor of a_0 and q is a factor of a_n.*

That is, the only possible rational zeroes are of the form $\frac{\text{factors of constant}}{\text{factors of leading coefficient}}$.

That's powerful! We can use this theorem to find a (relatively) short list of potential rational zeros.

Note: Henceforth, we will just write the final result of our synthetic division, rather than listing each step separately.

Example 9.5.4. Find all rational zeros of $f(x) = 4x^4 - 4x^3 + 5x^2 - 4x + 1$.

Solution: The leading coefficient is 4, which has factors $\pm 1, \pm 2$, and ± 4. The constant term is 1, which has factors ± 1. Therefore, the potential rational zeros are $\pm 1, \pm\frac{1}{2}, \pm\frac{1}{4}$. We now apply synthetic division to determine which are actual zeros.

First, notice that the signs alternate; this means that no negative number can be a zero. Consider: if $a > 0$, then $-a < 0$, and $4(-a)^4 - 4(-a)^3 + 6(-a)^2 - 4(-a) + 1 = 4a^4 + 4a^3 + 5a^2 + 4a + 1 > 0$. Therefore, we need only check $1, \frac{1}{2}$, and $\frac{1}{4}$. We begin with $x = 1$.

$$\begin{array}{r|rrrrr} 1 & 4 & -4 & 5 & -4 & 1 \\ & & 4 & 0 & 5 & 1 \\ \hline & 4 & 0 & 5 & 1 & 2 \end{array}$$

Thus $x = 1$ is not a zero (since division by $x - 1$ leaves a nonzero remainder). Next, we try $x = \frac{1}{2}$.

$$\begin{array}{r|rrrrr} \frac{1}{2} & 4 & -4 & 5 & -4 & 1 \\ & & 2 & -1 & 2 & -1 \\ \hline & 4 & -2 & 4 & -2 & 0 \end{array}$$

Therefore, $x = \frac{1}{2}$ is a zero, and $x - \frac{1}{2}$ is a factor of f. We have

$$f(x) = \left(x - \frac{1}{2}\right)(4x^3 - 2x^2 + 4x - 2).$$

We may factor the quotient by grouping, but it will be more instructive at this point to show how the method proceeds.

We are reduced to trying to find the zeros of $4x^3 - 2x^2 + 4x - 2$. Notice that any zero of $4x^3 - 2x^2 + 4x - 2$ is automatically a zero of f, so we need only look among the potential rational zeros we already have for f. It is not necessary to try $x = 1$ again, since we already know that $x = 1$ is not a zero of f, but we do need to try $x = \frac{1}{2}$ again, as it may be a repeated root.

$$\begin{array}{r|rrrr} \frac{1}{2} & 4 & -2 & 4 & -2 \\ & & 2 & 0 & 2 \\ \hline & 4 & 0 & 4 & 0 \end{array}$$

Thus we now have $4x^3 - 2x^2 + 4x - 2 = (x - \frac{1}{2})(4x^2 + 4)$, which means that

$$\begin{aligned} f(x) &= (x - \frac{1}{2})^2(4x^2 + 4) \\ &= (x - \frac{1}{2})^2(4)(x^2 + 1) \\ &= [(x - \frac{1}{2})^2(2^2)](x^2 + 1) \\ &= [(x - \frac{1}{2})(2)]^2(x^2 + 1) \\ &= (2x - 1)^2(x^2 + 1). \end{aligned}$$

The zeros of x^2+1 are $\pm i$, as can be seen by considering $x^2=-1$ or using the quadratic formula to solve $x^2+1=0$. Finally, we arrive at

$$f(x) = (2x-1)^2(x-i)(x+i).$$

The rational zero of f is $x=\frac{1}{2}$ (with multiplicity 2), and $x=i$ and $x=-i$ are nonreal complex zeros of f.

□

Notice that not all of the potential rational zeros actually were zeros, and that not all of the zeros were rational. Also, it is worth noting that synthetic division lends itself very nicely to this process; it takes far less time, effort, and space than polynomial long division. BE WARNED: synthetic division **only** works with divisors of the form $x-a$; if you try it with something else, it will fail.

Section 9.5 Exercises

Perform synthetic division where appropriate. If it is not appropriate, perform polynomial long division. In either case, give the remainder.

1. $(3x^3-5x^2-3x-1)\div(x-2)$
2. $(5x^5-3x^4+2x^2-x-2)\div(x+3)$
3. $(2x^4+x^3-x^2-5x-7)\div(2x+3)$
4. $(x^6-3x^4+3x^2-1)\div(x+1)$
5. $(-4x^5+5x^2+8)\div(x^2+1)$
6. $(6x^8-7x^5+4x^4+9x^3+5x-2)\div(x-4)$
7. Answer each of the following for the polynomial $f(x)=3x^3-2x^2+6x-4$. Give justification for **each** answer.
   (a) How many zeros does f have?
   (b) How many real zeros does f have?
   (c) List all the possible rational zeros of f.
   (d) Find all the zeros of f.
   (e) Classify each zero as rational, irrational, or nonreal complex.
   (f) Write f as a product of linear factors.
8. Answer each of the following for the polynomial $f(x)=3x^5-2x^4+6x^3-4x^2-24x+16$. Give justification for **each** answer.
   (a) How many zeros does f have?

(b) How many real zeros does f have?

(c) List all the possible rational zeros of f.

(d) Find all the zeros of f.

(e) Classify each zero as rational, irrational, or nonreal complex.

(f) Write f as a product of linear factors.

Find the zeros of each polynomial. Classify each zero as rational, irrational, or nonreal complex, and express f as a product of linear factors. Also give the multiplicity of each zero.

9. $x^4 + x^3 - 2x^2 + 4x - 24$

10. $x^6 - 4x^5 - x^4 + 20x^3 - 20x^2$

11. $x^7 + 2x^6 - 3x^5 - 6x^4 + 3x^3 + 6x^2 - x - 2$

12. $x^5 + 5x^4 + 11x^3 + 16x^2 + 18x + 12$

13. $2x^4 + 3x^3 - 21x^2 - 33x - 11$

14. $12x^4 - 46x^3 + 12x^2 + 20x - 8$

Chapter 10

Mathematical Induction

In this unit, we expound upon our earlier discussion of the Principle of Mathematical Induction. We provide a wide variety of applications of this principle. We utilize induction to prove both equalities and inequalities and investigate some limitations of inductive reasoning without proof. We also prove theorems concerning finite sums of arithmetic and geometric sequences as well as the Binomial Theorem.

Objectives

- To expound on the Principle of Mathematical Induction
- To introduce the Binomial Theorem

Terms

- Principle of Mathematical Induction
  - base case
  - inductive hypothesis
  - inductive step
- Binomial Theorem

10.1 Mathematical Induction

We begin this section with a review of the Principle of Mathematical Induction and proceed with examples using this principle.

Inductive reasoning is very natural to most people. We expect that certain things will happen "because they always have before." For example, if you let go of a book, you expect it to fall; this is what your experience tells you. You expect that if you turn your light switch, your light will come on; you expect that if a car is out of gas, it will not run. All of these are based on some sort of generalization of experiences and observations we have had, but, as much as we expect these things to happen, *none of them are guaranteed to happen.*

In mathematics, we can offer such a guarantee in the form of the Principle of Mathematical Induction.

Definition 10.1.1. Principle of Mathematical Induction: Let $S_1, S_2, S_3, \ldots$ represent statements, one statement for each natural number. Suppose that S_1 is known to be true, and that if S_k is known to be true for some $k \in \mathbb{N}$, S_{k+1} is also true. Then given any natural number n, S_n is true.

Recall that S_1 is called the **Base Case** and S_k implies S_{k+1} is called the **Induction Step**. We illustrate the principle by deriving some simple formulas.

Example 10.1.1. Prove that $1+2+3+\ldots+n = \dfrac{n(n+1)}{2}$ for all $n \in \mathbb{N}$.

Solution: Let S_n represent the statement $1+2+3+\ldots+n = \dfrac{n(n+1)}{2}$. Then S_1 is the statement $1 = \dfrac{1(1+1)}{2}$, which is true, establishing the base case.

Now assume that S_k is true for some $k \in \mathbb{N}$. Then $1+2+3+\ldots+k = \dfrac{k(k+1)}{2}$; this is called the **induction hypothesis**. Thus

$$\begin{aligned}
1+2+3+\ldots+k &= \frac{k(k+1)}{2} && \text{Induction Hypothesis} \\
1+2+3+\ldots+k+(k+1) &= \frac{k(k+1)}{2}+(k+1) && \text{Equals may be added to equals} \\
&= \frac{k^2+k}{2}+\frac{2k+2}{2} && \text{Fraction equivalence} \\
&= \frac{k^2+3k+2}{2} && \text{Fraction Addition I} \\
&= \frac{(k+1)(k+2)}{2} && \text{Factorization}
\end{aligned}$$

That is, $1+2+3+\ldots+(k+1) = \dfrac{(k+1)(k+2)}{2}$. But this is exactly the statement S_{k+1}; that is, $1+2+3+\ldots+(k+1) = \dfrac{(k+1)((k+1)+1)}{2}$. Therefore S_{k+1} is true.

We conclude (by the Principle of Mathematical Induction) that S_n is true for all $n \in \mathbb{N}$; that is, $1+2+3+\ldots+n = \dfrac{n(n+1)}{2}$.

□

Example 10.1.2. Determine all values of $n \in \mathbb{N}$ for which $2^n > n$.

Solution: Let's check several cases to see what's reasonable.

$$\begin{array}{ll} n=1: & 2^1 = 2 > 1 \\ n=2: & 2^2 = 4 > 2 \\ n=3: & 2^3 = 8 > 3 \\ n=4: & 2^4 = 16 > 4 \end{array}$$

It looks like $2^n > n$ for all $n \geq 1$, so let's try to prove this.

Let S_n be the statement $2^n > n$. We have already seen that S_1, S_2, S_3, and S_4 are all true, so we have more than taken care of the base case. Assume now that S_k is true for some $k \geq 1$, so that

$$\begin{array}{rcll} 2^k & > & k & \text{Induction Hypothesis} \\ 2(2^k) & > & 2k & \text{Multiply by 2} \\ 2^{k+1} & > & 2k & \text{Law of Exponents} \\ \vdots & & \vdots & \quad\vdots \\ \vdots & & \vdots & \quad\vdots \\ 2^{k+1} & > & k+1 & \text{Transitivity of Inequality} \end{array}$$

Obviously, we need to fill in the steps missing above. Note that $2k = k + k > k + 1$, since $k > 1$. By transitivity of inequaltiy, $2k > k + 1$. Combining two inequalities produces the compound inequality $2^{k+1} > 2k > k + 1$. Again by transitivity of inequaltiy, $2^{k+1} > k + 1$.

This shows that S_{k+1} is also a true statement. Therefore, by the principle of mathematical induction, $2^n > n$ for all $n \geq 1$.

□

Example 10.1.3. Determine all values of $n \in \mathbb{N}$ for which $2^n > n^2$.

Solution: Let's check several cases to see what's reasonable.

$$\begin{array}{ll} n=1: & 2^1 = 2 > 1^2 \\ n=2: & 2^2 \not> 2^2 \\ n=3: & 2^3 \not> 3^2 \\ n=4: & 2^4 \not> 4^2 \\ n=5: & 2^5 > 5^2 \\ n=6: & 2^6 > 6^2 \end{array}$$

It looks like for $n \geq 5$, $2^n > n^2$. This is what we will prove.

This induction proof is a little different; we start with $n = 5$ instead of $n = 1$. The principle of induction we gave above does not account for this; however, that version can be easily modified to start at any integer. We simply move the base case up to $n = 5$ here.

Accordingly, let S_n be the statement $2^n > n^2$. Then S_5 is the statement $2^5 > 5^2$, which is true. This establishes the base case. Now assume that S_k is true for some $k \geq 5$; this means that $2^k > k^2$.

Now we have to complete the following outlined argument:

$$\begin{array}{rcll} 2^k & > & k^2 & \text{Induction Hypothesis} \\ 2(2^k) & > & 2k^2 & \text{Multiply by 2} \\ 2^{k+1} & > & 2k^2 & \text{Law of Exponents} \\ \vdots & & \vdots & \vdots \\ \vdots & & \vdots & \vdots \\ 2^{k+1} & > & (k+1)^2 & \underline{\hspace{4cm}} \end{array}$$

We need to fill in the steps missing above. Since $2^{k+1} > 2k^2$ and our goal is to show that $2^{k+1} > (k+1)^2$ for $k \geq 5$, it is enough to show that $2k^2 \geq (k+1)^2$ for $k \geq 5$. In other words, it is enough to show that $2k^2 - (k+1)^2 > 0$. Simplification yields $2k^2 - (k+1)^2 = 2k^2 - k^2 - 2k - 1 = k^2 - 2k - 1$. We will now show that $k^2 - 2k - 1 \geq 0$ for $k \geq 5$.

The function $f(k) = k^2 - 2k - 1$ is an upward-opening parabola with vertex occurring at $k = \frac{-(-2)}{2(1)} = 1$. Therefore its graph rises to the right of 1. Now $f(5) = 14 > 0$, and, since the parabola is rising to the right, $f(k) > 0$ for any $k \geq 5$. This means that $2k^2 - (k+1)^2 > 0$ for $k \geq 5$, so $2k^2 > (k+1)^2$.

Combining $2^{k+1} > 2k^2$ with $2k^2 > (k+1)^2$, we conclude that $2^{k+1} > (k+1)^2$. Therefore S_{k+1} is true, and the principle of mathematical induction implies that $2^n > n^2$ for $n \geq 5$.

□

Example 10.1.4. Determine all values of $n \in \mathbb{N}$ for which $2^n > n!$.

Solution: Checking several cases we get that

$$\begin{array}{ll} n = 1: & 2^1 > 1! \\ n = 2: & 2^2 > 2! \\ n = 3: & 2^3 > 3! \\ n = 4: & 2^4 \not> 4! \\ n = 5: & 2^5 \not> 5! \\ n = 6: & 2^6 \not> 6! \end{array}$$

It looks like we actually have $2^n < n!$ most of the time! This is what we will prove.

We take as the base case $n = 4$, which gives $2^4 < 4!$, a true statement. Assume that for some $k \geq 4$, we have $2^k < k!$.

$$\begin{array}{rcll} 2^k & < & k! & \text{Induction Hypothesis} \\ (2^k) & < & (k+1)k! & \text{Multiply by } 0 < 2 < k+1 \\ 2^{k+1} & > & (k+1)! & \text{Law of Exponents, Defn. of Factorial} \end{array}$$

Therefore, $2^n < n!$ for $n \geq 4$.

□

Example 10.1.5. Show that $1 + 3 + 5 + 7 + ... + (2n - 1) = n^2$ for $n \in \mathbb{N}$.

Solution: The given statement is true for $n = 1$ since $1 = 1^2$. Assume that

$$1 + 3 + 5 + \ldots + 2k - 1 = k^2$$

for some $k \geq 1$. We will add $2(k+1)-1 = 2k+1$ to both sides of this equation. We have

$$\begin{aligned} 1+3+5+\ldots+(2k-1)+(2k+1) &= k^2+(2k+1) \\ &= (k+1)^2, \end{aligned}$$

which is exactly what we wanted to prove.

Therefore, $1+3+5+\ldots+(2n-1) = n^2$ for all $n \in \mathbb{N}$.

□

Notice that in none of the examples did we stop after examining just a few cases. There are many examples in which what you are trying to prove is true for several cases, but is not always true. To know for sure, you must go through a complete proof; checking examples **does not** constitute a proof.

Example 10.1.6. Determine whether $n^2 - n + 41$ is prime for all $n \geq 1$.

Solution: We begin by checking several values.

n	$n^2 - n + 41$	Prime?
1	41	Yes
2	43	Yes
3	47	Yes
4	53	Yes
5	61	Yes
6	71	Yes
7	83	Yes
8	97	Yes
9	113	Yes
10	131	Yes
20	421	Yes
30	911	Yes

It looks like $n^2 - n + 41$ will always yield a prime number. However, $41^2 - 41 + 41 = 41^2$ is not prime, and 41 is the first value of n for which $n^2 - n + 41$ is not prime! Lots of examples do not a proof make.

□

We now return to proving some theorems concerning sequences.

Theorem 105. *The nth term for an arithmetic sequence with first term a and difference d is* $a_n = a + (n-1)d$.

Proof. The proof is by induction on n. Let S_n represent the statement $a_n = a + (n-1)d$. Then S_1 is the statement $a_1 = a + (1-1)d = a$, which is true since we know that the first term is a. This establishes the base case.

Now assume that S_k is true for some $k \in \mathbb{N}$; this means that $a_k = a + (k-1)d$. We must show that S_{k+1} is true; that is, our goal is to show that the statement $a_{k+1} = a + ((k+1)-1)d$ is true also. (Notice that the right-hand side simplifies to $a + kd$.)

Since $a_k = a + (k-1)d$ and $a_{k+1} = a_k + d$ (by the definition of an arithmetic sequence), we have $a_{k+1} = a + (k-1)d + d = a + kd - d + d = a + kd$; thus $a_{k+1} = a + kd$, as desired. Therefore S_{k+1} is a true statement.

By the principle of mathematical induction, S_n is a true statement for all $n \in \mathbb{N}$; that is, $a_n = a + (n-1)d$. □

Sometimes we want to know more than just the terms in the sequence; we may also want to know what their sum is.

Theorem 106. *Given a general arithmetic sequence $a_1, \ldots, a_n$, the sum of the first n terms is $\dfrac{n(a_1 + a_n)}{2}$.*

Proof. The proof is by induction on n. If $n = 1$, then the "sum" is just a_1, and the formula gives $\dfrac{(1)(a_1 + a_1)}{2} = a_1$. Since these agree, we have established the base case.

Assume that the formula holds for some k, so that $a_1 + a_2 + \ldots + a_k = \dfrac{k(a_1 + a_k)}{2}$. Let d be the common difference for this arithmetic sequence. We have that

- $a_k = a + (k-1)d$,
- $a_{k+1} = a + kd$, and
- $a_1 = a$.

We want to show that $a_1 + a_2 + \ldots + a_{k+1} = \dfrac{(k+1)(a_1 + a_{k+1})}{2}$, which can also be written

$$\frac{(k+1)(a+a+kd)}{2} = \frac{(k+1)(2a+kd)}{2}.$$

Using substitution and arithmetic, we get

$$\begin{aligned}
a_1 + a_2 + \ldots + a_k + a_{k+1} &= \frac{k(a_1 + a_k)}{2} + a_{k+1} \\
&= \frac{k[a + (a + (k-1)d)]}{2} + a + kd \\
&= \frac{k(2a + kd - d)}{2} + \frac{2a + 2kd}{2} \\
&= \frac{2ka + k^2 d - kd + 2a + 2kd}{2} \\
&= \frac{2ka + k^2 d + 2a + kd}{2} \\
&= \frac{2a(k+1) + kd(k+1)}{2} \\
&= \frac{(k+1)(2a + kd)}{2},
\end{aligned}$$

as desired. Thus for any $n \in \mathbb{N}$, $a_1 + a_1 + \ldots + a_n = \dfrac{n(a_1 + a_n)}{2}$. □

Example 10.1.7. Find the sum of the following arithmetic sequence: 1,4,7, 10, ..., 109.

Solution: Notice that we need to know how many terms there are. We know that $109 = 1+3(n-1)$; solving for n we find $n = 37$. Now we can find the sum of the first 37 terms:

$$1+4+7+\ldots+109 = \frac{37(1+109)}{2} = 2035.$$

□

Theorem 107. *If $a_1, a_2, a_3, \ldots$ is a geometric sequence with common ratio r and initial term $a = a_1$, then $a_n = ar^{n-1}$.*

Proof. We see that $a_1 = ar^{1-1} = a$. Assume now that $a_k = ar^{k-1}$ for some k. Then $a_{k+1} = ra_k = r(ar^{k-1}) = ar^k$. This is exactly what we were hoping to show; therefore, $a_n = ar^{n-1}$ for all $n \in \mathbb{N}$. □

We may also want to add the terms in a geometric sequence.

Theorem 108. *If $n \in \mathbb{N}$ and $r \neq 1$, then $a+ar+ar^2+ar^3+\ldots+ar^{n-1} = a\cdot\left(\frac{1-r^n}{1-r}\right)$.*

Proof. For $n = 1$, we have $a = a\cdot\left(\frac{1-r^1}{1-r}\right)$, which is a true statement since $r \neq 1$. Now assume that $a+ar+\ldots+ar^{k-1} = a\cdot\left(\frac{1-r^k}{1-r}\right)$ for some $k \geq 1$. Then

$$\begin{aligned}
a+ar+\ldots+ar^{k-1}+ar^k &= a\cdot\left(\frac{1-r^k}{1-r}\right)+ar^k \\
&= a\cdot\left(\frac{1-r^k}{1-r}\right)+a\cdot r^k\left(\frac{1-r}{1-r}\right) \\
&= a\cdot\left(\frac{1-r^k}{1-r}+\frac{r^k-r^{k+1}}{1-r}\right) \\
&= a\cdot\left(\frac{1-r^{k+1}}{1-r}\right),
\end{aligned}$$

as desired. Thus $a+ar+ar^2+\ldots+ar^{n-1} = a\cdot\left(\frac{1-r^n}{1-r}\right)$ for all $n \geq 1$. □

Example 10.1.8. Find the sum of the first n terms and the first 8 terms in the sequence $\{1/2^n\}$.

Solution: The first term is $\frac{1}{2}$ and the ratio is $\frac{1}{2}$, so the sum of the first n terms is

$$\frac{1}{2}+\left(\frac{1}{2}\right)^2+\ldots+\left(\frac{1}{2}\right)^{n-1} = \frac{1}{2}\cdot\frac{1-\left(\frac{1}{2}\right)^n}{1-\left(\frac{1}{2}\right)} = \frac{1}{2}\cdot\left(\frac{\left(\frac{2^n-1}{2^n}\right)}{\left(\frac{1}{2}\right)}\right) = \frac{2^n-1}{2^n}.$$

Thus the sum of the first 8 terms is $\frac{2^8-1}{2^8} = \frac{255}{256}$.

□

A rather important theorem whose proof requires the Principle of Mathematical Induction is known as the Binomial Theorem. This theorem gives a formula for expanding powers of binomials; *i.e.*, $(a+b)^n$ and is the topic of the next section.

Section 10.1 Exercises

1. Prove that for all $n \in \mathbb{N}$, $4n+6 < 3n^2+3n+5$.

2. Prove that for all $n \in \mathbb{N}$, $2n^2+1 < n^3+4$.

3. Prove that for all $n \in \mathbb{N}$, $3n^2+n < 2n^3+3$.

4. Prove that for all $n \in \mathbb{N}$, $1+4+7+\cdot+(3n-2) = \dfrac{n(3n-1)}{2}$.

5. Prove that for all $n \in \mathbb{N}$, $1+5+9+\cdot+(4n-3) = 2n^2-n$.

Notation: If $a_1, a_2, a_3, \ldots, a_n$ is a finite list of numbers (field elements), then **sigma notation** is defined as follows:

$$\sum_{i=1}^{n} a_i = a_1 + a_2 + \cdots + a_n.$$

The variable i is called the **index of summation** and runs through the values indicated, beginning at the value below the sigma and ending at the value above the sigma.

6. Prove that for all $n \in \mathbb{N}$, $\sum_{i=1}^{n} i^2 = \dfrac{n(n+1)(2n+1)}{6}$.

7. Prove that for all $n \in \mathbb{N}$, $\sum_{i=1}^{n} i^3 = \dfrac{n^2(n+1)^2}{4}$.

8. Prove that for all $n \in \mathbb{N}$, $\sum_{i=1}^{n} 2i^2+i = \dfrac{n(n+1)(4n+5)}{6}$.

9. Prove that for all $n \in \mathbb{N}$, $\sum_{i=1}^{n} 2i^3-i = \dfrac{n(n+1)(n^2+n-1)}{2}$.

10. Prove that for all $n \in \mathbb{N}$, $\sum_{i=1}^{n} 4i^3+6i = n(n+1)(n^2+n+3)$.

11. Prove that for all $n \in \mathbb{N}$, $\sum_{i=1}^{n} 2i^2+i = \dfrac{n(n+1)(4n+5)}{6}$.

12. Prove that for all $n \in \mathbb{N}$, $\sum_{i=1}^{n} i^3+3i = \dfrac{n(n+1)(n^2+5n+2)}{4}$.

13. Prove that the following properties hold:

    (a) $c\sum_{i=1}^{n} a_i = \sum_{i=1}^{n} ca_i$

(b) $\sum_{i=1}^{n}(a_i+b_i)=\sum_{i=1}^{n}a_i+\sum_{i=1}^{n}b_i$

(c) $\sum_{i=1}^{n}c=nc$

14. Define $\bar{x}=\frac{1}{n}\sum_{i=1}^{n}x_i$. Show that $\sum_{i=1}^{n}(x_i-\bar{x})^2=\sum_{i=1}^{n}x_i^2-n\bar{x}^2$ using the above problem.

15. (a) Show that if $2+2^2+\ldots+2^k=2^{k+1}$, then $2+2^2+\ldots+2^{k+1}=2^{k+2}$.

    (b) Does this prove that $2+2^2+\ldots+2^n=2^{n+1}$ for all $n\in\mathbb{N}$? Why or why not?

16. A **triomino** is an L-shaped figure made up of three congruent squares (each one unit on a side). Show that if a chessboard is $2^n\times 2^n$ for some $n\in\mathbb{N}$, then the chessboard can be tiled by triominos if one corner is removed. ("Tiling" means that you may not cut or overlap triominos, and none may hang over the edge.)

10.2 The Binomial Theorem

The Binomial Theorem is an important result which gives gives a formula for expanding powers of binomials. Since the proof of this theorem requires mathematical induction, the theorem is typically deferred until after induction has been covered in detail. However, the Binomial Theorem could be stated early in a College Algebra course (without proof). The statement does, however, require some preliminary definitions and notation.

Definition 10.2.1. Let $n, k \in \mathbb{N}$ with $k \leq n$. The **binomial coefficient**

$$\binom{n}{k} = \frac{n!}{k!(n-k)!}$$

(pronounced "n choose k") is the number of different ways to choose k objects out of a pool of n distinct objects, where the order is irrelevant.

Example 10.2.1. $\binom{7}{3} = \dfrac{7!}{3!(7-3)!} = 35.$

□

Example 10.2.2. $\binom{n}{0} = \dfrac{n!}{0!n!} = \dfrac{n!}{n!} = 1$ regardless of the value of $n \in \mathbb{N}$.

□

Example 10.2.3. How many ways are there to select a committee of 5 people from a class of 38 people?

Solution: There are $\binom{38}{5} = \dfrac{38!}{5!(38-5)!} = 501942$ ways. (Most calculators will handle these calculations.)

□

We will not prove that binomial coefficients count what we claim they do; that is a topic for another course.

One fascinating fact is that the binomial coefficients are the entries in Pascal's triangle! Pascal's triangle begins:

$$\begin{array}{lccccccccccccc}
\text{Row 0} & & & & & & & 1 & & & & & & \\
\text{Row 1} & & & & & & 1 & & 1 & & & & & \\
\text{Row 2} & & & & & 1 & & 2 & & 1 & & & & \\
\text{Row 3} & & & & 1 & & 3 & & 3 & & 1 & & & \\
\text{Row 4} & & & 1 & & 4 & & 6 & & \boxed{4} & & 1 & & \\
\text{Row 5} & & 1 & & 5 & & 10 & & 10 & & 5 & & 1 & \\
\text{Row 6} & 1 & & 6 & & 15 & & 20 & & 15 & & 6 & & 1
\end{array}$$

The top row is row 0, and the leftmost diagonal (all 1's) is diagonal 0. The binomial coefficient $\binom{n}{k}$ is the entry in row n and diagonal k in Pascal's triangle. For example, $\binom{4}{3}$ is the entry in row 4 (start counting at the top with row 0) and diagonal 3; it is boxed in the triangle above.

Examining Pascal's triangle gives us the following theorem, that each entry is the sum of the two above it, which we will not prove.

Theorem 109. *For $n, k \in \mathbb{N}$ with $k \leq n$,* $\binom{n}{k} + \binom{n}{k+1} = \binom{n+1}{k+1}$.

Example 10.2.4. Compute $\binom{8}{3} + \binom{8}{4}$ and $\binom{9}{4}$.

Solution. We have that

$$\begin{aligned} \binom{9}{4} &= \frac{9!}{4!(9-4)!} \\ &= \frac{9!}{4!5!} \\ &= \frac{9 \cdot 8 \cdot 7 \cdot 6}{4!} \\ &= \frac{3 \cdot 3 \cdot 8 \cdot 7 \cdot 6}{24} \\ &= \frac{3 \cdot 7 \cdot 6}{1} \\ &= 126 \end{aligned}$$

and also

$$\begin{aligned} \binom{8}{3} + \binom{8}{4} &= \frac{8!}{3!(8-3)!} + \frac{8!}{4!(8-4)!} \\ &= \frac{8!}{3!5!} + \frac{8!}{4!4!} \\ &= \frac{8 \cdot 7 \cdot 6 \cdot 5!}{3!5!} + \frac{8 \cdot 7 \cdot 6 \cdot 5 \cdot 4!}{4!4!} \\ &= \frac{8 \cdot 7 \cdot 6}{3!} + \frac{8 \cdot 7 \cdot 6 \cdot 5}{4!} \\ &= \frac{56}{1} + \frac{70}{1} \\ &= 125 \end{aligned}$$

□

It is also apparent from Pascal's triangle that $\binom{n}{k} = \binom{n}{n-k}$ since the triangle is symmetric. This result is also easily obtained from the definition of the binomial coefficients. For example, $\binom{4}{1} = \frac{4!}{1!(4-1)!} = \frac{4!}{(4-1)!1!} = \binom{4}{4-1}$.

We have been referring to $\binom{n}{k}$ as a binomial "coefficient;" this terminology is made clear in the following extremely important theorem.

Theorem 110 (Binomial Theorem). *Let a and b be elements of a field and $n \in \mathbb{N}$. Then*

$$(a+b)^n = \binom{n}{0}a^n b^0 + \binom{n}{1}a^{n-1}b^1 + \binom{n}{2}a^{n-2}b^2 + \binom{n}{3}a^{n-2}b^3 + \ldots + \binom{n}{k}a^{n-k}b^k + \ldots + \binom{n}{n-2}a^2 b^{n-2} + \binom{n}{n-1}a^1 b^{n-1} + \binom{n}{n}a^0 b^n.$$

The coefficient of $a^k b^{n-k}$ is the binomial coefficient $\binom{n}{k}$. Notice the pattern in the terms: the exponents on a and b always add up to n, and the exponent on b always matches the number in the bottom of $\binom{n}{k}$. Also, reading from left to right, the exponents on a descend while the exponents on b ascend.

Example 10.2.5. Determine the coefficient of x^7 in $(3x-2)^{12}$.

Solution: Let $a = 3x$ and $b = -2$, so that we have $(a+b)^{12}$. We need to have a^7 in order to get x^7, so we are looking at the term $\binom{12}{7}a^7 b^5 = \binom{12}{7}3^7(-2)^5 x^7$. Therefore the coefficient of x^7 is $\binom{12}{7}3^7(-2)^5$.

□

Corollary 10.2.1. *For a field element x and $n \in \mathbb{N}$,*

$$(x+1)^n = x^n + nx^{n-1} + \binom{n}{2}x^{n-2} + \ldots + \binom{n}{n-2}x^2 + nx + 1.$$

Proof. Apply the binomial theorem with $a = x$ and $b = 1$. □

Now we finally see the relationship between the entries in Pascal's triangle and the coefficients in $(x+1)^n$. Notice that it is far easier to remember "1-3-3-1" and know that $(x+1)^3 = x^3 + 3x^2 + 3x + 1$ than it is to compute $(x+1)(x+1)(x+1)$.

Example 10.2.6. Compute $(x+1)^6$.

Solution: We can simply read the coefficients out from Pascal's triangle:

$$(x+1)^6 = x^6 + 6x^5 + 15x^4 + 20x^3 + 15x^2 + 6x + 1.$$

□

Example 10.2.7. Compute $(2x+1)^4$.

Solution: $(2x+1)^4 = (2x)^4 + 4(2x)^3 + 6(2x)^2 + 4(2x) + 1 = 16x^4 + 32x^3 + 24x^2 + 8x + 1$.

□

Section 10.2 Exercises

Expand the given expression.

1. $(x+y)^4$
2. $(x+4y)^3$
3. $(2x+y)^5$
4. $(3x-2y)^3$
5. $(7x-5y)^3$
6. $\left(2x-\frac{1}{2}y\right)^3$
7. $(ix+y)^2$
8. $(ix+2y)^4$
9. $\left(2ix-\frac{3}{2}y\right)^5$
10. $(\sqrt{x}+\sqrt{y})^3$
11. $(-4+i)^3$
12. $(7+4i)^5$
13. $(2+3i)^4$
14. $(x^2+y^2)^2$
15. $(x^2+y^2)^3$
16. $((2-i)x+3y)^3$
17. $((-2+3i)x^3+(1+2i)y^2)^3$

Find the requested term by making use of the Binomial Theorem.

18. Find the fourth term in the expansion of $(2x+y)^6$.
19. Find the third term in the expansion of $(x-3y)^7$.
20. Find the fifth term in the expansion of $(4x+y)^6$.
21. Find the fourth term in the expansion of $(2x+3y)^8$.
22. Find the sixth term in the expansion of $(\sqrt{x}+\sqrt{y})^8$.
23. Find the fifth term in the expansion of $(\sqrt{x}+2\sqrt{y})^6$.
24. Find the third term in the expansion of $(4\sqrt{x}-3\sqrt{y})^5$.
25. Find the third term in the expansion of $(x^{2/3}+y)^6$.
26. Find the fourth term in the expansion of $(2x^{2/3}+y^{3/5})^7$.
27. Find the third term in the expansion of $(ix+y)^4$.
28. Find the fifth term in the expansion of $(ix^{2/3}+y)^8$.

Find the coefficient of the requested term using the Binomial Theorem.

29. Find the coefficient of x^2y^4 in the expansion of $(3x-y)^6$.
30. Find the coefficient of x^3y^2 in the expansion of $(x+\sqrt{3}y)^5$.
31. Find the coefficient of x^2y in the expansion of $(ix+y)^7$.

32. Find the coefficient of $\frac{y^5}{x^3}$ in the expansion of $\left(\frac{3}{x}+2y\right)^8$.

33. Find the coefficient of $\frac{x^3}{y^3}$ in the expansion of $\left(\frac{2y}{x}+\frac{x}{y}\right)^8$.

34. Prove that $\binom{n}{k}+\binom{n}{k+1}=\binom{n+1}{k+1}$

35. Prove that $\binom{n}{k}=\binom{n}{n-k}$

Named Theorems

Solutions to Selected Exercises

Logic (statements), page 3

1. Statement; true.

3. Statement; false.

5. Statement; true.

7. Hypothesis: $x = 3$. Conclusion: $x^2 - 1 = 8$.

9. Hypothesis: I stay up too late tonight. Conclusion: I won't feel well tomorrow.

11. First rewrite: If the candidate is able to pass the driver's test, then the candidate is able to parallel park. Hypothesis: the candidate is able to pass the driver's test. Conclusion: the candidate is able to parallel park. Other variants of the "if-then" form are possible, but the ability to parallel park should be in the conclusion.

13. This is a conditional equation; it is true for $x = 2$ and false for $x = 1$.

15. This is an identity; it is an example of the distributive law.

17. This is a contradiction; it is never true.

19. The conjunction (**and**) is false since plastic is not made from iron. The disjunction (**or**) is true since oranges do contain Vitamin C.

21. Converse: if $x^2 = 4$, then $x = 2$. This is false; $x = -2$ is an equally valid conclusion. Contrapositive: if $x^2 \neq 4$, then $x \neq 2$. This is true; the contrapositive is logic01ally equivalent to the original conditional statement.

23. Converse: if I pass this course, then I studied appropriately. We cannot determine the truth value. Contrapositive: If I do not pass this course, then I did not study appropriately. This is true (given that the original conditional statement was true).

25. Hypothesis: I stay up too late tonight. Conclusion: I won't feel well tomorrow.

27. I won't feel well tomorrow.

29. Hypothesis: $x = 2$. Conclusion: $(x-2)(x+3)(x-1) = 0$.

31. Nothing.

Logic (arguments), page 7

1. This is valid.

3. This is not valid; the hypotheses don't tell us about what happens to non-dogs.

5. Assume that x and y are both even integers. Then $x = 2a$ and $y = 2b$ for some integers a and b. Therefore, $x + y = 2a + 2b = 2(a + b)$ using the distributive law. That is, $x + y$ equals 2 times another integer, so $x + y$ is even.

7. Tameka likes to read.

9. This argument is not valid. We don't know what Company X cars are being used for, how often they've been repaired, how many are still in use, or what percentage are still in use. It is also possible that Company Y has only been around for 14 years, but ALL of their cars are still functioning perfectly. We just don't have enough information to draw any conclusions.

11. This argument is invalid even though the conclusion is true. If the argument were valid, then the following (absurd) argument would be, too: all dogs pant; all lions pant; therefore, all dogs are lions.

13. This argument is not valid. All we are told is that rainy days are cloudy; we are told nothing about non-rainy days. Therefore, on non-rainy days we can't conclude anything.

15. $x = -2$ is part of the collection since $(-2)^2 = 4 > 3$.

Sets, page 15

1. This is a set; an object either is a cat, or it isn't.

3. This is not a set; "fast" is subjective.

5. This is not a set; "rich" is subjective.

7. {Ronald Reagan, Jimmy Carter, Bill Clinton, Gerald Ford, George Bush} The universal set could be the set of all people in the world.

9. $\{2, 4, 6, 8, 10, 12, 14, 16, 18, 20\}$. The universal set could be the set of all integers.

11. True, since 2 is explicitly listed as belonging to A.

13. False, $4 \in A$.

15. $\{2n | n \text{ is a natural number}\}$

17. $\{2n | n \text{ is an integer and } 2n < 15\}$.

19. A and C both contain exactly one element. B and D both contain exactly two elements.

21. $B = D$ since both contain exactly the elements 2 and -2.

23. 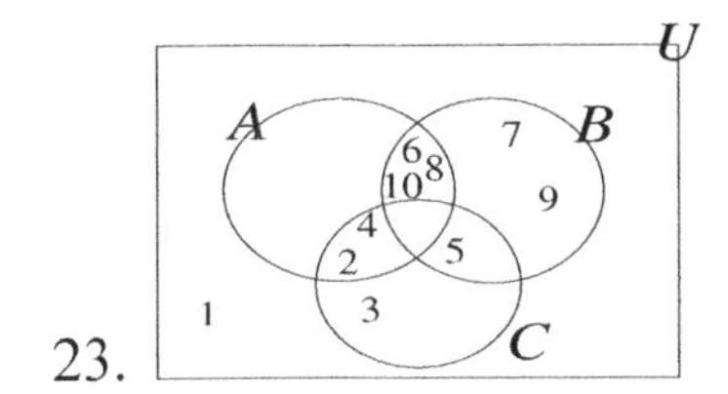

25. Since only 20 need new tires, at most 20 can need new tires *and* something else.

27. Since 30 need gear repairs, there are at most 15 that do not. If the 20 needing new tires all need gear repairs as well, then there are exactly 15 needing neither, and that is the most possible.

29. Let U be the universal set of all things, P be the set of all things made by people, M the set of all machines, and C the set of all computers. The given hypotheses lead to the Venn diagram shown, and we are forced to the conclusion given.

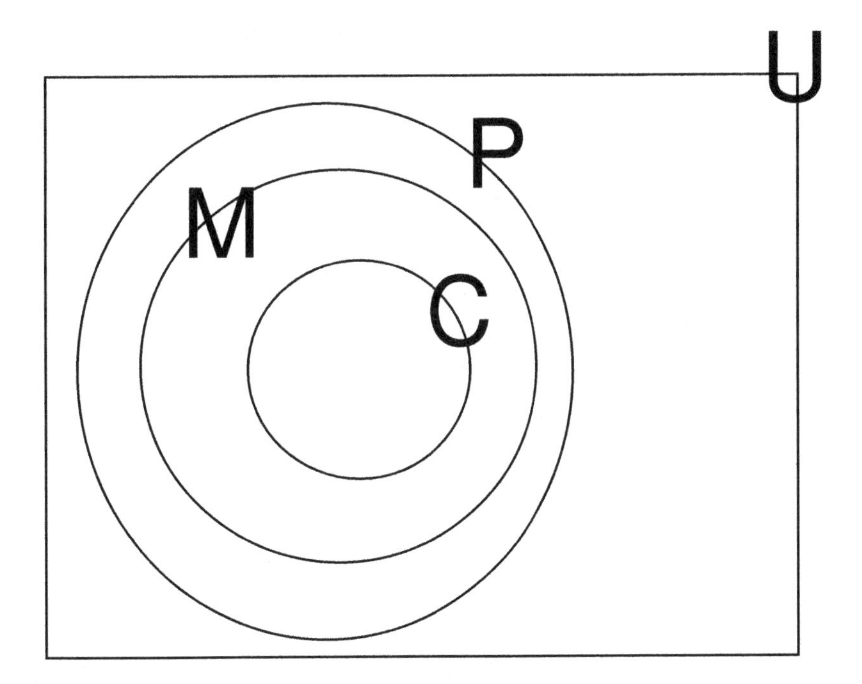

31. (a) See below.

(b) $V = \{a, b, c, d, e, i, o, u\}$. V is made up of everything in the two circles combined.

(c) $R = \{a, e\}$. R is made up of the region shared by the two circles.

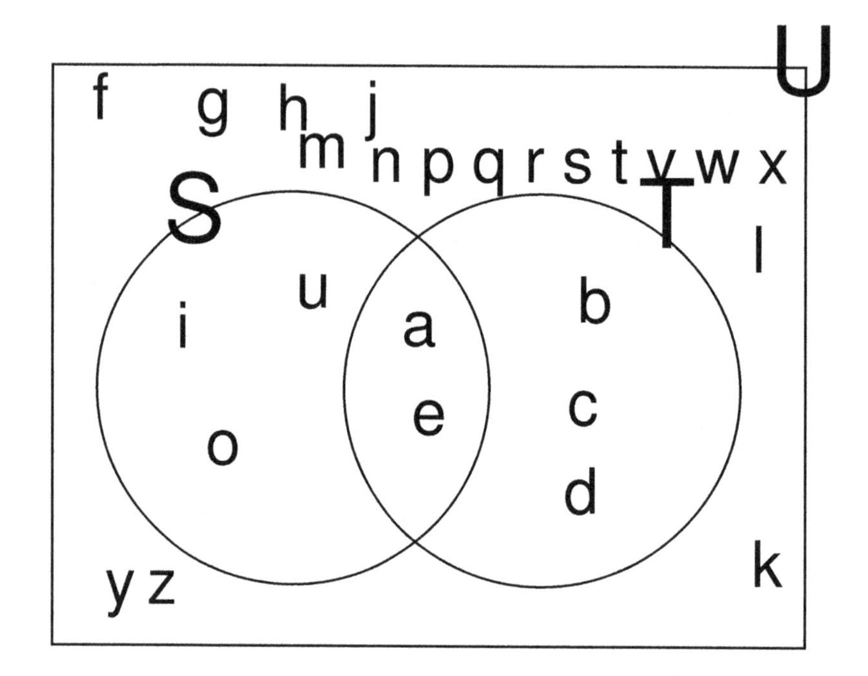

33. Recall that $\mathbb{Z}$ stands for the set of all integers, and $\mathbb{Q}$ stands for the set of all fractions. Since any integer n can be written as $\frac{n}{1}$, each integer is also a fraction.

35. True.

37. True.

39. False.

41. True. (Assuming no deformed horses.)

42. If $x \in A$, then $x \in A$. Since this is the definition of subset, $A \subseteq A$.

43. The sets in Exercises 39 are such an example.

44a. If $x \in C$, then $x \in A$ and $x \in B$, so, in particular, $x \in A$. Therefore $C \subseteq A$. Similarly, $C \subseteq B$.

45. If $x \in \{x|x^2 = 1\}$, then $x = \pm 1$, so $x \in \{1, -1\}$. Conversely, if $x \in \{1, -1\}$, then $x^2 = 1$, so $x \in \{x|x^2 = 1\}$.

46. Let A be any set. Every element of the empty set (of which there are none) is a member of A. Thus, $\emptyset \subseteq A$.

Sets, page 21

1. $A \cap B = \{-2, 2\}$

3. $B \cup C = \{-10, -8, -6, -4, -2, -1, 0, 1, 2, 4, 6, 8, 10\}$

5. $A' = \{-1, 0, 1\}$

7. $(A \cup B) \cap C = C$. (Notice that $A \cup B = U$, and $U \cap X = X$ for any $X \subseteq U$.)

9. Suppose that $x \in A \cap B$. That means that x is a member of both A and B, so it is certainly a member of A. Therefore, $A \cap B \subseteq A$. Similarly, $A \cap B \subseteq B$.

11. If $x \in A \cap B$, then x is in both A and B. Therefore, x is in both B and A, so $A \cap B \subseteq B \cap A$. On the other hand, if $x \in B \cap A$, then x is in both B and A, so x is in both A and B. Thus, $B \cap A \subseteq A \cap B$. Since each set is contained in the other, we have $A \cap B = B \cap A$.

13. Let $A = \{1,2\}$ and let $B = \{a\}$. Then $A \times B = \{(1,a),(2,a)\}$ and $B \times A = \{(a,1),(b,1)\}$.

15. Let $x \in (A \cup B)'$. Then $x \notin (A \cup B)$. That is, x is not in $A \cup B$, so $x \notin A$ and $x \notin B$. Therefore, $x \in A'$ and $x \in B'$, so $x \in A' \cap B'$.

    Now let $x \in A' \cap B'$. Then $x \in A'$ and $x \in B'$, so $x \notin A$ and $x \notin B$. Thus, x belongs to neither A nor B, so x cannot be in $A \cup B$. Therefore, $x \in (A \cup B)'$.

17. Let $A = \{H_1, H_2, H_3\}$ and let $B = \{S_1, S_2\}$. Then

$$A \times B = \{(H_1,S_1),(H_1,S_2),(H_2,S_1),(H_2,S_2),(H_3,S_1),(H_3,S_2)\},$$

    and she has 6 hat-scarf combinations.

Sets, page 24

1. (a) The cardinality of A is 4.
   (b) The cardinality of V is 5.

3. If $A = B$, then A and B have exactly the same elements. That is, $x \in A$ if and only if $x \in B$. This means that the pairings (x,x) for all $x \in A$ give an explicit one-to-one correspondence between A and B.

5. Probably the simplest one-to-one correspondence is the one defined by $(n, 2n+2)$ that matches each number with two more than its double. Thus $(0,2),(2,6),(3,8),(143,288)$, etc. are all pairs.

7. (a) $A = B$ because they contain exactly the same elements.
   (b) $A \neq B$ since $-1 \in B$ but $-1 \notin A$.
   (c) $S = T$ because they contain exactly the same elements.
   (d) $A \neq B$ since $0 \in B$ but $0 \notin A$.

9. (a) Let $A = \{1,2,3\}$ and $B = \{4,5\}$.
   (b) Let $A = \{1,2,3\}$ and $B = \{3,4,5\}$.

Natural Numbers, page 28

1. Let $A = \{1,2,3,4\}$ and let $B = \{a,b,c\}$. Notice that $A \cap B = \emptyset$, $n(A) = 4$, and $n(B) = 3$. Also, $A \cup B = \{1,2,3,4,a,b,c\}$, so $n(A \cup B) = 7$. Therefore, $4+3=7$.

3. Let $C = \{x,y\}$. Using sets A and B from above, we illustrate $2 \cdot (3+4)$ with

$$\begin{gathered} C \times (B \cup A) \\ = \{(x,a),(y,a),(x,b),(y,b),(x,c),(y,c),(x,1),(y,1),(x,2),(y,2),(x,3),(y,3), \\ (x,4),(y,4)\}. \end{gathered}$$

We can also illustrate $2 \cdot 3 + 2 \cdot 4$ with

$$\begin{gathered} (C \times B) \cup (C \times A) \\ = \{(x,a),(y,a),(x,b),(y,b),(x,c),(y,c)\} \cup \{(x,1),(y,1),(x,2),(y,2),(x,3),(y,3), \\ (x,4),(y,4)\}, \end{gathered}$$

which is the same set! Therefore, they are equivalent, so $2(3+4) = 2 \cdot 3 + 2 \cdot 4$.

5. Using sets A and C from above, we have $2 \cdot 4 = n(C \times A) = n(\{(x,1),(y,1),(x,2),(y,2),(x,3),(y,3),(x,4),(y,4)\})$. We also have $4 \cdot 2 = n(A \times C) = n(\{(1,x),(1,y),(2,x),(2,y),(3,x),(3,y),(4,x),(4,y)\})$. There is a one-to-one correspondence between these two sets; namely, the element $(a,b) \in A \times C$ corresponds to the element $(b,a) \in C \times A$.

7. (a) $A \times B = \{(1,a),(1,b),(1,c),(2,a),(2,b),(2,c),(3,a),(3,b),(3,c)\}$.
   (b) $B \times A = \{(a,1),(b,1),(c,1),(a,2),(b,2),(c,2),(a,3),(b,3),(c,3)\}$.
   (c) Let each element of $A \times B$ correspond to the element of $B \times A$ written directly below it; the rule for this correspondence is $(x,y) \leftrightarrow (y,x)$. (Just switch coordinates.) This illustrates the fact that multiplication of natural numbers is commutative.

9. Let $S = \{1,2\}$. If a is a natural number, then a is the cardinal number of some set A. Let $x \in A$. Then $\{(1,x),(2,x)\} \subseteq S \times A$. Thus, there can be no one-to-one correspondence between $\{1\}$ and $S \times A$!

11. Let $X, Y \in \mathscr{P}(A)$. Then $X, Y \subseteq A$, so $X \cup Y \subseteq A$, since if a belongs to X or Y, then x belongs to A.

Field Properties, page 31

1. 0 is the additive identity. (Additive identity)

3. Multiplication distributes over addition. (Distributive law)

5. -7 is the additive inverse of 7. (Additive inverse)

7. 1 is the multiplicative identity. (Multiplicative identity)

9. Multiplication distributes over addition. (Distributive law)

11. The additive inverse of a is $-a$. (Additive inverse)

13. Not all nonzero elements of $\mathbb{Z}$ have a multiplicative inverse in $\mathbb{Z}$.

15. The distributive law says $7 \cdot 1.52 + 3 \cdot 1.52 = (7+3) \cdot 1.52 = 10 \cdot 1.52 = 15.20$.

17.
$$\begin{aligned}(b+c)a &= a(b+c) && \text{Commutativity of Multiplication}\\ &= ab+ac && \text{Distributive Law}\\ &= ba+ca && \text{Commutativity of Multiplication.}\end{aligned}$$

Extending, page 34

1. $-(-3.14159) = 3.14159$.

3. $-(-a) = a$.

5. $-(-12) = 12$.

7. $(4x+7)^{-1} = \dfrac{1}{4x+7}$, provided $x \neq -7/4$.

9. $\left(\dfrac{13}{4}\right)^{-1} = \dfrac{4}{13}$.

11. $(3+4)^{-1} = \dfrac{1}{7}$.

13. $13 - 9 = 13 + (-9)$.

15. $x - y = x + (-y)$.

17. $2 - 3 + 6 = (2-3) + 6$.

19. $a \cdot b + c \div d - e \cdot f = [(a \cdot b) + (c \div d)] - (e \cdot f)$.

Rational Numbers, page 38

1. $\dfrac{4}{3}$

3. $\dfrac{19}{12}$

5. $\dfrac{239}{168}$

7. $\dfrac{41}{108}$

9. $-\dfrac{9}{5}$

11. $\dfrac{18}{35}$

13. $-\dfrac{8}{27}$

15. $\dfrac{18}{17}$

17. $-\dfrac{26}{21}$

19. $-\dfrac{1}{9}$

21. $\dfrac{2}{63}$

23. $\dfrac{1}{3}$

25. $\dfrac{10}{1} = 10$

27. $\dfrac{248}{999}$

29. $\dfrac{1}{5}$

31. $\dfrac{10577}{495}$

33. $\dfrac{2891}{9999}$

35. This terminates since $20 = 2^2 \cdot 5$.

37. This repeats since $120 = 2^3 \cdot 3 \cdot 5$.

39. This terminates since $64000 = 2^9 \cdot 5^3$.

41. $0.\overline{4} - 0.444 = 0.000\overline{4}$.

43. The rational number $\left(\dfrac{37}{15} + \dfrac{38}{17}\right) \cdot \dfrac{1}{2} = \dfrac{1199}{510}$ is between the two given rational numbers.

45. $5.711, 5.712, 5.713$.

47. $\dfrac{5}{8} + \dfrac{3}{7} = \dfrac{5 \cdot 7}{8 \cdot 7} + \dfrac{3 \cdot 8}{7 \cdot 8} = \dfrac{35}{56} + \dfrac{24}{56} = \dfrac{59}{56} \in \mathbb{Q}$.

49. Let $\dfrac{a}{b} \in \mathbb{Q}$ and $\dfrac{c}{d} \in \mathbb{Q}$, where $a, b, c, d \in \mathbb{Z}$ and $b \neq 0$ and $d \neq 0$. Thus, $\dfrac{a}{b} \cdot \dfrac{c}{d} = \dfrac{ac}{bd} \in \mathbb{Q}$ since $ac \in \mathbb{Z}$, $bd \in \mathbb{Z}$, and $bd \neq 0$.

51. $\frac{a}{b} \div \frac{c}{d} = \frac{ad}{bc} = \frac{a}{b} \cdot \frac{d}{c}$. Since $\mathbb{Q}$ is closed under multiplication, it is enough to show that $\frac{d}{c} \in \mathbb{Q}$. Since $\frac{c}{d} \neq 0$, $c \neq 0$, so $\frac{d}{c} \in \mathbb{Q}$.

53.

$$\begin{aligned}\frac{a}{b} \cdot \frac{c}{d} &= \frac{ac}{bd}\\ &= \frac{ca}{db}\\ &= \frac{c}{d} \cdot \frac{a}{b}.\end{aligned}$$

55.

$$\begin{aligned}\left(\frac{a}{b} \cdot \frac{c}{d}\right) \cdot \frac{e}{f} &= \frac{ac}{bd} \cdot \frac{e}{f}\\ &= \frac{(ac)e}{(bd)f}\\ &= \frac{a(ce)}{b(df)}\\ &= \frac{a}{b} \cdot \frac{ce}{df}\\ &= \frac{a}{b} \cdot \left(\frac{c}{d} \cdot \frac{e}{f}\right).\end{aligned}$$

57.

$$\begin{aligned}\frac{1}{1} \cdot \frac{a}{b} &= \frac{1 \cdot a}{1 \cdot b}\\ &= \frac{a}{b}.\end{aligned}$$

59.

$$\begin{aligned}\frac{a}{b} \cdot \frac{b}{a} &= \frac{ab}{ba}\\ &= \frac{ab}{ab}\\ &= \frac{1}{1}.\end{aligned}$$

Irrational Numbers, page 42

1. $\frac{17}{31}$ is rational.

3. $(\sqrt{2}+13)-(1+\sqrt{2}) = 12$ is rational.

5. $\sqrt{2} \cdot \sqrt{2} = 2$ is rational even though $\sqrt{2}$ is not.

7. If $\sqrt{3}$ is rational, then $\sqrt{3} = \frac{a}{b}$ for some integers a and b, with $b \neq 0$ and not both a and b divisible by 3. Therefore, $b\sqrt{3} = a$, so $(b\sqrt{3})^2 = a^2$. That is, $3b^2 = a^2$. However, if 3 is a factor of a^2, then 3 must also be a factor of a, so $a = 3c$ for some integer c.

   Substituting for a gives $3b^2 = 9c^2$, so $b^2 = 3c^2$. However, if 3 is a factor of b^2, then 3 is a factor of b, contradicting the fact that a and b do not both have a factor of 3. Therefore, our original assumption (that $\sqrt{3}$ is rational) must have been in error, so $\sqrt{3}$ is irrational.

9. The given number is irrational because its decimal expansion can never terminate or repeat – the pattern keeps inserting more zeros!

11. Suppose that $\sqrt{2}+3 \in \mathbb{Q}$. Since $\mathbb{Q}$ is closed under subtraction, $\sqrt{2}+3-3 = \sqrt{2} \in \mathbb{Q}$, a contradiction. Therefore, $\sqrt{2}+3 \notin \mathbb{Q}$.

13. The pattern is the same as in 11: Since $\mathbb{Q}$ is closed under subtraction, if $a+b \in \mathbb{Q}$, then $a+b-b = a \in \mathbb{Q}$, a contradiction. Therefore, $a+b \notin \mathbb{Q}$.

15. Your **own** words!

Field Properties, page 48

1. Subtraction is not commutative on the integers; $5-3=2$, but $3-5=-2$. Division is not commutative on the nonzero rational numbers: $\frac{2}{3} \div \frac{3}{4} = \frac{8}{9}$, but $\frac{3}{4} \div \frac{2}{3} = \frac{9}{8}$. There are many other examples.

3. Just about any two numbers a and b will work. For example, $(2+1)^2 = 3^2 = 9$, but $2^2+1^2 = 4+1 = 5$.

5. Zero is the additive identity.

7. Multiplication is commutative.

9. This is the definition of subtraction.

11. This is the definition of division.

13. This is the distributive law.

15. Show that $(x+1)(x+2) = x^2+3x+2$. (Note: x^2 means $x \cdot x$.)

$$
\begin{aligned}
(x+1)(x+2) &= [x(x+2)]+[1(x+2)] && \text{Distributive Law of Multiplication over Addition} \\
&= [x \cdot x + x \cdot 2] + [1 \cdot x + 1 \cdot 2] && \text{Distributive Law of Multiplication over Addition} \\
&= [x^2+2x]+[1x+2] && \text{Commutative Law of Multiplication; Definition of } x^2 \\
&= [(x^2+2x)+1x]+2 && \text{Associative Law of Addition} \\
&= [x^2+(2x+1x)]+2 && \text{Associative Law of Addition} \\
&= [x^2+(2+1)x]+2 && \text{Distributive Law of Multiplication over Addition} \\
&= [x^2+3x]+2 && 2+1=3 \\
&= x^2+3x+2 && \text{Order of Operations}
\end{aligned}
$$

17. The distributive law implies that $7x+5x = (7+5)x = 12x$.

19. $0 \cdot 1 = 0$ since 1 is the multiplicative identity.

21.

$$
\begin{aligned}
(a+b)+(c+d) &= [(a+b)+c]+d && \text{Associativity of Addition} \\
&= [a+(b+c)]+d && \text{Associativity of Addition} \\
&= [a+(c+b)]+d && \text{Commutativity of Addition} \\
&= [(a+c)+b]+d && \text{Associativity of Addition} \\
&= (a+c)+(b+d) && \text{Associativity of Addition}
\end{aligned}
$$

Therefore, $(a+b)+(c+d) = (a+c)+(b+d)$ by transitivity of equality. The other part is similar.

Equality Axioms, page 50

1. Equality is reflexive.

3. Substitution.

5. If $a = b$, then $b = a$ by symmetry of equality. Thus we have $b = a$ and $a = c$, so $b = c$ by transitivity of equality.

7. If $a = b$, then $a + c = b + c$ by Theorem 6. Now by substitution, since $c = d$, $a + c = b + d$.

9. **Commutativity of addition** refers to the fact that the **operation** of addition commutes: $a + b = b + a$ for all field elements a and b. **Symmetry of equality** refers to the fact that the **relation** of equality is symmetric: if $a = b$, then $b = a$.

11. By Exercise 10, we know that $ac = bc$. Substituting d for c (since they are equal) gives $ac = bd$.

Uniqueness, page 51

1. Suppose that a is another identity for multiplication (aside from 1). Then $a \cdot 1 = 1$ (since a is an identity), and $a \cdot 1 = a$ (since 1 is an identity). Therefore, using Exercise 3 from the previous section, $a = 1$. Thus, there is only one identity for multiplication.

3. Since multiplicative inverses are unique, x must be 6^{-1}.

5. Additive inverses are unique.

Multiplication with Additive Identities and Inverses, page 55

1. Since $(x+1)(x-3) = 0$, the Zero Product Theorem implies that either $x + 1 = 0$ or $x - 3 = 0$. Therefore, $x = -1$ or $x = 3$ by the uniqueness of additive inverses.

3. Since $4(x+2) = 0$, the Zero Product Theorem implies that either $4 = 0$ or $x + 2 = 0$. Since $4 \neq 0$, $x + 2 = 0$, and $x = -2$ by the uniqueness of additive inverses.

5. $x = \frac{1}{2}$ or $x = \frac{1}{3}$ or $x = \frac{1}{4}$.

7. We compute:

$$
\begin{aligned}
(x+a)(x+b) &= x \cdot x + xb + ax + ab && \text{Exercise 6} \\
&= x^2 + bx + ax + ab && \text{Commutativity of multiplication} \\
&= x^2 + (b+a)x + ab && \text{Distributive Property/Summand Permutation} \\
&= x^2 + (a+b)x + ab && \text{Commutativity of Addition}
\end{aligned}
$$

Therefore, $(x+a)(x+b) = x^2 + (a+b)x + ab$ by the transitivity of equality.

9. $(x+a)(x+a) = x^2 + (a+a) + a^2$ by Exercise 7. Therefore, $(x+a)(x+a) = x^2 + 2ax + a^2$.

11.

$$
\begin{aligned}
x^2 - 8x + 12 &= x^2 + (-6 + (-2))x + (-6)(-2) \\
&= (x + (-6))(x + (-2)) \\
&= (x-6)(x-2)
\end{aligned}
$$

using Exercise 7 (in reverse).

13. $x^2 - 64 = x^2 - 8^2 = (x-8)(x+8)$ using Exercise 8.

15.

$$\begin{aligned} x^2+2x-15 &= x^2+(5+(-3))x+(5)(-3) \\ &= (x+5)(x+(-3)) \\ &= (x+5)(x-3) \end{aligned}$$

using Exercise 7.

16.

$(ax+b)(cx+d)$	$=$	$(ax)(cx)+(ax)d+b(cx)+bd$	Exercise 6
	$=$	$acx^2+adx+bcx+bd$	Factor Permutation, Associativity, and Commutativity
	$=$	$acx^2+(ad+bc)x+bd$	Distributive Law

Therefore, $(ax+b)(cx+d) = acx^2+(ad+bc)x+bd$ by transitivity of equality.

17. $(2x-3)(x+1)$

19. $(3x-1)(x-4)$

21. $(6x+5)(2x-1)$

23. $x^2-3x+2=0$ implies $(x-2)(x-1)=0$ by Exercise 7. Therefore, by the Zero Product Theorem, $x-2=0$ or $x-1=0$, so $x=2$ or $x=1$.

25. $x=-5$ or $x=3$.

27. $x=-4/3$ or $x=-1/2$.

29. $x=-1/2$ or $5/6$.

31.

$-(a-b)$	$=$	$(-1)(a-b)$	$(-1)a=-a$
	$=$	$(-1)a-(-1)b$	Generalized Distributive Law (first part)
	$=$	$-a+[-(-1)b]$	$(-1)a=-a$ and Definition of Subtraction
	$=$	$-a+[-(-(1\cdot b)]$	$(-a)b=-(ab)$
	$=$	$-a+(1\cdot b)$	$-(-a)=a$
	$=$	$-a+b$	1 is the Multiplicative Identity

Therefore, $-(a-b)=-a+b$ by transitivity of equality.

33. If $abc=0$, then, by definition, $(ab)c=0$. Therefore, either $ab=0$ or $c=0$ by the Zero Product Theorem. If $ab=0$, then $a=0$ or $b=0$ by the Zero Product Theorem. Thus, in any case, $a=0$ or $b=0$ or $c=0$.

35. Let w represent the whole number. One more than w is $w+1$, and two more than w is $w+2$. We are told that the product of these two numbers is 2, so $(w+1)(w+2)=2$. Using Exercise 3, we see that $w^2+3w+2=2$, so $w^2+3w=0$. This is equivalent to $w(w+3)=0$, and now the Zero Product Theorem implies that $w=0$ or $w=-3$. Since -3 is not a whole number, $w=0$.

Fractions, page 62

1. $\frac{1}{b}=1\cdot b^{-1}$ by the definition of division, and $1\cdot b^{-1}=b^{-1}$ since 1 is the multiplicative identity. Therefore, $\frac{1}{b}=b^{-1}$ by transitivity of equality.

3. $\frac{b}{b}=b\cdot b^{-1}$ by the definition of division. Since b and b^{-1} are multiplicative inverses, $b\cdot b^{-1}=1$. Therefore, $\frac{b}{b}=1$ by transitivity of equality.

5. $x-(a-b)=x+[-(a-b)]=x+[-a+b]$ using Exercise 1. This is equal to $x-a+b$ by the definition of subtraction.

7. $\left(-\frac{1}{b}\right)\cdot(-b)=\frac{1}{b}\cdot b=1$, using the facts that $(-a)(-b)=ab$ and $\frac{a}{b}\cdot b=a$. Therefore, $(-b)^{-1}=-\frac{1}{b}$ by the uniqueness of multiplicative inverses. But by Exercise 5, $(-b)^{-1}=\frac{1}{-b}$, so $-\frac{1}{b}=\frac{1}{-b}$ by transitivity of equality.

9. Since $\frac{a}{b}=0$, $b\cdot\frac{a}{b}=b\cdot 0$. Therefore, $a=0$, using the facts that $b\cdot\frac{a}{b}=a$ and $b\cdot 0=0$.

11. $\frac{3}{x}+\frac{2x+3}{5}=\frac{3\cdot 5+x(2x+3)}{x\cdot 5}$ by Fraction Addition II. This is just $\frac{15+2x^2+3x}{5x}$, or $\frac{2x^2+3x+15}{5x}$.

13. $x-\frac{3-x}{2}=\frac{x}{1}+\frac{-(3-x)}{2}=\frac{x\cdot 2+1(-3+x)}{1\cdot 2}$ by Fraction Addition II and Exercise 1. This equals $\frac{2x-3+x}{2}=\frac{3x-3}{2}$.

15. $\frac{x^2-9x+20}{x-6}\cdot\frac{x-3}{x^2-7x+12}=\frac{(x-4)(x-5)(x-3)}{(x-6)(x-3)(x-4)}=\frac{x-5}{x-6}$ by Fraction Simplification.

17. $\frac{62x^3-5x+52x^2-4}{(2x^2-2x-1)(16x^2-1)}$

19. $\frac{(2x+5)(x+5)}{(2x-3)(x+2)}$

21. $\dfrac{2x^2+1}{2x^2+x+3}$

23. Try just about any value for x: $\dfrac{1^2+1}{1^2+2} = \dfrac{2}{3} \neq \dfrac{1}{2}$.

25. $x = 2$

27. $x = -1$

Exponents and Induction, page 69

1. $5^6 x^x = (5x)^6$

3. $z^5 z^{-3} = z^{5-3} = z^2$.

5. x^8

7. $\dfrac{1}{(x+2)^3}$

9. ≈ -1.007.

11. 1874.173.

13. $(4x^5)(5x^3) = 20x^{5+3} = 20x^8$.

15. $\dfrac{x^2 y}{729}$

17. $2\dfrac{-x^2+2xy+y^2}{(x+y)(-x+y)} \dfrac{2x}{x+y} - \dfrac{2y}{x-y} = \dfrac{2x(x-y)-2y(x+y)}{(x+y)(x-y)} = \dfrac{2x^2-2xy-2xy-2y^2}{(x+y)(x-y)}$
$= \dfrac{2x^2-4xy-2y^2}{(x+y)(x-y)}$

19. $(2t^3)^{-2} = \dfrac{1}{(2t^3)^2} = \dfrac{1}{2^2 t^{3\cdot 2}} = \dfrac{1}{4t^6}$.

21. $\dfrac{x^3}{x+1}$

23. $\dfrac{(1-y^2x^2)y^2}{(x^3y^2-1)x^2}$

25. $3\dfrac{y^2-x^3}{x(3y^2+x^2)}$

27. For the base case, we have $(a^m)^1 = a^m = a^{m\cdot 1}$, so the theorem holds. Assume now that the theorem holds for some $k \in \mathbb{N}$, so that, for that k, $(a^m)^k = a^{mk}$. Then

$$\begin{aligned} (a^m)^{k+1} &= (a^m)^k(a^m) && \text{Definition of exponent} \\ &= (a^{mk})a^m && \text{Raising a power to a power} \\ &= a^{mk+m} && \text{Multiplication of like bases} \\ &= a^{mk+m\cdot 1} && \text{1 is the multiplicative identity} \\ &= a^{m(k+1)} && \text{Distributive Law} \end{aligned}$$

Therefore, $(a^m)^{k+1} = a^{m(k+1)}$ by transitivity. By the principle of mathematical induction, $(a^m)^n = a^{mn}$ for all $m, n \in \mathbb{N}$.

29. We will prove part 2 here.

If $n > 0$, then this is a theorem we have already proven.

If $n = 0$, then $(ab)^n = (ab)^0 = 1 = 1 \cdot 1 = a^0 b^0 = a^n b^n$.

If $n < 0$, we have $(ab)^n = \frac{1}{(ab)^{-n}}$, where $-n > 0$. Now by previous results, we get $\frac{1}{a^{-n}b^{-n}} = \frac{1}{a^{-n}}\frac{1}{b^{-n}} = a^n b^n$.

Complex Numbers, page 78

1. $(-\sqrt{2}+1,4) - (\sqrt{2},6) = (-\sqrt{2}+1,4) + (-\sqrt{2},-6) = (-2\sqrt{2}+1,-2)$

3. $(2-4i) - (12-i) = 2-4i-12+i = -10-3i$

5. $(2,-0.5)\cdot(8,-2) = (16-1,-4+(-4)) = (15,-8)$

7. $(3.1,-7)^{-1} = \left(\frac{3.1}{3.1^2+(-7)^2}, -f-(-7)3.1^2+(-7)^2\right) = \left(\frac{3.1}{58.61}, \frac{7}{58.61}\right)$
$= \frac{1}{58.61}(3.1,7)$

9. $-1(4,2) = (-4,-2)$

11. $(2.2,5) \div (7,13) = (2.2,5)\cdot(7,13)^{-1} = (2.2,5)\cdot\left(\frac{7}{7^2+13^2}, \frac{-13}{7^2+13^2}\right)$
$= \frac{1}{218}(2.2,5)\cdot(7,-13) = \frac{1}{218}(2.2(7)-5(-13), 2.2(-13)+5(7)) = \frac{1}{218}(80.4,6.4)$

13. $|(-6,10)| = \sqrt{(-6)^2+10^2} = \sqrt{136} = 2\sqrt{34}$.

15. $\overline{(2,2)} = (2,-2)$.

17. $\frac{3-4i}{2+i} = \frac{3-4i}{2+i}\cdot\frac{2-i}{2-i} = \frac{6-3i-8i+4i^2}{4-i^2} = \frac{2-11i}{5} = \frac{2}{5} - \frac{11}{5}i$

19. $-i$
21. i
23. 1
25. i
27. i
29. 1

31. $(3,1)z-(1,-1)=(4,2) \implies (3,1)z=(4,2)+(1,-1)=(5,1) \implies z=(3,1)^{-1}(5,1)=$ $\left(\frac{3}{10},\frac{-1}{10}\right)(5,1)=\frac{1}{10}(3,-1)(5,1)=\frac{1}{10}(16,-2)=\frac{1}{5}(8,-1)$

33. Although this doesn't seem at first glance to be a difference of squares, it is. $x^2+4=$ $z^2-(2i)^2=(z-2i)(z+2i)$. Therefore, $z=\pm 2i$.

35. $(2,0)z+(2,-1)z=(4,3) \implies (4,-1)z=(4,3) \implies z=(4,-1)^{-1}(4,3)=\frac{1}{17}(4,1)(4,3)=$ $\frac{1}{17}(13,16)$

37. $(5,0)z=(4,-12) \implies z=\frac{1}{5}(4,-12)$.

39. $(2,1)z+(13,-3)=(-4,2)z+(7,2) \implies [(2,1)-(-4,2)]z=(7,2)-(13,-3) \implies$ $(6,-1)z=(-6,5) \implies z=(6,-1)^{-1}(6,-5) \implies z=\frac{1}{37}(6,1)(6,-5)=\frac{1}{37}(41,-24)$

41. $a(b,c)=(ab,ac)$ and $(a,0)(b,c)=(ab-0\cdot c, ac+0\cdot b)=(ab,ac)$.

43. Let $(a,b)\in\mathbb{C}$ and $(c,d)\in\mathbb{C}$. Then

$$\begin{array}{rcll} (a,b)(c,d) & = & (ac-bd, ad+bc) & \text{Definition of multiplication in } \mathbb{C} \\ & = & (ca-db, cb+da) & \text{Commutativity of multiplication in } \mathbb{R} \\ & = & (c,d)(a,b) & \text{Definition of multiplication in } \mathbb{C} \end{array}$$

Therefore, $(a,b)(c,d)=(c,d)(a,b)$ by transitivity.

45.

$$\begin{aligned} (a+bi)(c+di) &= ac+adi+bic+bidi \\ &= ac+(ad+bc)i-bd \\ &= (ac-bd)+(ad+bc)i. \end{aligned}$$

It does match our original definition.

Order Axioms, page 87

1. $-(-3)=3$ is positive since $3\in\mathbb{N}$, and all natural numbers are positive by theorem 54.

3. $(-2)^{-1}=\frac{1}{-2}=-\frac{1}{2}$ is negative.

5. It cannot be determined whether $-a$ is positive, negative, or zero.

7. $-3^2 = -9$ is negative.

9. $(-x) \cdot x = -x^2$ is negative since x^2 is positive.

11. $3x - 7 < 5x + 4 \implies 3x - 5x < 4 + 7 \implies -2x < 11 \implies x > 11/2$. The solution set is $(11/2, \infty)$.

13. We must have $x + 1 > 0$ and $x - 5 > 0$ or $x + 1 < 0$ and $x - 5 < 0$. In the first case, we require that $x > -1$ and $x > 5$; these are both satisfied if $x > 5$. In the second case, we require that $x < -1$ and $x < 5$; these are both satisfied if $x < -1$. Therefore, our solution set is $(-\infty, -1) \cup (5, \infty)$.

15. $2x^2 - 3x > x^2 - 2 \implies x^2 - 3x + 2 > 0 \implies (x-2)(x-1) > 0$. Therefore, either

    (a) $x - 2 > 0$ and $x - 1 > 0$, in which case $x > 2$ (and $x > 1$, but that's redundant),
    OR

    (b) $x - 2 < 0$ and $x - 1 < 0$, in which case $x < 1$ (and $x < 2$, but that's redundant).

    Therefore, the solution set is $(-\infty, 1) \cup (2, \infty)$.

    We can also analyze this on a number line:

		1		2	
$x-2$:	+		−		+
$x-1$:	+		−		+
Overall	+		−		+

17. $[3/4, \infty)$.

19. Let x be the number of minutes of a call. Then we want to know under what circumstances it is the case that $0.99 + 0.07(x - 20) < 0.10x$ (if $x \leq 20$), and $0.99 < .10x$ (if $x \geq 10$). We solve: $-0.41 + 0.07x < 0.10x \implies 0.03x > -0.41 \implies x > 0$ (since we are only interested in calls of a positive duration). This is insufficient since it includes values of x that are less than 20. We also have $x > 9.9$; thus, if a call lasts longer than 9.9 minutes, it is cheaper with the 0.99 plan.

21. If $a + c < b + c$, then $(a+c) + (-c) < (b+c) + (-c)$ by Theorem 48. Therefore $a < b$.

23. Let $\frac{a}{b} \in \mathbb{Q}^+$ and $\frac{c}{d} \in \mathbb{Q}^+$. Then a and b have the same sign, and c and d have the same sign. Therefore, by multiplying both numerator and denominator by -1 if necessary, we may assume that a, b, c, d are all positive. Now $\frac{a}{b} + \frac{c}{d} = \frac{ad + bc}{bd}$ and $\frac{a}{b} \cdot \frac{c}{d} = \frac{ac}{bd}$ both have positive numerators and denominators, so the product of a numerator and denominator is positive, as required. Therefore, they are also in $\mathbb{Q}^+$ and Q^+ is closed under addition and multiplication.

    Let $\frac{a}{b} \in \mathbb{Q}$. Then ab is positive, negative, or zero by the trichotomy property in $\mathbb{Z}$. In the first case, $\frac{a}{b} \in \mathbb{Q}^+$; in the second case, $-\frac{a}{b} \in \mathbb{Q}^+$, and in the third case, $\frac{a}{b} = 0$. (Note that $b \neq 0$, so $a = 0$ if $ab = 0$.) Therefore, the trichotomy property holds and $\mathbb{Q}$ is an ordered field.

One-Dimensional Coordinates, page 99

1. $|3| = 3$ since $3 \geq 0$.

3. $\left|\frac{-2}{3}\right| = -\frac{-2}{3} = \frac{2}{3}$ since $-2/3 < 0$. (We know $3 > 0$, so $1/3 > 0$; and since $2 > 0 \implies -2 < 0$, we get $(-2)(1/3) < 0$.)

5. The most we can simplify is $|a|$ unless we know what a is.

7. $|-8-(-2)| = |-6| = 6$.

9. $|8-(-1.7)| = |9.7| = 9.7$.

11. $|1.41-\sqrt{2}| = \sqrt{2}-1.41 \approx 0.0042$.

13. $|x-3| = 5 \implies x-3 = 5$ or $x-3 = -5$. Therefore, $x = 8$ or $x = -2$.

15. $x = -55/24$ or $x = -19/8$.

17. $[-7/3, 1]$

19. $[-3, 5/3]$

21. $(-51/2, 3/2)$

23. $|x^2| < 16 \implies x^2 < 16 \implies x^2 - 16 < 0 \implies (x-4)(x+4) < 0$. Therefore, either $x-4 > 0$ and $x+4 < 0$ (impossible), or $x-4 < 0$ and $x+4 > 0$, in which case $x < 4$ and $x > -4$. The solution set is therefore $(-4, 4)$.

25. We have $|x-y| = 1$. Choosing a direction on the road to be "positive", the gas station is at either $x+1$ or $x-1$. It fits, since the directions did not indication which way down the road to proceed.

27. Assume that $x, y > 0$. If $x > y$, then $x^2 > xy > y^2$, so $x^2 > y^2$.

    If $x^2 > y^2$, then $x^2 - y^2 > 0$, so $(x-y)(x+y) > 0$. This means that either $x-y > 0$ and $x+y > 0$, or $x-y < 0$ and $x+y < 0$. Since $x > 0$ and $y > 0$ by assumption, the second case cannot hold; therefore, $x-y > 0$, so $x > y$.

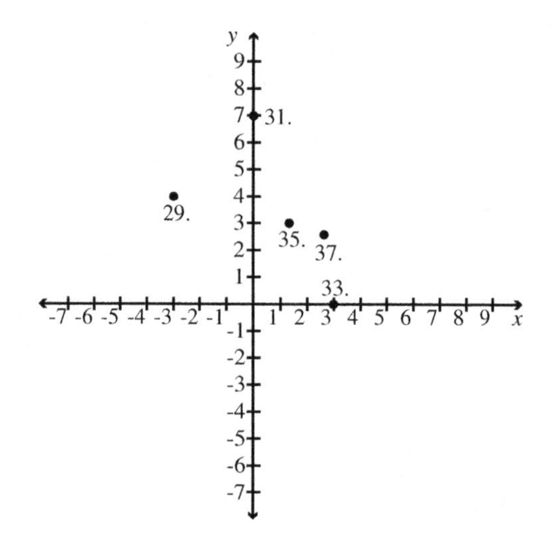

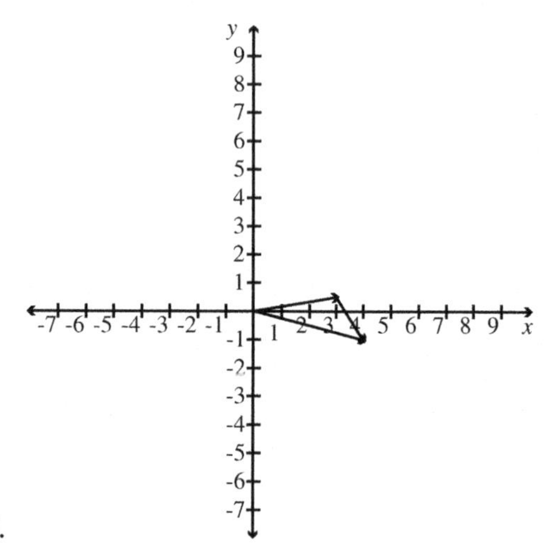

39.

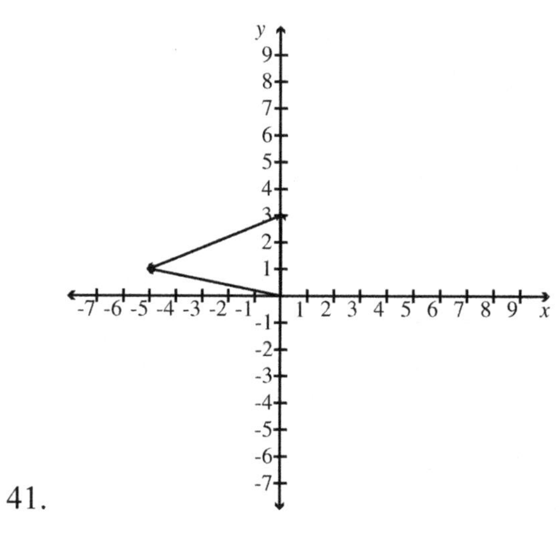

41.

43. $(1,1)(1,3) = (1-3, 3+1) = (-2,4)$.

45. $(3,1)(3,-1) = (9+1, -3+3) = (10,0)$.

The modulus of the product is the product of the moduli. Also, the angle the product makes with the positive x-axis is the sum of the angles the factors make with the positive x-axis.

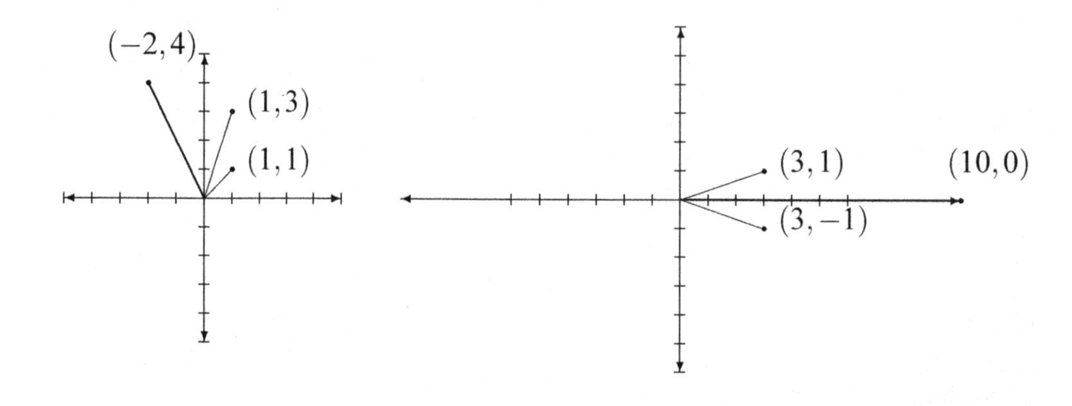

47. From the parallelogram law, we have the picture below. Analytically, we have $(a,b)+\overline{(a,b)}=(a,b)+(a,-b)=(2a,0)=2a\in\mathbb{R}$.

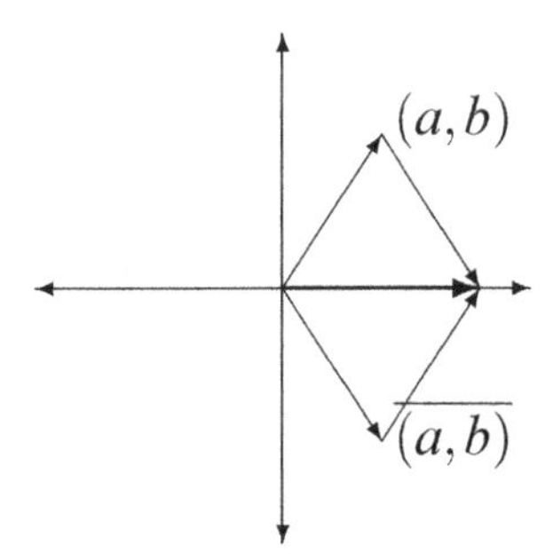

48. (a) $(-4,7)$

    (c) $(1,-2)$

    (e) $(-6,3)$

    (g) $(-3,-3)$

49. $\sqrt{(15-(-2))^2+(7-5)^2}=\sqrt{293}$.

51. $\sqrt{65}$

53. $\frac{\sqrt{2018}}{5}$

55. $(x-2)^2+(y+0.8)^2=4$

57. The center of the circle is $(3.5,-0.5)$ since it must be halfway between the points, and the radius of the circle must be $\frac{1}{2}\sqrt{(5-2)^2+(3-(-4))^2}=\frac{1}{2}\sqrt{58}$. Therefore, an equation of the circle is $(x-3.5)^2+(y+0.5)^2=\frac{58}{4}=\frac{29}{2}$.

59. Let P be the constant; then $\sqrt{(x-a)^2+(y-b)^2}+\sqrt{(x-c)^2+(y-d)^2}=P$.

61. (a) It is reasonable to make Houston the origin $(0,0)$ and make the road you are driving along the positive x-axis. In this case, the coordinates would be $(143,0)$ and then $(78,0)$. Other choices are also possible.

    (b) The distance would be $\sqrt{(143-78)^2+(0-0)^2}=65$.

    (c) Because of the way we set up the coordinate system, we can just consider our points as points on a line that are associated with the real numbers 143 and 78. These are $|143-78|=65$ miles apart.

63. If two points lie on the same vertical line, they have coordinates (a,b) and (a,c) (same x-coordinate). Therefore, the distance between them is $\sqrt{(a-a)^2+(b-c)^2}=\sqrt{(b-c)^2}=|b-c|$.

65. $(x-h)^2+(y-k)^2=r^2 \implies x^2-2hx+h^2+y^2-2ky+k^2-r^2=0$, or $x^2+y^2+(-2h)x+(-2k)y+(h^2+k^2-r^2)=0$. Thus, $A=-2h, B=-2k$, and $C=h^2+k^2-r^2$.

Matrices, page 110

1. A is a 3×3 matrix.

3. The $1,3$ entry of A is $\sqrt{5}$.

5. The $3,1$ entry of A is $-\dfrac{5}{7}$.

7. $\begin{bmatrix} 2 & 7 & -3 \\ 4 & -1 & 6.2 \end{bmatrix} + \begin{bmatrix} 5 & -2 & 2 \\ -4 & 2.5 & 1 \end{bmatrix} = \begin{bmatrix} 2+5 & 7+(-2) & -3+2 \\ 4+(-4) & -1+2.5 & 6.2+1 \end{bmatrix} = \begin{bmatrix} 7 & 5 & -1 \\ 0 & 1.5 & 7.2 \end{bmatrix}$

9. $\begin{bmatrix} 1 & 4 \\ -2 & 0 \\ 3 & 1 \end{bmatrix} \begin{bmatrix} 4 & 8 \\ 2 & -3 \end{bmatrix} = \begin{bmatrix} 1\cdot4+4\cdot2 & 1\cdot8+4\cdot(-3) \\ -2\cdot4+0\cdot2 & -2\cdot8+0\cdot(-3) \\ 3\cdot4+1\cdot2 & 3\cdot8+1\cdot(-3) \end{bmatrix} = \begin{bmatrix} 12 & -4 \\ -8 & -16 \\ 14 & 21 \end{bmatrix}$

11. $4\begin{bmatrix} 2 & -3 \\ 5 & 3.4 \end{bmatrix} = \begin{bmatrix} 4(2) & 4(-3) \\ 4(5) & 4(3.4) \end{bmatrix} = \begin{bmatrix} 8 & -12 \\ 20 & 13.6 \end{bmatrix}$.

13. $AB, AAB, AAAB$, etc. are defined.

15. No products are defined.

17. $X = \begin{bmatrix} -2 & 17 \\ -11 & 6 \end{bmatrix}$

19. We have $2\begin{bmatrix} x & y \\ z & w \end{bmatrix} = B - A$, so $\begin{bmatrix} x & y \\ z & w \end{bmatrix} = \frac{1}{2}(B-A) = \frac{1}{2}\begin{bmatrix} -4 & -1 \\ -2 & 2 \end{bmatrix}$. Therefore, $x = -2, y = -1/2, z = -1$, and $w = 1$.

21. $X = \begin{bmatrix} 7/3 & 0 \\ 3/2 & -4/3 \end{bmatrix}$

23. $x = \begin{bmatrix} 20 & -6 \\ -4 & 3 \end{bmatrix}$

25. $x = \begin{bmatrix} 9 & -22 \\ -10 & -9 \end{bmatrix}$

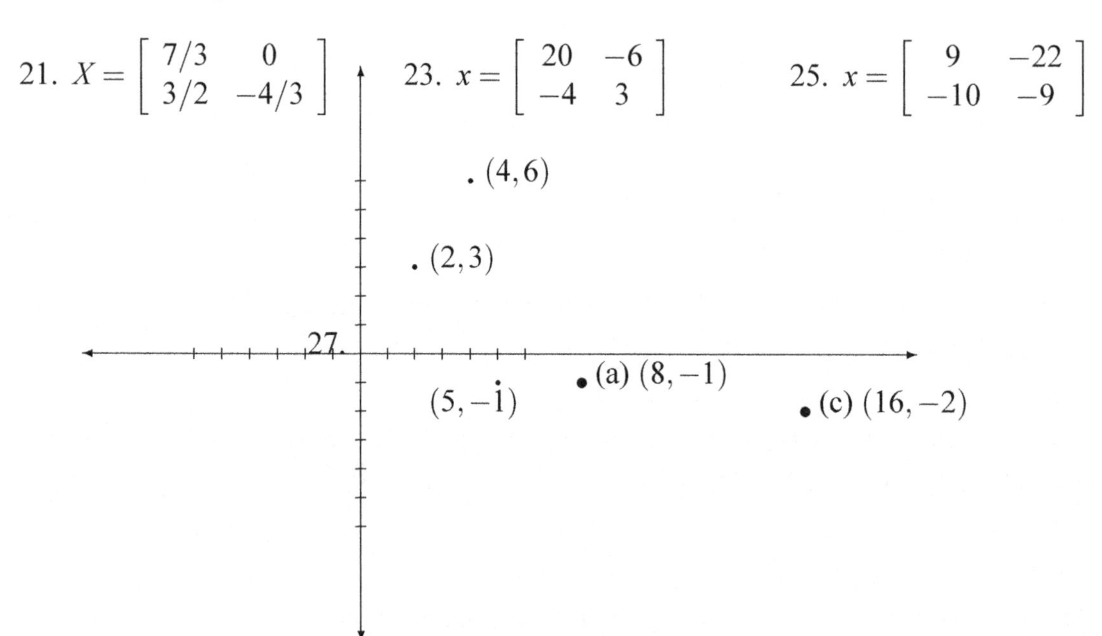

There appears to be some sort of magnification combined with a rotation about the origin.

29. Let $A=[a_{ij}]$ and $B=[b_{ij}]$ be $m\times n$ matrices. Then

$$\begin{aligned} A+B &= [a_{ij}]+[b_{ij}] && \text{Substitution} \\ &= [a_{ij}+b_{ij}] && \text{Definition of Matrix Addition} \\ &= [b_{ij}+a_{ij}] && \text{Addition of Real Numbers is Commutative} \\ &= [b_{ij}]+[a_{ij}] && \text{Definition of Matrix Addition} \\ &= B+A && \text{Substitution} \end{aligned}$$

Therefore, $A+B=B+A$ by transitivity of equality. Notice that, as with complex numbers, we reduced the question about matrix addition into multiple questions about real number addition, about which we know a great deal.

Matrix Inverses, page 116

1. $\begin{vmatrix} 4 & 3 \\ 3 & 2 \end{vmatrix} = 4(2)-3(3)=-1.$ $\begin{bmatrix} 4 & 3 \\ 3 & 2 \end{bmatrix}^{-1} = \frac{1}{-1}\begin{bmatrix} 2 & -3 \\ -3 & 4 \end{bmatrix} = \begin{bmatrix} -2 & 3 \\ 3 & -4 \end{bmatrix}.$

3. $\begin{vmatrix} 0 & -4 \\ -2 & 3 \end{vmatrix} = 0(3)-(-2)(-4)=-8.$

5. $-341/336$. The inverse is $\begin{bmatrix} -1/8 & -2/3 \\ -5/7 & 1/2 \end{bmatrix}$

7. $X=\begin{bmatrix} 2 & -3 \\ 4 & 1 \end{bmatrix}^{-1}\begin{bmatrix} 2 & -1 \\ 6 & -9 \end{bmatrix} = \frac{1}{2(1)-4(-3)}\begin{bmatrix} 1 & 3 \\ -4 & 2 \end{bmatrix}\begin{bmatrix} 2 & -1 \\ 6 & -9 \end{bmatrix} = \frac{1}{14}\begin{bmatrix} 20 & -28 \\ 4 & -14 \end{bmatrix}.$

9. $X=\begin{bmatrix} -10 & 14 & -14 \\ 15 & 52 & -6 \end{bmatrix}$

11.

$$\begin{aligned} & 3\begin{bmatrix} 28 & -46 \\ 51 & 13 \end{bmatrix} + \begin{bmatrix} 15 & 12 \\ -10 & 7 \end{bmatrix} X = \begin{bmatrix} -6 & 14 \\ 20 & 48 \end{bmatrix} \\ \Longrightarrow & \begin{bmatrix} 15 & 12 \\ -10 & 7 \end{bmatrix} X = \begin{bmatrix} -6 & 14 \\ 20 & 48 \end{bmatrix} - \begin{bmatrix} 84 & -138 \\ 153 & 39 \end{bmatrix} \\ \Longrightarrow & \begin{bmatrix} 15 & 12 \\ -10 & 7 \end{bmatrix} X = \begin{bmatrix} -90 & 152 \\ -133 & 9 \end{bmatrix} \\ \Longrightarrow & X = \begin{bmatrix} 15 & 12 \\ -10 & 7 \end{bmatrix}^{-1} \begin{bmatrix} -90 & 152 \\ -133 & 9 \end{bmatrix} \\ \Longrightarrow & X = \frac{1}{225}\begin{bmatrix} 7 & -12 \\ 10 & 15 \end{bmatrix} \begin{bmatrix} -90 & 152 \\ -133 & 9 \end{bmatrix} \\ \Longrightarrow & X = \frac{1}{225}\begin{bmatrix} 966 & 956 \\ -2895 & 1655 \end{bmatrix} \end{aligned}$$

13. We have $|A|=a(kb)-(ka)b=0.$

15. This is $\begin{bmatrix} 2x+5y \\ -x+3y \end{bmatrix} = \begin{bmatrix} -4 \\ 2 \end{bmatrix}$. Therefore, we need $2x+5y=-4$ and $-x+3y=2$.

Systems of Equations, page 121

1. $2(8)-12(1)=4$ and $3(8)-(1)=23$. Since the first equation is not satisfied, we do not have a solution.

3. $\begin{bmatrix} 3 & -2 \\ -7 & 3 \end{bmatrix}\begin{bmatrix} x \\ y \end{bmatrix} = \begin{bmatrix} 4 \\ 5 \end{bmatrix}$.

   $\begin{bmatrix} 3 & -2 \\ -7 & 3 \end{bmatrix}\begin{bmatrix} x \\ y \end{bmatrix} = \begin{bmatrix} 4 \\ 5 \end{bmatrix} \Longrightarrow \begin{bmatrix} x \\ y \end{bmatrix} = \frac{1}{-5}\begin{bmatrix} 3 & 2 \\ 7 & 3 \end{bmatrix}\begin{bmatrix} 4 \\ 5 \end{bmatrix} = \frac{1}{-5}\begin{bmatrix} 22 \\ 43 \end{bmatrix}$. Therefore, $x=-4.4$ and $y=-8.6$.

5. $\begin{bmatrix} -2 & 4 & -2 \\ 3 & -2 & 2 \\ 1 & 5 & -1 \end{bmatrix}\begin{bmatrix} x \\ y \\ z \end{bmatrix} = \begin{bmatrix} -3 \\ 1 \\ 0 \end{bmatrix}$.

   $\begin{bmatrix} -2 & 4 & -2 \\ 3 & -2 & 2 \\ 1 & 5 & -1 \end{bmatrix}\begin{bmatrix} x \\ y \\ z \end{bmatrix} = \begin{bmatrix} -3 \\ 1 \\ 0 \end{bmatrix} \Longrightarrow \begin{bmatrix} x \\ y \\ z \end{bmatrix} = \begin{bmatrix} 9 \\ -5.5 \\ -18.5 \end{bmatrix}$ (from a graphing calculator).

7. Let x be Papa Joe's age, and let y be his daughter's age. Then $x=2y$ and $x-10=3(y-10)$.

9. Let x represent the number of solid color neckwarmers sold, and let y be the number of print neckwarmers sold. Then $x+y=40$ and $9.9x+12.75x=421.65$.

11. Your calculator will give you an error; since these three points are collinear, there is no circle through them. This is reflected in the lack of an inverse for the matrix you find.

13. The general form is $x^2+y^2+Ax+By+C=0$, so we have the following three equations:

$$1^2+2^2+A(1)+B(2)+C=0 \Longrightarrow A+2B+C=-5$$

$$3^2+(-1)^2+A(3)+B(-1)+C=0 \Longrightarrow 3A-B+C=-10$$

$$4^2+4^2+A(4)+B(4)+C=0 \Longrightarrow 4A+4B+C=-32$$

   These give the matrix equation $\begin{bmatrix} 1 & 2 & 1 \\ 3 & -1 & 1 \\ 4 & 4 & 1 \end{bmatrix}\begin{bmatrix} A \\ B \\ C \end{bmatrix} = \begin{bmatrix} -5 \\ -10 \\ -32 \end{bmatrix}$. Solving this on a graphing calculator gives $\begin{bmatrix} A \\ B \\ C \end{bmatrix} = \begin{bmatrix} -7 \\ -3 \\ 8 \end{bmatrix}$. Therefore, the desired equation is $x^2+y^2-7A-3B+8=0$.

15. A parabola has the equation $y = Ax^2 + Bx + C$, so we get the three equations

$$1 = 16A - 4B + C$$

$$-3 = 4A + 2B + C$$

$$5 = C.$$

These give the matrix equation $\begin{bmatrix} 16 & -4 & 1 \\ 4 & 2 & 1 \\ 0 & 0 & 1 \end{bmatrix} \begin{bmatrix} A \\ B \\ C \end{bmatrix} = \begin{bmatrix} 1 \\ -3 \\ 5 \end{bmatrix}$. A graphing calculator gives $\begin{bmatrix} A \\ B \\ C \end{bmatrix} = \begin{bmatrix} -5/6 \\ -7/3 \\ 5 \end{bmatrix}$. Therefore, the desired equation is $y = -\frac{5}{6}x^2 - \frac{7}{3}x + 5$.

Substitution and Elimination, 128

1. From $x + y = 12$, we get $y = 12 - x$. Therefore, $3x - (12 - x) = 4$, so $4x = 16$ and $x = 4$. This gives $y = 8$.

3. $-2(2x + 5y) = -2(10)$, so $-4x - 10y = -20$. Combining this with $4x + 3y = -3$ gives $-7y = -23$, so $y = 23/7$. Now $2x + 5(23/7) = 10$, so $2x = -45/7$, and $x = -45/14$.

5. We have $-2(6x + 9y) = -2(12)$ or $-12x - 18y = -24$, so $-13y = -22$. Therefore, $y = 22/13$. Now $6x + 9(22/13) = 12$, so $6x = -42/13$, and $x = -7/13$.

7. We have $y = 17 - 5x$, so $-15x + 13(17 - 5x) = 22$, and $-80x = -199$. Therefore, $x = 199/80$, and $y = 17 - 5(199/80) = 73/16$.

9. Since $I_1 = I_2 + I_3$, we have $5(I_2 + I_3) + 2I_3 = 12$ and $5(I_2 + I_3) + 3I_2 = 12$. Therefore, $5I_2 + 7I_3 = 12$ and $8I_2 + 5I_3 = 12$. These are equivalent to $8(5I_2 + 7I_3) = 8(12)$, or $40I_2 + 56I_3 = 96$, and $-5(8I_2 + 5I_3) = -5(12)$, or $-40I_2 - 25I_3 = -60$. These give $31I_3 = 36$, so $I_3 = 36/31$. Now $I_2 = \frac{(12 - 5(36/31))}{8} = 24/31$. Finally, $I_1 = 60/31$.

11. If x represents the number of apple trees and y represents the number of pear trees, then we have $x + y = 150$ and $8x + 5y = 885$. This gives $y = 150 - x$, so $8x + 5(150 - x) = 885$. That is, $3x = 135$, so $x = 45$. Thus, $y = 105$.

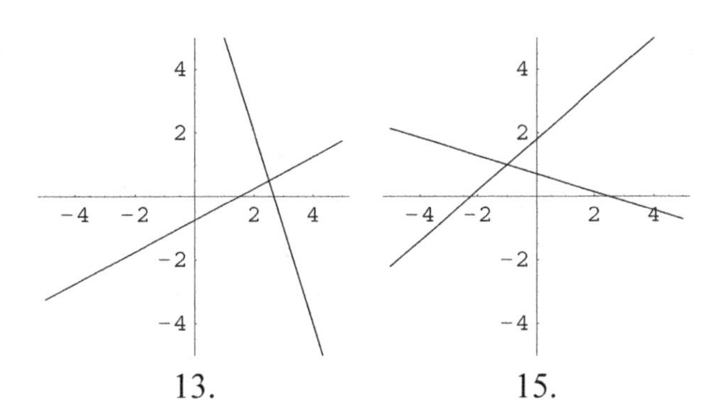

13. 15.

From the graphs, the point of intersection for 13 appears to be around $(2.5, 0.5)$. The point of intersection for 15 appears to be around $(-1, 1)$.

For 13, we can solve by substitution since $y = 8 - 3x$. We get $2x - 4(8 - 3x) = 3$, so $14x = 35$, and $x = 5/2$. This gives $y = 1/2$, which agrees with our graphical solution.

For 15, we have $\begin{bmatrix} -4 & 5 \\ 2 & 7 \end{bmatrix}\begin{bmatrix} x \\ y \end{bmatrix} = \begin{bmatrix} 9 \\ 5 \end{bmatrix}$. Therefore, $\begin{bmatrix} x \\ y \end{bmatrix} = -\frac{1}{38}\begin{bmatrix} 7 & -5 \\ -2 & -4 \end{bmatrix}\begin{bmatrix} 9 \\ 5 \end{bmatrix} = -\frac{1}{38}\begin{bmatrix} 38 \\ -38 \end{bmatrix}$. This gives $x = -1$ and $y = -1$, again in agreement with our graphical solution.

17. We get $x^2 - 2x = 2$, so $x^2 - 2x + 1 = 3$. This implies that $(x-1)^2 = 3$, so $x - 1 = \pm\sqrt{3}$, and $x = 1 \pm \sqrt{3}$.

19. We have $y^2 + y = 6$, so $y^2 + y - 6 = 0$. Therefore, $(y-2)(y+3) = 0$, and $y = 2$ or $y = -3$. If $y = 2$, then $x^2 + 4 = 13$, so $x^2 = 9$. Therefore, $x = \pm 3$. Both of these satisfy the second equation as well, so $(3, 2)$ and $(-3, 2)$ are both solutions. If $y = -3$, then $x = \pm 2$, and these both satisfy the given equations. Thus, $(2, -3)$ and $(-2, -3)$ are also solutions.

    The graph of this system is the intersection of a circle and a parabola, as shown.

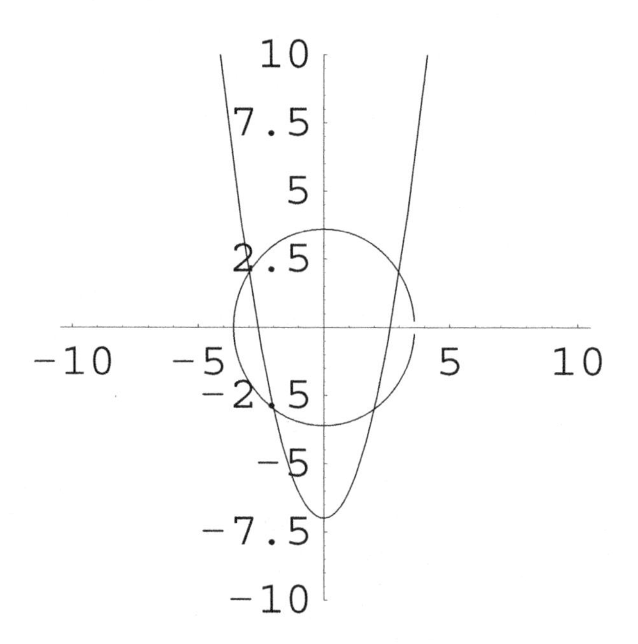

Functions, page 137

1. The domain of f is $\{x | 2x + 5 \neq 0\} = \{x | x \neq -5/2\} = (-\infty, 5/2) \cup (5/2, \infty)$.

3. The domain is $\{x | x^2 - 5x - 6 \neq 0\}$. We solve: $x^2 - 5x - 6 = 0$ implies $(x-6)(x+1) = 0$, so $x = 6$ or $x = -1$ by the zero product theorem. Therefore, the domain is $\{x | x \neq 6$ and $x \neq -1\} = (-\infty, -1) \cup (-1, 6) \cup (6, \infty)$.

5. The domain is $\{x | x^2 - 9 \geq 0\}$. We need to solve this inequality: $x^2 - 9 \geq 0 \implies (x-3)(x+3) \geq 0$. We use the number line below to determine the sign of $x^2 - 9$.

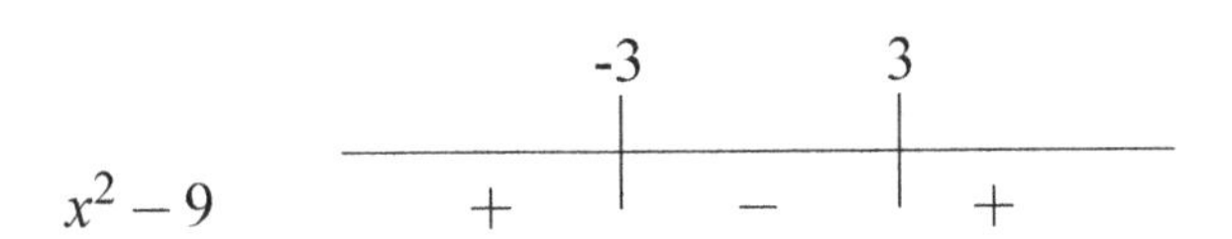

Thus, the domain is $(-\infty, -3] \cup [3, \infty)$.

7. $f(0) = \sqrt{(0)^2 - 2(0) + 2} = \sqrt{2}$. $f(-2) = \sqrt{(-2)^2 - 2(-2) + 2} = \sqrt{10}$. $f(3) = \sqrt{3^2 - 2(3) + 2} = \sqrt{5}$.

9. $x(1) = 1^1 = 1$. $x(-3) = (-3)^{-3} = \dfrac{1}{(-3)^3} = \dfrac{1}{-27}$. $x(2) = 2^2 = 4$.

11.

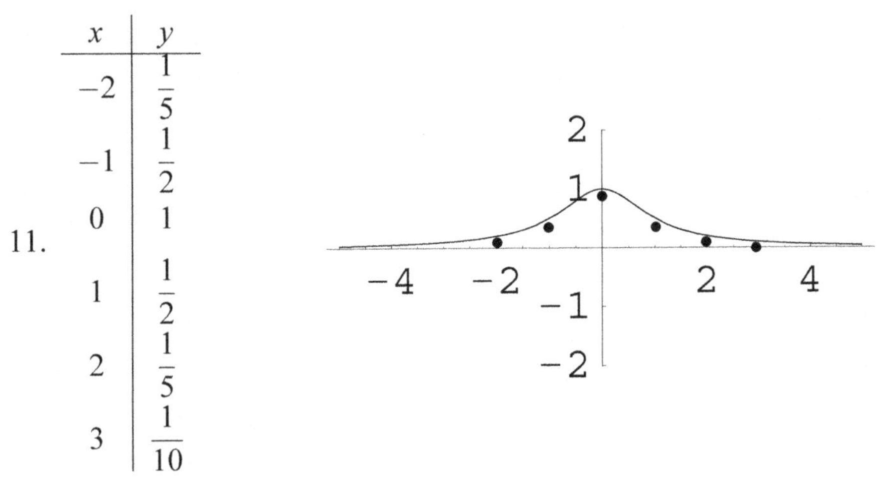

x	y
-2	$\frac{1}{5}$
-1	$\frac{1}{2}$
0	1
1	$\frac{1}{2}$
2	$\frac{1}{5}$
3	$\frac{1}{10}$

13. Compute $h(0.5) = 16(0.5)(1-0.5) = 4$ feet, or 48 inches. One second is a LONG time to be in the air.

Sequences, page 142

1. $3, 6, 9, 12, 15$

3. $\frac{4}{3}, \frac{4}{3}, \frac{4}{3}, \frac{4}{3}, \frac{4}{3}$

5. $\frac{3}{5}, \frac{9}{7}, 3, \frac{81}{11}, \frac{243}{13}$

7. $2n - 1$

9. $-3n + 8$

11. $(-2)^n \cdot 3$

13. $\dfrac{n+1}{n+2}$

15. $(-1)^{n-1} n$

17. $\dfrac{n(n+1)}{2}$

19. 23

21. $4\pi + 1$

23. 35.5

25. -6

27. 2

29. -18

31. $x = 2$. (Solve $\frac{x}{x-1} = \frac{x+2}{x}$.)

Operations on Functions, page 149

1. $(f+g)(x) = \frac{1}{x^2+1} + \frac{1}{x-1} = \frac{x-1+x^2+1}{(x^2+1)(x-1)} = \frac{x(x+1)}{(x^2+1)(x-1)}$. (Notice that this does not simplify.) The domain of $f+g$ is $\{x|x \neq 1\}$.

   $\left(\frac{f}{g}\right)(x) = \frac{x-1}{x^2+1}$. The domain of $\frac{f}{g}$ is $\{x|x \neq 1\}$.

   $(f-g)(4) = f(4) - g(4) = \frac{1}{17} - \frac{1}{3} = \frac{-14}{51}$.

3. $(fg)(x) = (2x-3)(x^2+2)$. $(f \circ g)(x) = f(g(x)) = f(x^2+2) = 2(x^2+2) - 3 = 2x^2+1$. Both domains are $\mathbb{R}$.

5. Since A is 2×2, if X is 2×1, the product will be 2×1.

7. $f\left(\begin{bmatrix} 12 \\ -8 \end{bmatrix}\right) = \begin{bmatrix} 1 & 2 \\ -2 & 3 \end{bmatrix}\begin{bmatrix} 12 \\ -8 \end{bmatrix} = \begin{bmatrix} -4 \\ -48 \end{bmatrix}$.

9. (a) $(f \circ g)(x) = f(\sqrt{x}) = \frac{1}{\sqrt{x}-1}$. The domain is $\{x|x \geq 0 \text{ and } x \neq 1\} = [0,1) \cup (1,\infty)$.

   (b) $(g \circ f)(x) = g\left(\frac{1}{x-1}\right) = \sqrt{\frac{1}{x-1}}$. The domain is $\left\{x|x \neq 1 \text{ and } \frac{1}{x-1} \geq 0\right\}$. This gives $\{x|x > 1\} = (1,\infty)$.

   (c) $(f \circ g)(4) = f(\sqrt{4}) = f(2) = \frac{1}{2-1} = 1$.

   (d) $(g \circ f)(3) = g\left(\frac{1}{3-1}\right) = g(1/2) = \sqrt{1/2}$.

11. f is neither even nor odd.

13. g is odd since $g(-x) = \frac{1}{-x} = -\frac{1}{x} = -g(x)$.

15. Neither.

17. f is odd.

19. (a) $f(150) = 25$ dollars.

    (b) $f(245) = 25 + 0.35(245 - 200) = 40.75$ dollars.

    (c) The charge is \$0.35 for each mile *beyond* 200.

21. An even function is symmetric about the y-axis.

23. Assume that f and g are both even functions. Then $(f+g)(-x) = f(-x) + g(-x) = f(x) + g(x) = (f+g)(x)$ using the definition of function addition and the fact that f and g are both even. Therefore, $f+g$ is even.

25. Assume that f and g are both even functions. Then $(fg)(-x) = f(-x)g(-x) = f(x)g(x) = (fg)(x)$. Therefore, fg is even.

27. Assume that f is even and g is odd. Then $(fg)(-x) = f(-x)g(-x) = f(x)(-g(x)) = -f(x)g(x) = -(fg)(x)$. Therefore, fg is odd.

29. f is a function, a list of ordered pairs, a set. $f(x)$ is a *number*, an output, a single value. For example, $f = \{(1,3),(2,5)\}$ is a function, and $f(2)$ is the number 5.

31. Let f and g be functions. Then

$$\begin{aligned}(fg)(x) &= f(x)g(x) && \text{Definition of function multiplication}\\ &= g(x)f(x) && \text{Multiplication of real numbers is commutative}\\ &= (gf)(x) && \text{Definition of function multiplication}\end{aligned}$$

Therefore, $(fg)(x) = (gf)(x)$ for all $x \in D(fg) = D(f) \cap D(g) = D(gf)$, so $fg = gf$.

Graphs of Functions, page 158

1. Not a function.

3. Function.

5. Not a function.

7. Function.

9. 2, 3, and 7 appear to have domain $\mathbb{R}$.

10. In 2, $f(2) = 5$ (it appears). In 3, $f(2) = -1$ (it appears). In 7, $f(2) = 1$ (it appears).

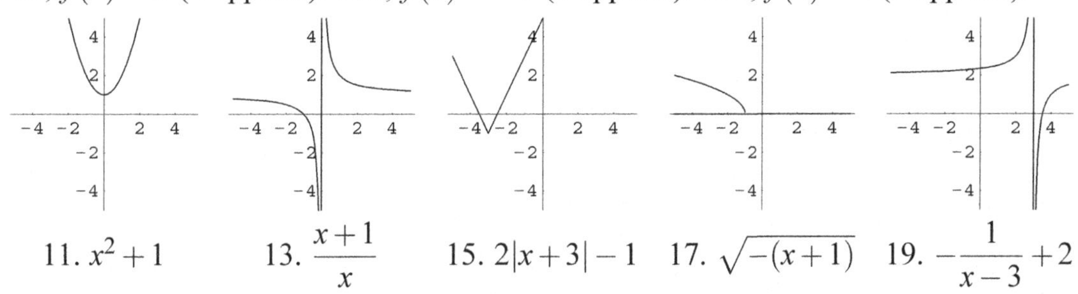

11. x^2+1 13. $\dfrac{x+1}{x}$ 15. $2|x+3|-1$ 17. $\sqrt{-(x+1)}$ 19. $-\dfrac{1}{x-3}+2$

In 11, the graph of f is the same as the graph of x^2, but translated up 1 unit. In 13, $f(x) = 1+\dfrac{1}{x}$, so we have the graph of $\dfrac{1}{x}$ translated up 1 unit. In 15, We translate the graph of $|x|$ left 3 units, then perform a vertical stretch by a factor of 2, and finally translate the result down 1 unit. In 17, we have $f(x) = g(x+1)$, where $g(x) = \sqrt{-x}$. Thus, g is a reflection of $\sqrt{x}$ about the y-axis, and then f translates this to the left by 1 unit. Finally, in 19, we have a translation of $\dfrac{1}{x}$ to the right by 3, then a reflection across the x-axis, and then a translation up 2 units.

21. 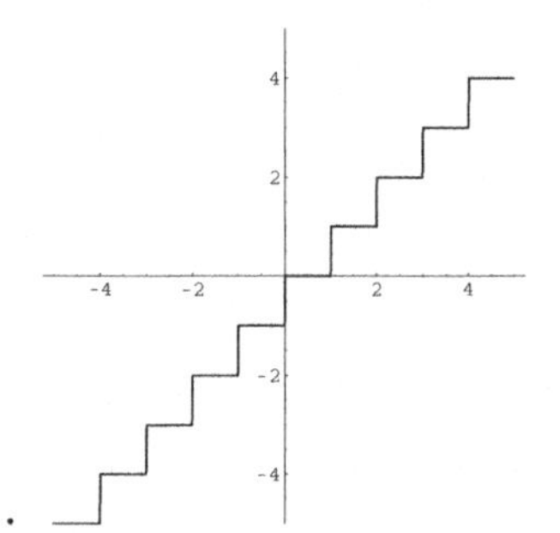

The vertical segments are just an artifact of the graphing program; they are not actually part of the graph.

23. This is a transformation of the square root function.

25. This is a transformation of the cubing function.

27. Let $x \in \mathbb{R} - Z$. Then x is between two integers, say $a < x < a+1$ and a is an integer. Since there are no integers between a and $a+1$, we must have $a = \lfloor x \rfloor$, $a+1 = lceilx\rceil$. Also, $a+1 < x+1 < a+2$, so $\lfloor x+1 \rfloor = a+1$. Therefore, $\lceil x \rceil = a+1 = \lfloor x+1 \rfloor$.

Inverse Functions, page 166

1. $f(g(x)) = \sqrt{2\left(\frac{(x+3)^2-1}{2}\right)+1} - 3 = \sqrt{(x+3)^2} - 3 = |x+3| - 3 = x+3-3 = x$ (since $x \geq 3$ due to the domain of g). Also, $g(f(x)) = \frac{(\sqrt{2x+1}-3+3)^2-1}{2} = \frac{(\sqrt{2x+1})^2-1}{2} = \frac{2x+1-1}{2} = x$. Therefore, f and g are inverses.

3. $f(g(x)) = (x+5) - 5 = x$. $g(f(x)) = (x-5) + 5 = x$.

5. f is one-to-one since if $(a+2)^3 = (b+2)^3$, then $a+2 = b+2$, so $a = b$. We need to solve $x = (y+2)^3$ for y. This gives $\sqrt[3]{x} = y+2$, so $y = \sqrt[3]{x} - 2$. Therefore, $f^{-1}(x) = \sqrt[3]{x} - 2$.

7. f is not one-to-one since $f(2) = 5 = f(5)$.

9. g is one-to-one. $g^{-1} = \{(2,1), (3,2), (4,0), (-3,-2), (5,5)\}$.

11. f is not one-to-one since $f(1) = \frac{|1|}{1} = 1 = \frac{|2|}{2} = f(2)$.

13. If $g(a) = g(b)$, then $\frac{-5}{2a-3} = \frac{-5}{2b-3}$, so $2a-3 = 2b-2$ and $a = b$. Therefore, g is one-to-one. $g^{-1}(x) = \frac{-3x+5}{-2x}$.

15. If $f(a) = f(b)$, then $\frac{3a+5}{6a-1} = \frac{3b+5}{6b-1}$, so $(3a+5)(6b-1) = (3b+5)(6a-1)$. This gives

$$\begin{aligned} 18ab - 3a + 30b - 5 &= 18ab - 3b + 30a - 5 \\ 27b &= 27a \\ b &= a \end{aligned}$$

Therefore, f is one-to-one. Now we must solve $x = \frac{3y+5}{6y-1}$ for y. We have

$$\begin{aligned} 3y+5 &= x(6y-1) = 6xy - x \\ 3y - 6xy &= -x-5 \\ (3-6x)y &= -x-5 \\ y &= \frac{-x-5}{3-6x} \end{aligned}$$

Therefore, $f^{-1}(x) = \dfrac{x+5}{6x-3}$.

17. This function is not one-to-one; it fails the horizontal line test.

19. This function is one-to-one; its inverse is graphed to the right (along with the original function).

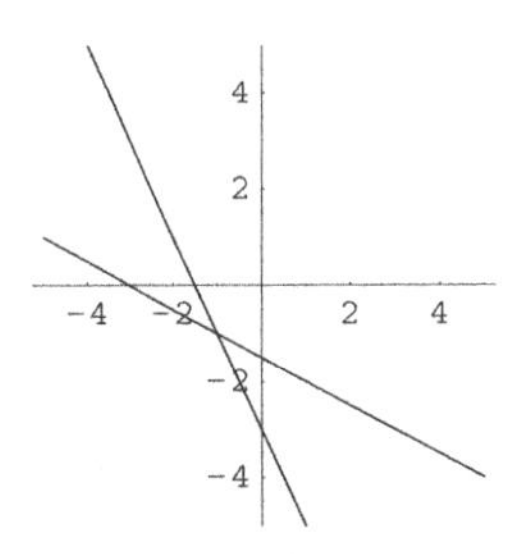

21. A function is defined as a set of ordered pairs such that no first element is paired with more than one second element. A one-to-one function satisfies that *as well as* the condition that no second element is paired with more than one first element.

Exponential Functions, page 175

1. 64.3634

3. 15.1602

5. 0.1013

7. −4.7288

9. -7.072×10^{18}

11. 1.7743

13. The calculator gives 1.

15. $f(-2) = 4^2 = 16; f(-1) = 4^1 = 4; f(0) = 4^0 = 1; f(2) = 4^{-2} = \dfrac{1}{16}$.

17. $f(-2) = (2/3)^2 = 4/9$. $f(-1) = (2/3)^1 = 2/3$. $f(0) = (2/3)^0 = 1$. $f(1) = (2/3)^{-1} = 3/2$.

19. $f(0) = 1, f(-2) = \dfrac{25}{16}, f(-1) = \dfrac{5}{4}, f(1) = \dfrac{4}{5}$.

21. $f(-4)=2^0-2=-1$. $f(-3)=2^1-2=0$. $f(-2)=2^2-2=2$. $f(-1)=2^3-2=6$.

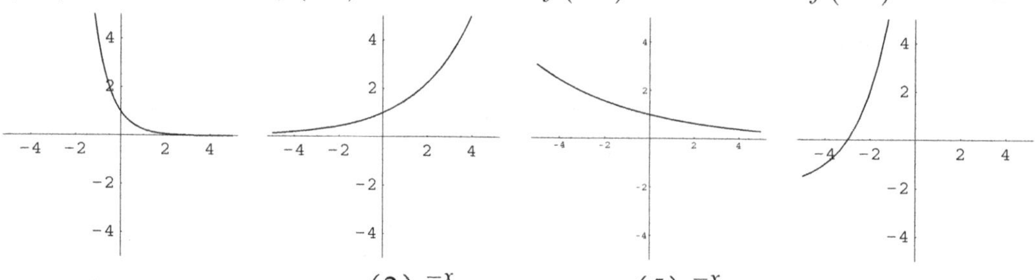

15. 4^{-x} 17. $\left(\frac{2}{3}\right)^{-x}$ 19. $\left(\frac{5}{4}\right)^{-x}$ 21. $2^{x+4}-2$

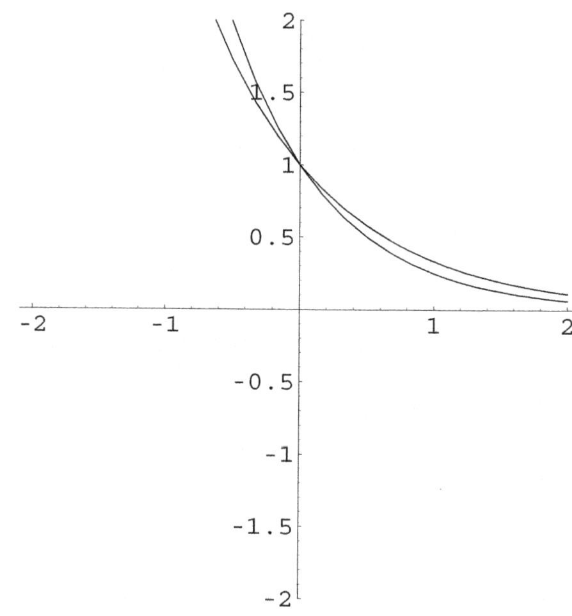

23.

$\left(\frac{1}{4}\right)^x \le \left(\frac{1}{3}\right)^x$ for $x \le 0$.

25. The value of the investment will be $1500(1+0.08/12)^{18} \approx 1690.57$ dollars, so the interest is \$190.57.

27. Since the principal (\$5000) and the term (4 years) are the same for all three, we only need to compute the values

(a) $\left(1+\dfrac{0.072}{4}\right)^4 \approx 1.07397$ (for 7.2% compounded quarterly),

(b) $\left(1+\dfrac{0.0715}{365}\right)^{365} \approx 1.07411$ (for 7.15% compounded daily),

(c) $\left(1+\dfrac{0.071}{8760}\right)^{8760} \approx 1.07358$ (for 7.1% compounded hourly).

Of these, 7.15% compounded daily will give the best return.

29. We will have $1000\left(\dfrac{1}{2}\right)^{10000/1200} \approx 3.1$ grams left.

31. (a) 2
 (b) 2.59374
 (c) 2.70481
 (d) 2.71828
 (e) The value seems to be getting close to a particular number, about 2.71828.

Logarithmic Functions, page 180

1. $\log_8 512 = \log_8 8^3 = 3.$

3. $\log_5 625 = \log_5 5^4 = 4.$

5. $\log_6 \frac{1}{36} = \log_6 6^{-2} = -2.$

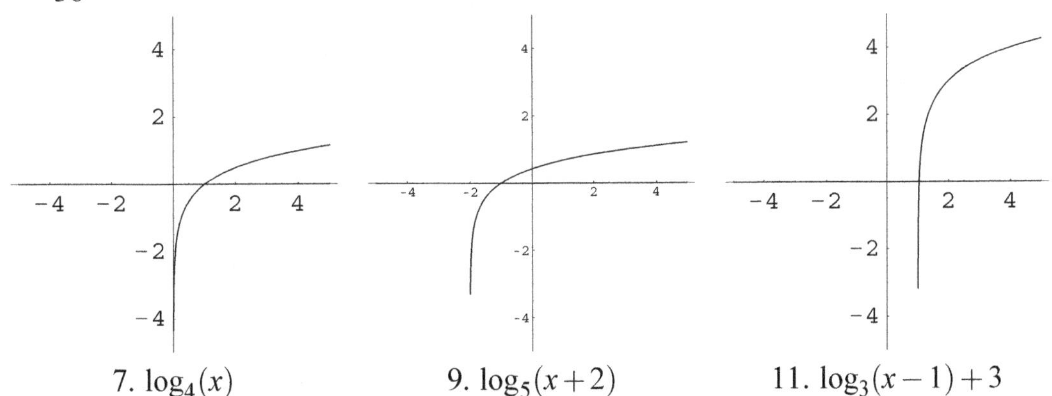

7. $\log_4(x)$ 9. $\log_5(x+2)$ 11. $\log_3(x-1)+3$

13. $\log_4(3x^4) = \log_4(3) + \log_4(x^4) = \log_4(3) + 4\log_4(x).$

15. $\log_6(x) - \log_6(3)$

17. $\log_5(x) + 3\log_5(x+1)$

19. $\log_2(1) - \log_2(2) = -1$

21. $\log_3\left(\frac{2x^2(x-4)^3}{(x+1)^4}\right) = \log_3(2x^2(x-4)^3) - \log_3((x+1)^4) = \log_3(2) + 2\log_3(x) + 3\log_3(x-4) - 4\log_3(x+1)$

23. $\log_3(2x)$

25. $\log_2(x+y)^3$

27. $\log_7(x-5)^{3/2}$

29. $\log_2(3x) - \log_2(x+3) + \log_2(x) = \log_2\left(\frac{3x^2}{x+3}\right)$

31. $2\log_{10}(3x+2y) - 2\log_{10}(6x+4y) = 2\log_{10}\left(\frac{3x+2y}{6x+4y}\right) = 2\log_{10}(1/2) = -2\log_{10}(2).$

33. $\log_a(80) = \log_a(8) + \log_a(10) \approx 2.3026 + 2.0794 = 4.3820.$

35. $\log_a(640) = \log_a(8^2) + \log_a(10) \approx 2(2.0794) + 2.3026 = 6.4614.$

37. $\log_a(1.25) = \log_a(10/8) = \log_a(10) - \log_a(8) \approx 0.2232.$

39. 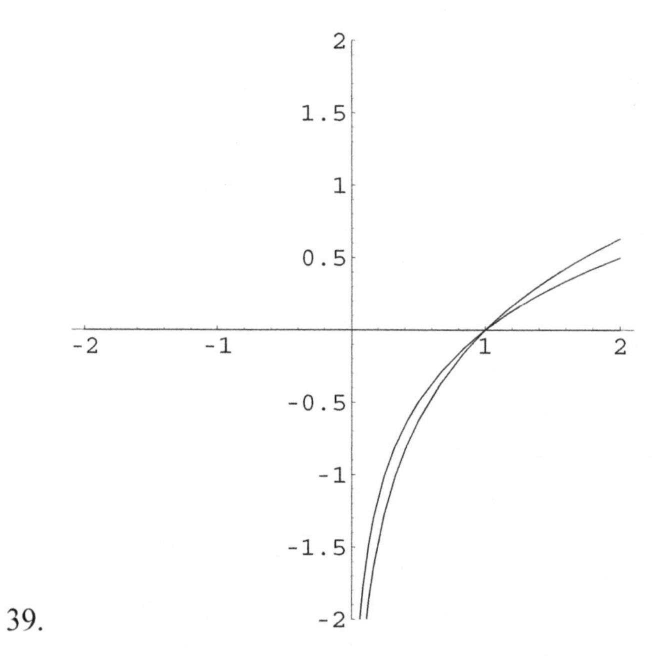

$\log_3 x < \log_4 x$ for $x < 1$.

41. Let $x = \log_a(u)$ and $y = \log_a(v)$. Then $a^x = u$ and $a^y = v$, so $\frac{u}{v} = a^{x-y}$. Therefore, $\log_a \frac{u}{v} = x - y = \log_a u - \log_a v$.

The Natural Exponential and Logarithmic Function, page 184

1. $\log 1000 = \log 10^3 = 3.$

3. $\ln 13 \approx 2.56495.$

5. $\log_8 14 = \dfrac{\ln 14}{\ln 8} \approx 1.26912$

7. $\log_{1/3} 42 = \dfrac{\ln 42}{\ln(1/3)} \approx -3.40217.$

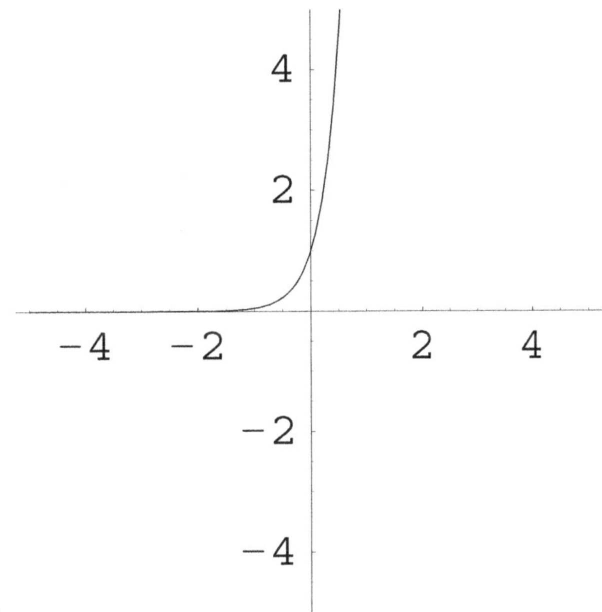

9.

11. $\log(2)+\log(x)$

13. $\log(5)+2\log(x)+\log(y)$

15. $\log(x)+3\log(x+1)$

17. $\ln(1)-\ln(e)=-1$

19. $\log(2)+2\log(x)+3\log(x-4)-4\log(x+1)$

21. $\log(6x)$

23. $\log(x+y)^3$

25. $\ln(x-5)^{3/2}$

27. $\log\left(\dfrac{3x^2}{x+2}\right)$

29. $-\log(4)$

31. (a) $\left(1+\dfrac{1}{1}\right)^1=2.$

    (b) 2.59374

    (c) $\left(1+\dfrac{1}{100}\right)^{100}\approx 2.70481.$

    (d) 2.71828

    (e) The calculator gives 1. When it calculates $1+\dfrac{1}{10^{14}}$, it gets 1 because of its level of precision.

(f) The y-values should be very close to 2.71828.

Exponential and Logarithmic Equations, page 191

1. $3^{x+2} = 27 \implies 3^{x+2} = 3^3 \implies x+2 = 3$. Therefore, $x = 1$.

3. $x = -6$

5. $e^{3x+2} = 5 \implies \ln\left(e^{3x+2}\right) = \ln 5 \implies 3x+2 = \ln 5 \implies x = \dfrac{\ln(5)-2}{3}$.

7. $x \approx 8.693$.

9. $x = 0$ or $x \approx 0.387$.

11. $x \approx 1339.431$.

13. $x \approx -1.926$

15. $x \approx 9.833$.

17. $\log(4x^2) = 2 \implies 10^{\log(4x^2)} = 10^2 \implies 4x^2 = 100 \implies x^2 = 25 \implies x = \pm 5$. Since -5 is not in the domain of the common logarithm function, it is not a solution of the original equation. Therefore, the only solution is $x = 5$.

19. No solution.

21. We know that $A(t) = A_0(1/2)^{t/5700} < 0.0001A_0$. Therefore, $(1/2)^{t/5700} < 0.0001$, so $\ln\left[(1/2)^{t/5700}\right] < \ln(0.0001)$, and thus $\dfrac{t}{5700}\ln(1/2) < \ln 0.0001$. Notice that $\ln(1/2) < 0$, so when we divide by it, we must reverse the direction of the inequality. Thus, $t > 5700\dfrac{\ln(0.0001)}{\ln(1/2)} \approx 75,700$ years.

23. About 4 and a half years.

Linear Functions, page 198

1. Parallel

3. Perpendicular.

5. 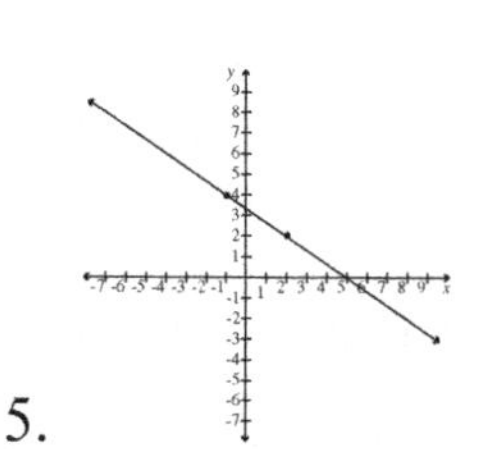

7. 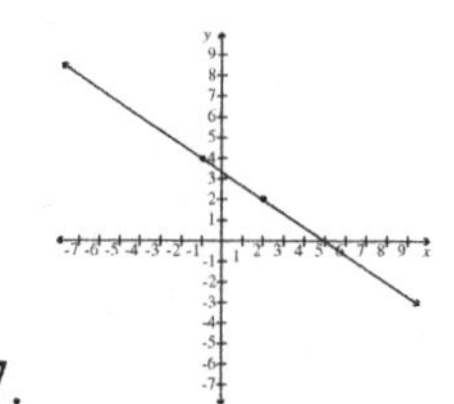

9. $y = \dfrac{3}{8}(x-4)$

11. $y = \dfrac{1}{9}x + 3$

13. $y + 7 = \dfrac{5}{2}(x-12)$

15. $y+7=-\frac{2}{5}(x-12)$

17. $y=-2500(x-0.55)+2400$.

Quadratic Functions, page 203

Complete the square to write each quadratic function in standard form.

1. $f(x)=(x+3)^2-21$

3. $f(x)=-2(x+2)^2+23$

5. $(-4,-1)$

7. $(3/2,116)$

9. $f(x)=-1.5x^2+7.5x-4$

11. $f(x)=-0.003x^2+50$.

13. A punted football follows a parabolic path. Its height at $t=0$ s is 3 feet, its height at $t=2$ s is 95 feet, and its height at $t=4$ s is 59 feet.

   (a) $h(t)=-16t^2+78t+3$.

   (b) 98.0625 feet.

   (c) About 4.9 seconds.

Solving Quadratic Equations, page 214

1. $x=-3\pm\sqrt{6}$

3. $x=-\frac{4}{3}\pm\frac{\sqrt{2}}{3}i$

5. $x=\frac{-5\pm\sqrt{57}}{2}$

7. $x=-\frac{\pm\sqrt{13}}{3}$

9. $x^2=\pm 4$

11. $x=-6$ or $x=-2$

13. $x=-1\pm\sqrt{5}$

15. $x=5$ or $x=4$.

17. $x=0$ or $x=-1\pm\frac{\sqrt{2}}{2}$ or $x=1\pm\frac{\sqrt{2}}{2}$

19. $x=-\frac{1}{2}$ or $x=4$.

21. 14 hours.

23. $(1+\sqrt{2})^2=3+2\sqrt{2}$.

Polynomial Functions and Equations, page 220

1. At most 2 zeros. $(\nwarrow,\nearrow)$

3. At most 5 zeros. $(\nwarrow,\searrow)$

5. $(x^5-3x^4-2x^3+x^2+4x+1)=(x^3-2)(x^2-3x)+(x^2-2x+1)$

7. $(3x^3-4x^2+2x+2)=(3x+3)\left(x^2-\frac{7}{3}x+3\right)-7$

9. 0.8844

11. 1216

13. 0

15. 2 has multiplicity 1, -3 has multiplicity 4, and $\frac{5}{3}$ has multiplicity 3.

17. -1 has multiplicity 1, -2 has multiplicity 2, -3 has multiplicity 3, and -4 has multiplicity 4.

Rational Roots, page 227

1.
$$\begin{array}{r|rrrr} 2 & 3 & -5 & -3 & -1 \\ & & 6 & 2 & -2 \\ \hline & 3 & 1 & -1 & \boxed{-3} \end{array}$$

3.
$$\begin{array}{r|rrrrr} \frac{-3}{2} & 2 & 1 & -1 & -5 & -7 \\ & & -3 & 3 & -3 & 12 \\ \hline & 2 & -2 & 2 & -8 & \boxed{5} \end{array}$$

5. $(-4x^5+5x^2+8)=(x^2+1)(-4x^3+4x+5)+\boxed{(-4x)}$

7. (a) 3.

   (b) 1

   (c) $\pm1,\pm\frac{1}{3},\pm2,\pm\frac{2}{3},\pm4,\pm\frac{4}{3}$.

   (d) $x=\frac{2}{3},\pm\sqrt{2}i$.

   (e) $\frac{2}{3}$ is rational and $\pm\sqrt{2}i$ are both nonreal complex.

   (f) $f(x)=(3x-2)(x-\sqrt{2}i)(x+\sqrt{2}i)$.

   Find the zeros of each polynomial. Classify each zero as rational, irrational, or nonreal complex, and express f as a product of linear factors. Also give the multiplicity of each zero.

9. Zeros: $x=2$ (rational), $x=-3$ (rational), $x=\pm2i$ (nonreal complex). All have multiplicity 1. $f(x)=(x-2)(x+3)(x+2i)(x-2i)$.

11. $x=2$ (rational, multiplicity 1), $x=\pm1$ (rational, multiplicity 3 each). $f(x)=(x-2)(x+1)^3(x-1)^3$.

13. $x=-\frac{1}{2},-1$ (rational, multiplicity 1), $x=\pm\sqrt{11}$ (irrational, multiplicity 1). $f(x)=(2x+1)(x+1)(x-\sqrt{11})(x+\sqrt{11})$.

Mathematical Induction, page 236

1. If $n=1$, then $4(1)+6=10<11=3(1)^2+3(1)+5$, so the base case holds. Suppose that the inequality holds for some $k \geq 1$, so that $4k+6<3k^2+3k+5$. Then

$$\begin{aligned} 4(k+1)+6 &= (4k+6)+4 \\ &< 3k^2+3k+5+4 \\ &= (3k^2+1)+(3k+3)+5 \\ &< 3(k+1)^2+3(k+1)+5 \end{aligned}$$

since $3k^2+1<3k^2+6k+3=3(k+1)^2$. Therefore, the inequality holds for $k+1$, so it also holds for all natural numbers n.

3. If $n=1$, then $3(1)^2+(1)=4<5=2(1)^3+3$. If $n=2$, the inequality also holds. Suppose that the inequality holds for some $k \geq 2$, so that $3k^2+k<2k^3+3$. Then

$$\begin{aligned} 3(k+1)^2+(k+1) &= (3k^2+k)+(7k+4) \\ &\leq 2k^3+3 \end{aligned}$$

since $4>3$ and $3k^2<2k^3$ for $k>\frac{3}{2}$. Prove that for all $n \in \mathbb{N}$, $3n^2+n<2n^3+3$.

5. If $n=1$, we have $1=2-1$, so the formula holds. Now suppose that the formula holds for some $k \geq 1$, so that $1+5+9+\ldots+(4k-3)=2k^2-k$. Then

$$\begin{aligned} 1+5+9+\ldots+(4k-3)+(4(k+1)-3) &= 2k^2-k+(4k+1) \\ &= 2k^2+3k+1 \\ &= 2(k+1)^2-k, \end{aligned}$$

as desired. Therefore, the formula holds for all $n \in \mathbb{N}$.

7. If $n=1$, then $\sum_{i=1}^{n} i^3 = 1^3 = 1 = \dfrac{1^2(1+1)^2)}{4}$. Now suppose that the formula holds for some $k \geq 1$, so that $\displaystyle\sum_{i=1}^{k} i^3 = \frac{k^2(k+1)^2}{4}$. Then

$$\begin{aligned} \sum_{i=1}^{k+1} i^3 &= 1^3+2^3+\ldots+k^3+(k+1)^3 \\ &= \frac{k^2(k+1)^2}{4}+(k+1)^3 \\ &= (k+1)^2\left(\frac{k^2}{4}+k+1\right) \\ &= (k+1)^2\left(\frac{k^2+4k+4}{4}\right) \\ &= \frac{(k+1)^2(k+2)^2}{4} \\ &= \frac{(k+1)^2((k+1)+1)^2}{4}, \end{aligned}$$

as desired. Therefore, the formula holds for all $n \in \mathbb{N}$.

The Binomial Theorem, page 241

1. $x^4+4x^3y+6x^2y^2+4xy^3+y^4$

3. $32x^5+80x^4y+80x^3y^2+40x^2y^3+10xy^4+y^5$

5. $343x^3-735x^2y+525xy^2-125y^3$

7. $-x^2+2ixy+y^2$

9. $32ix^5-120x^4y-180ix^3y^2+135x^2y^3+\frac{405}{8}ixy^4-\frac{243}{32}y^5$

11. $-52+47i$

13. $-119-120i$

15. $x^6+3x^4y^2+3x^2y^4+y^6$

17. $46x^9+9ix^9+57x^6y^2-66ix^6y^2-18x^3y^4-51ix^3y^4-11y^6-2y^6$

19. $189x^5y^2$.

21. $48384x^5y^3$

23. $240xy^2$

25. $15x^{8/3}y^2$

27. $-6x^2y^2$

29. 135

31. 0

33. 0

35. $\binom{n}{k}=\frac{n!}{k!(n-k)!}$ and $\binom{n}{n-k}=\frac{n!}{(n-k)!(n-(n-k))!}=\frac{n!}{(n-k)!k!}$.

Index

Made in the USA
Las Vegas, NV
21 August 2022